Supra-materials Nanoarchitectonics

Supra-materials Nanoarchitectonics

Edited by

Katsuhiko Ariga

Masakazu Aono

ELSEVIER

AMSTERDAM • BOSTON • HEIDELBERG • LONDON
NEW YORK • OXFORD • PARIS • SAN DIEGO
SAN FRANCISCO • SINGAPORE • SYDNEY • TOKYO

William Andrew is an imprint of Elsevier

William Andrew is an imprint of Elsevier
The Boulevard, Langford Lane, Kidlington, Oxford, OX5 1GB, United Kingdom
50 Hampshire Street, 5th Floor, Cambridge, MA 02139, United States

Notices
Knowledge and best practice in this field are constantly changing. As new research and experience broaden our understanding, changes in research methods, professional practices, or medical treatment may become necessary.

Practitioners and researchers must always rely on their own experience and knowledge in evaluating and using any information, methods, compounds, or experiments described herein. In using such information or methods they should be mindful of their own safety and the safety of others, including parties for whom they have a professional responsibility.

To the fullest extent of the law, neither the Publisher nor the authors, contributors, or editors, assume any liability for any injury and/or damage to persons or property as a matter of products liability, negligence or otherwise, or from any use or operation of any methods, products, instructions, or ideas contained in the material herein.

Library of Congress Cataloging-in-Publication Data
A catalog record for this book is available from the Library of Congress

British Library Cataloguing-in-Publication Data
A catalogue record for this book is available from the British Library

ISBN: 978-0-323-37829-1

Publisher: Matthew Deans
Acquisition Editor: Simon Holt
Editorial Project Manager: Sabrina Webber
Production Project Manager: Susan Li
Designer: Greg Harris

Typeset by TNQ Books and Journals

Contents

Contributors

K. Cervantes-Salguero
Tohoku University, Sendai, Miyagi, Japan

M. Ebara
National Institute for Materials Science (NIMS), Tsukuba, Japan

K. Eid
Zhejiang University of Technology, Zhejiang, P.R. China; Changchun Institute of Applied Chemistry, Changchun, P.R. China; University of Chinese Academy of Sciences, Beijing, P.R. China

M.W. England
National Institute of Advanced Industrial Science and Technology (AIST), Nagoya, Japan

T. Funaki
Research Center for Photovoltaics, National Institute of Advanced Industrial Science and Technology (AIST), Tsukuba, Ibaraki, Japan

Y. Hoshino
Kyushu University, Fukuoka, Japan

A. Hozumi
National Institute of Advanced Industrial Science and Technology (AIST), Nagoya, Japan

S. Ishihara
National Institute for Materials Science, Tsukuba, Japan

N. Kameta
Research Institute for Sustainable Chemistry, National Institute of Advanced Industrial Science and Technology (AIST), Tsukuba, Ibaraki, Japan

K. Katagiri
Hiroshima University, Higashihiroshima, Japan

W. Kurashige
Tokyo University of Science, Shinjuku-ku, Tokyo, Japan

J. Labuta
National Institute for Materials Science, Tsukuba, Japan

Y. Manaka
Renewable Energy Research Center, National Institute of Advanced Industrial Science and Technology (AIST), Koriyama, Fukushima, Japan

Yoshitake Masuda
National Institute of Advanced Industrial Science and Technology (AIST), Nagoya, Japan

N. Miyamoto
Fukuoka Institute of Technology, Fukuoka, Japan

T.N. Murakami
Research Center for Photovoltaics, National Institute of Advanced Industrial Science and Technology (AIST), Tsukuba, Ibaraki, Japan

S. Murata
Tohoku University, Sendai, Miyagi, Japan

M. Naito
National Institute for Materials Science, Tsukuba, Japan

Y. Negishi
Tokyo University of Science, Shinjuku-ku, Tokyo, Japan; Tokyo University of Science, Noda, Chiba, Japan; Institute for Molecular Science, Okazaki, Aichi, Japan

Y. Niihori
Tokyo University of Science, Shinjuku-ku, Tokyo, Japan

K. Nobusawa
Osaka University, Osaka, Japan

Y. Oaki
Keio University, Kohoku-ku, Yokohama, Japan

S. Sharma
Tokyo University of Science, Shinjuku-ku, Tokyo, Japan

H.-T. Sun
Soochow University, Suzhou, People's Republic of China

H. Wang
Zhejiang University of Technology, Zhejiang, P.R. China

L. Wang
Zhejiang University of Technology, Zhejiang, P.R. China

S. Yamamoto
Fukuoka Institute of Technology, Fukuoka, Japan

Preface

How can we create functional materials? We do not always have to create totally new stuffs from zero. In our history, a telescope was invented through wise combinations and arrangements of simple glasses and a light bulb was fabricated with bamboo fibers. Materials themselves are important but architecting them into functional systems and materials are much more important. Supra-materials, which are expected to have advanced functions as compared to conventional materials, would be created through architecting various nanomaterials.

However, fabrication of systems and materials is not always easy in nanoscale regions where uncontrollable fluctuations and unexpected disturbances significantly affect structural formation. Influence of thermal/statistical fluctuations and mutual interactions unavoidably occurs between component atoms, molecules, and materials. Outputs such as materials structures and properties are not simply determined by known inputs. In the nanoscale region, what we try to prepare cannot be simply attained through conventional one-by-one construction strategy. Therefore, synthesis and fabrication of functional materials using nanoscale components have to be a novel concept, nanoarchitectonics.

A novel terminology, *nanoarchitectonics*, which means nano + architecto + nics, was invented in 2000 as a new paradigm of materials science and technology by Masakazu Aono that was introduced at the *1st International Symposium on Nanoarchitectonics Using Suprainteractions* in Tsukuba, Japan. Nanoarchitectonics aims at opening a new paradigm of development of materials design and fabrication with the following working principles:

1. Nanomaterials or nanosystems with functional reliable functions can be constructed by organizing nanoscale parts even with some unavoidable unreliability.
2. Functions are not always originated only from the individual nanoparts but their interactions may create new functionalities.
3. Unexpected functionalities can emerge from assembling or organizing a huge number of nanoparts.

Nanoarchitectonics has begun to spread into many fields including nanostructured materials synthesis, supramolecular assembly, nanoscale structural fabrications, materials hybridizations, materials and structures for energy and environmental sciences, device and physical application, and bio and medical applications.

In this book entitled, *Supra-Materials Nanoarchitectonics*, basic materials to various recent hot examples are assembled with classification of (1) Nanoarchitectonics of Basics Materials; (2) Nanoarchitectonics of Inorganic Materials; (3) Nanoarchitectonics of Biomaterials and for Biomedicals; and (4) Nanoarchitectonics for Energy/Environment. Current frontiers and the future of materials science and technology are also discussed in this book.

Katsuhiko Ariga
Masakazu Aono

Overview

This short chapter provides some overview comments on nanoarchitectonics of materials design and synthesis. It can be used as a certain guidance for this book through showing tiny essences of this book's contents.

Recent advances and new directions for nanoarchitectonics in small scale regions will be described in the former several sections. Functional nanomaterials including colloidal nanoparticles of metals, carbons, semiconductors, and ceramics, which have already become an important class of materials in the current nanotechnology, serve as the minimum objects for building nanoarchitectures.

Topics in some chapters show the fundamental aspects and amazing properties of nanoclusters of metals or metal-ion-doped molecular crystals. It has been long accepted that representative noble metals, gold and platinum, are metallic, shiny, conductive, and nontarnish materials, but we see the emergent features totally different from their bulk crystals as their sizes are going down into nanoscale. Most of the metal nanoclusters of single nanometers are no longer jewelry; they become high-tech materials with the catalytic, paramagnetic, plasmonic, and photoluminescence properties that are independent of size. Although nucleation and growth of the nanoclusters had been limited to manipulate in special environments such as in vacuo or under high temperature and pressure, the evolution of synthetic techniques allows us to produce a quantity of nanoclusters, and to manipulate them by hand even under ambient conditions. Relationships between their equilibrium structures being dependent on sizes and properties are also demonstrated.

Nanoarchitectonics can create the free-standing nanocrystals of semiconductors exhibiting the quantum confinement effect. These nanocrystals are frequently denoted as "artificial atoms" because of their electronic states being described by atomic-like wave functions. The energy quantization enables us to tune optical absorption and emission spectra that are dependent on size of the nanocrystals for each semiconductor. Combination of size-dependent absorption-emission properties and the unique electronic structures makes the nanocrystals key components to fabricate various quantum dot devices including lasers, solar cells, memories, and light-emitting diodes. A fourth topic focuses on the synthesis of well-controlled shapes of ceramics nanocrystals. The nanocrystals are prepared in liquid phase from the viewpoint of green and sustainable technology. The precise control over the nucleation and growth conditions in aqueous media encourages anisotropic growth of nanocrystals to give the unique morphological crystals with acicular, cylindrical, multineedle, sheet-like shapes.

Further self-assembly of morphologically controlled nanocrystals for their film formation is still one of the exciting and attractive fields in surface science. The prototypes of devices by taking advantage of immense surface area of the nanocrystals are also demonstrated. The fifth topic demonstrates the supramolecular complexes. The final topic showcases the supramolecular self-assembly at surfaces. Since the initial study on surface chemistry focusing the absorption behaviors of small molecules, including oxygen or carbon dioxide, the choice of adsorbates employed has broadened to organic molecules with multiple functionalities over several decades, spawning a field of host–guest chemistry. Diverse substrates of conductors, semiconductors, or insulators are available. Applications of such supramolecular assemblies working as platforms include repellent painting, oil/water separation devices, and molecular sensors by taking advantage of specific molecular recognitions. Again, the present subsection highlights various tiny objects of organics and inorganics with unique properties that serve as the fundamental building blocks for various supramolecular nanoarchitectonics applications.

The later parts of this book discuss two additional topics: (1) biomaterials and biomedicals; and (2) energy/environmental sciences. Of course these two areas are hot and very important in recent technologies and have attracted attention on the basis of high social demands. Biorelated sections includes DNA, smart polymers, bioinspired materials, and biomedical usages. On the other hand, energy and environmental sections showcase recent hot examples on materials conversion (mainly catalysis), sensing, solar cells, hydrogen storage, and CO_2 separation. Although their details are not described in this overview, one can understand the high importance of nanoarchitectonics to practically useful materials fields.

Naoto Shirahata
Katsuhiko Ariga

Nanoarchitectonics of Basics Materials

Nanocluster Science

S. Sharma[1], W. Kurashige[1], Y. Niihori[1] and Y. Negishi[1,2,3]

[1]*Tokyo University of Science, Shinjuku-ku, Tokyo, Japan;* [2]*Tokyo University of Science, Noda, Chiba, Japan;* [3]*Institute for Molecular Science, Okazaki, Aichi, Japan*

CHAPTER OUTLINE

1.1 INTRODUCTION

Small metal clusters up to 2 nm ($< \sim 100$ atoms) in size have different geometric and electronic structures from their corresponding bulk metals (Fig. 1.1) [1]. With regard to their geometry, these clusters adopt special atomic arrangements, such as icosahedral structures, to reduce their surface energy, instead of the close packing found in bulk metals. In addition, their electronic structures are not continuous, as in bulk metals, but rather discretized. Owing to these characteristics, small metal clusters exhibit different

Supra-materials Nanoarchitectonics. http://dx.doi.org/10.1016/B978-0-323-37829-1.00001-8

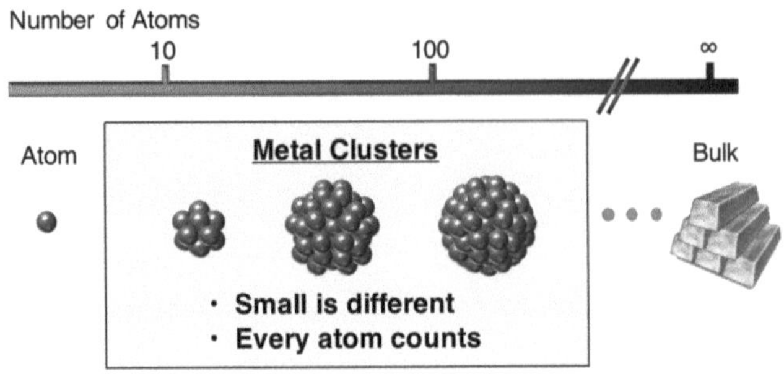

■ **FIGURE 1.1** Metal clusters discussed in this chapter.

physical and chemical properties and functionalities from the bulk metal. Furthermore, these properties and functions undergo significant changes depending on the number of constituent atoms in the metal cluster (Fig. 1.1). Metal clusters with these characteristics have attracted much attention as new functional nanomaterials in a wide range of fields, from basic science to applied chemistry.

When working with these small metal clusters, it is essential to synthesize clusters with an atomic level of precision so as to gain an understanding of their fundamental nature and to allow for practical applications in nanoscale devices or as catalysts by controlling their functionalities. Numerous studies regarding such precise synthesis have been reported in the case of phosphine-protected metal clusters, beginning in the end of 1960s [2–18]. However, phosphine-protected metal clusters are susceptible to degradation in solution and in the solid state, and this poor stability restricts the applications of these materials [19]. In contrast, thiolate-protected metal clusters [20–61] show high stability both in solution and in the solid state owing to the formation of strong bonds between the metal and the thiolate group. Therefore, in recent years, intensive research has been conducted on thiolate-protected metal clusters. As a result, various precise methods have been established for the synthesis of these clusters. In addition, much knowledge has been gained with respect to their stability, geometrical and electronic structures, and physical and chemical properties. Furthermore, certain guidelines have been established for the functionalization and application of these metal clusters. Thus, the techniques and knowledge necessary to use thiolate-protected metal clusters in various applications have gradually been obtained in recent years. In this chapter, we focus on gold clusters, which are the most-studied type of cluster, and introduce the latest information regarding their precise synthesis as well as their stability, electronic and geometrical structures, physical and chemical properties, and functionalization.

1.2 PRECISE SYNTHESIS

Thiolate-protected gold clusters [$Au_n(SR)_m$] having a defined number of constituent atoms are typically synthesized by one of the following four methods:

1. High-resolution separation of a mixture composed of various sizes according to the number of constituent atoms (Fig. 1.2A);
2. Exposure of a mixture composed of various sizes to extreme conditions followed by the collection of only those clusters stable under such conditions (Fig. 1.2B);
3. Controlling the growth rate of the clusters so as to obtain a uniform chemical composition (Fig. 1.2C);
4. Replacing the ligands of the cluster with ligands having different bulkiness so as to render clusters with specific chemical composition stable (Fig. 1.2D).

Hereafter, each of these methods is explained in detail.

1.2.1 Fractionation

$Au_n(SR)_m$ clusters are generally prepared by adding a reducing agent to a solution containing thiol and gold ions [20]. The aggregation of the resultant gold atoms leads to the formation of $Au_n(SR)_m$ clusters, and the products obtained by this method normally have a distribution in the number of constituent atoms [21,29,62–80]. The high-resolution separation of such a mixture according to the number of atoms is one of the most efficient methods of obtaining $Au_n(SR)_m$ clusters with well-defined numbers of constituent atoms (Fig. 1.2A) [21,29,62–80].

Polyacrylamide gel electrophoresis is a highly effective technique for separating hydrophilic thiolate-protected $Au_n(SR)_m$ clusters [21,62–64,68–72]. Using this method, clusters such as $Au_{10}(SG)_{10}$, $Au_{15}(SG)_{13}$, $Au_{18}(SG)_{14}$, $Au_{22}(SG)_{16}$, $Au_{22}(SG)_{17}$, $Au_{25}(SG)_{18}$, $Au_{29}(SG)_{20}$, $Au_{33}(SG)_{22}$, and $Au_{39}(SG)_{24}$ (SG = glutathiolate) have been isolated with high purity (Fig. 1.3A) [62]. On the contrary, fractional precipitation [73,74] and fractional extraction [75–78] have been primarily employed for the separation of hydrophobic ligand-protected $Au_n(SR)_m$ clusters. In addition to these frequently used conventional methods, in recent years, reverse-phase high-performance liquid chromatography has also proven very effective for the high-resolution separation of both types of $Au_n(SR)_m$ clusters [65–67,79,80]. $Au_n(SC_{12}H_{25})_m$ clusters have been systematically isolated by this method over a wide range of sizes, from $Au_{38}(SC_{12}H_{25})_{24}$ to $Au_{\sim 520}(SC_{12}H_{25})_{\sim 130}$ (Fig. 1.3B) [66]. Thus, several methodologies have

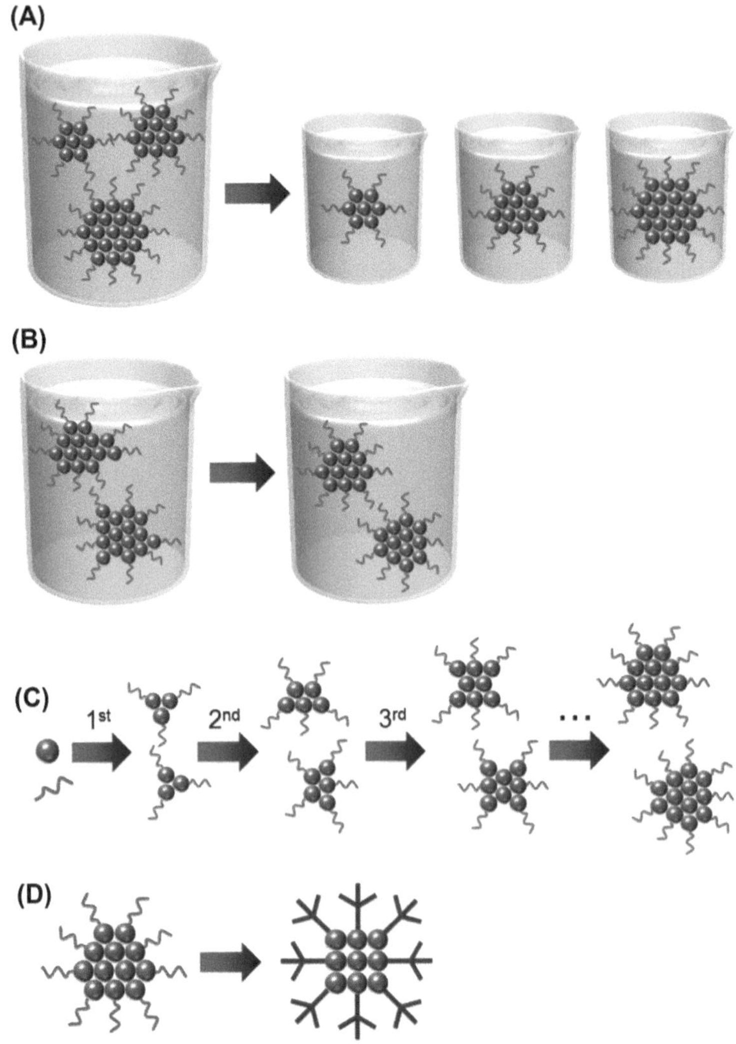

■ **FIGURE 1.2** Typical methods for the precise synthesis of $Au_n(SR)_m$ clusters: (A) fractionation, (B) size focusing, (C) slow reduction, and (D) transformation from one stable size to another.

been established to date for the fractionation of $Au_n(SR)_m$ clusters, and precise systematic isolation can be now carried out for $Au_n(SR)_m$ clusters protected either with hydrophilic or hydrophobic thiolates using these techniques.

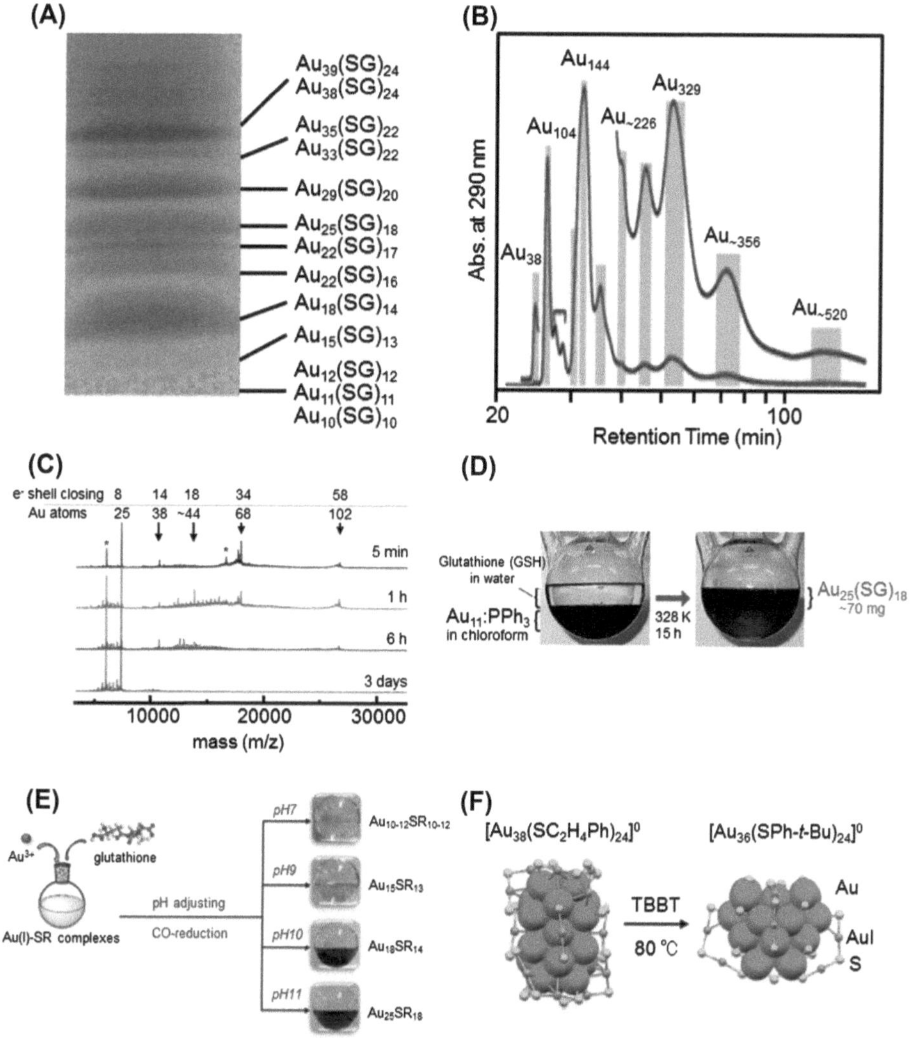

■ **FIGURE 1.3** Examples of the precise synthesis of $Au_n(SR)_m$ clusters: (A) separation of $Au_n(SG)_m$ clusters by PAGE, (B) separation of $Au_n(SC_{12}H_{25})_m$ clusters by reverse-phase high-performance liquid chromatography, (C) size focusing into $Au_{25}(SC_2H_4Ph)_{18}$, (D) size focusing into $Au_{25}(SG)_{18}$, (E) precise synthesis of each $Au_n(SG)_m$ cluster by slow reduction, and (F) transformation from $Au_{38}(SC_2H_4Ph)_{24}$ to $Au_{36}(SPh\text{-}^tBu)_{24}$. *The figures were adapted from Negishi Y, Nobusada K, Tsukuda T. Glutathione-protected gold clusters revisited: bridging the gap between gold(I)-thiolate complexes and thiolate-protected gold nanocrystals. J Am Chem Soc 2005;127:5261—70 with permission. Copyright 2005 American Chemical Society; Negishi Y, Nakazaki T, Malola S, Takano S, Niihori Y, Kurashige W, Yamazoe S, Tsukuda T, Häkkinen H. A critical size for emergence of nonbulk electronic and geometric structures in dodecanethiolate-protected Au clusters. J Am Chem Soc 2015; 137:1206—12 with permission. Copyright 2015 American Chemical Society; Dharmaratne AC, Krick T, Dass A. Nanocluster size evolution studied by mass spectrometry in room temperature $Au_{25}(SR)_{18}$ synthesis. J Am Chem Soc 2009;131:13604—05 with permission. Copyright 2009 American Chemical Society; Shichibu Y, Negishi Y, Tsukuda T, Teranishi T. Large-scale synthesis of thiolated Au_{25} clusters via ligand exchange reactions of phosphine-stabilized Au_{11} clusters. J Am Chem Soc 2005;127: 13464—65 with permission. Copyright 2005 American Chemical Society; Yu Y, Chen X, Yao Q, Yu Y, Yan N, Xie J. Scalable and precise synthesis of thiolated Au_{10-12}, Au_{15}, Au_{18}, and Au_{25} nanoclusters via pH controlled CO reduction. Chem Mater 2013;25:946—52 with permission. Copyright 2013 American Chemical Society; Zeng C, Liu C, Pei Y, Jin R. Thiol ligand-induced transformation of $Au_{38}(SC_2H_4Ph)_{24}$ to $Au_{36}(SPh\text{-}t\text{-}Bu)_{24}$. ACS Nano 2013;7:6138—45 with permission. Copyright 2013 American Chemical Society.*

1.2.2 Size Focusing

The fractionation methods noted herein are suitable for the systematic isolation of a series of $Au_n(SR)_m$ clusters. However, these methods are not conducive to mass production. To synthesize $Au_n(SR)_m$ clusters with a specific chemical composition on a large scale (~100 mg), it is necessary to instead achieve a selective synthesis, and size focusing is an effective means of doing so (Fig. 1.2B). Several $Au_n(SR)_m$ clusters show higher resistance to deterioration in solution [62] or by thiol etching [81–86] than other clusters of the same type. As such, the less stable clusters could be converted into stable clusters when exposed to extreme conditions [81–86]. As an example, when $Au_n(SR)_m$ clusters (SR = SG[86] or SC_2H_4Ph[83]; $25 \leq n < 38$) are exposed to extreme conditions, unstable $Au_n(SR)_m$ clusters are transformed into stable $Au_{25}(SR)_{18}$ clusters. Stable clusters can be precisely synthesized on a large scale with this size-focusing method (Fig. 1.3C and D).

1.2.3 Slow Reduction

$NaBH_4$ is typically employed as the reducing agent for the generation of gold atoms, although the mild agent CO can also be used [30,72,87,88]. In the latter case, the gold atoms are generated more slowly and so the $Au_n(SR)_m$ clusters are also formed at a reduced pace [87,88]. This lower synthesis rate tends to produce more uniform clusters (Fig. 1.2C) and the precise and size-selective synthesis of $Au_n(SR)_m$ clusters up to $Au_{25}(SG)_{18}$ has been realized by this method (Fig. 1.3E) [88]. Even when using $NaBH_4$ as the reducing agent, slow reduction can be achieved by controlling the pH of the solution. Au_{76}(4-MEBA)$_{44}$ (4-MEBA = 4-(2-mercaptoethyl)benzoic acid) has been synthesized in this manner [89]. However, this method is only applicable to small $Au_n(SR)_m$ clusters protected by hydrophilic thiolates. It is expected that the size-selective synthetic methods based on this principle are established even for $Au_n(SR)_m$ clusters protected by hydrophobic thiolates, as well as for large hydrophilic $Au_n(SR)_m$ clusters.

1.2.4 Transformation From One Stable Size to Another

The chemical composition of stable clusters varies depending on the bulkiness of the thiolate functional group [52,90]. Therefore, when the ligand of a stable cluster is replaced with a bulkier thiolate, distortion of the metallic core is induced, resulting in the formation of clusters with a different composition (Fig. 1.2D) [91]. An example is the reaction of phenylethanethiolate (SC_2H_4Ph)-protected $Au_{38}(SC_2H_4Ph)_{24}$ with 4-tert-butylbenzenthiol (SPh-^{t}Bu) in solution, which generates Au_{36}(SPh-^{t}Bu)$_{24}$ as the primary product (yield ~90%) (Fig. 1.3F) [92,93]. This technique enables the synthesis of stable clusters having different chemical compositions from $Au_n(SC_2H_4Ph)_m$ or $Au_n(SC_xH_{2x+1})_m$ (SC_xH_{2x+1} = alkanethiolate), and $Au_{28}(SPh\text{-}^tBu)_{20}$, $Au_{36}(SPh\text{-}^tBu)_{24}$, and $Au_{133}(SPh\text{-}^tBu)_{52}$ have all been synthesized by this method [92–96].

1.3 BASIC PROPERTIES

A number of $Au_n(SR)_m$ clusters have been precisely synthesized to date by the methods described in the previous section [97]. The subsequent study of these clusters has assessed their stability and also their geometrical and electronic structures and physical and chemical properties. In the following sections, we summarize the most recent findings regarding the basic properties of $Au_n(SR)_m$ clusters.

1.3.1 Stability

Upon allowing a series of $Au_n(SG)_m$ clusters ($n < 25$) (Fig. 1.3A) to react with an excess of glutathione, size reduction occurs via etching to generate Au(I)-thiolate oligomers. Although a similar etching reaction will take place in the case of larger $Au_n(SG)_m$ clusters ($25 \leqq n \leqq 39$), the size reduction stops at the $Au_{25}(SG)_{18}$ stage (Fig. 1.4) rather than proceeding to the oligomer [81]. This phenomenon is due to the fact that $Au_{25}(SG)_{18}$ has especially high stability against etching among the series of $Au_n(SG)_m$ clusters

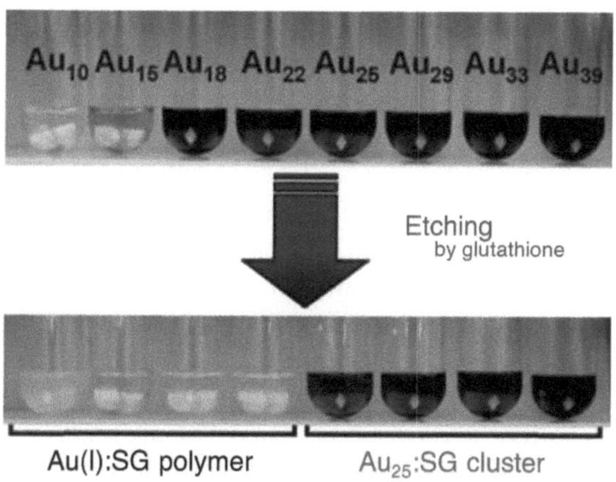

■ **FIGURE 1.4** Comparison of the stability against etching among a series of $Au_n(SG)_m$ clusters (n = 10–39). *This figure was adapted from Shichibu Y, Negishi Y, Tsunoyama H, Kanehara M, Teranishi T, Tsukuda T. Extremely high stability of glutathionate-protected Au_{25} clusters against core etching. Small 2007;3:835–39 with permission. Copyright 2007 Wiley-VCH Publishing.*

shown in Fig. 1.3A. Thus, the stability of $Au_n(SR)_m$ clusters differs greatly depending on their chemical composition. The size-selective synthesis of stable clusters based on the size-focusing technique described in the previous section utilizes this difference in stability among a series of $Au_n(SR)_m$ clusters.

The chemical composition of such stable clusters also changes depending on the thiolate functional group employed. As an example, the clusters $[Au_{25}(SR)_{18}]^-$, $Au_{38}(SR)_{24}$, $Au_{130}(SR)_{50}$, $Au_{144}(SR)_{60}$, and $Au_{329}(SR)_{84}$ are all stable when the chemical composition is $Au_n(SC_2H_4Ph)_m$ or $Au_n(SC_xH_{2x+1})_m$ [62,65,76,77,98,99]. In contrast, clusters such as $Au_{28}(SPh\text{-}^tBu)_{20}$, $Au_{36}(SPh\text{-}^tBu)_{24}$, and $Au_{133}(SPh\text{-}^tBu)_{52}$ are stable when SPh-^{t}Bu is used as the ligand [92–96]. Others, such as $[Au_{23}(Sc\text{-}C6)_{16}]^-$, are stable when the ligand is *c*-hexanethiolate (S*c*-C6) [100]. These clusters are stabilized by geometric factors, electronic factors, or both, even though the number of Au atoms and thiolate ligands varies significantly for each [97,101].

1.3.2 Geometrical Structures

Until 2007, it had been believed that $Au_n(SR)_m$ clusters have a geometrical structure in which the gold core is covered by thiolates. Since then, however, several $Au_n(SR)_m$ clusters have been shown by single crystal X-ray structural analysis to have a gold core covered by the multiple -S(R)[-Au-S(R)-]$_x$ staples ($x = 1\text{-}4$) (Fig. 1.5) [37,92–96,100,102–105]. Based on the geometrical structures determined for $Au_n(SR)_m$ clusters to date, it can be assumed that almost all $Au_n(SR)_m$ clusters have this type of core–shell structure in small cluster regions [97].

In addition, the gold core structure has been found to vary depending on the bulkiness of the ligand. In the case of $Au_n(SC_2H_4Ph)_m$ and $Au_n(SC_xH_{2x+1})_m$ clusters, clusters with sizes below $Au_{144}(SR)_{60}$ have a gold core based on an icosahedral or a Marks decahedral structure (Fig. 1.5A and B) [102–105]. When the cluster size exceeds $Au_{144}(SR)_{60}$, the core transitions to one based on a face-centered cubic (fcc) structure, just as in bulk Au [66,67]. In contrast, small $Au_n(SR)_m$ clusters having a gold core with an fcc structure can be synthesized in the case of $Au_n(SPh\text{-}^tBu)_m$ and $Au_n(Sc\text{-}C6)_m$, both of which employ ligands having bulky functional groups (Fig. 1.5C and D) [92–96,100]. The use of the adamantine thiolate ligand (S-Adm), which has a functional group of even greater bulkiness, results in the production of a $Au_{38}S_2(S\text{-}Adm)_{20}$ cluster with a gold core based on a bcc structure [106]. It is likely that the continued design of new thiolates for use as ligands

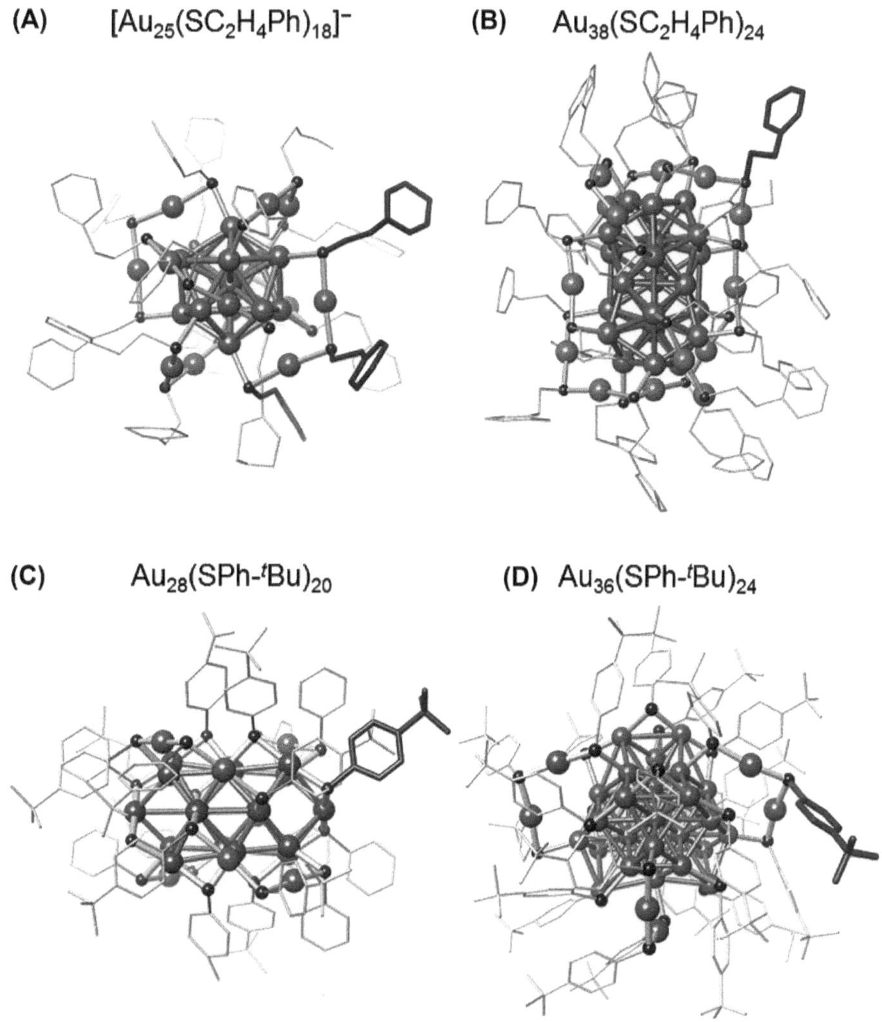

■ **FIGURE 1.5** Crystal structures of $[Au_{25}(SC_2H_4Ph)_{18}]^-$, $Au_{38}(SC_2H_4Ph)_{24}$, $Au_{28}(SPh\text{-}^tBu)_{20}$, and $Au_{36}(SPh\text{-}^tBu)_{24}$. Light gray and dark gray spheres represent Au and S atoms, respectively, while gray lines show the ligands. *The figures were adapted from Zeng C, Qian H, Li T, Li G, Rosi, NL, Yoon B, et al. Total Structure and electronic properties of the gold nanocrystal $Au_{36}(SR)_{24}$. Angew Chem Int Ed 2012;51:13114–18 with permission. Copyright 2012 Wiley-VCH Publishing; Zeng C, Li T, Das A, Rosi NL, Jin R. Chiral structure of thiolate-protected 28-gold-atom nanocluster determined by X-ray crystallography. J Am Chem Soc 2013;135: 10011–13 with permission. Copyright 2013 American Chemical Society; Heaven MW, Dass A, White PS, Holt KM, Murray RW. Crystal structure of the gold nanoparticle $[N(C_8H_{17})_4][Au_{25}(SCH_2CH_2Ph)_{18}]$. J Am Chem Soc 2008;130:3754–55 with permission. Copyright 2008 American Chemical Society; Qian H, Eckenhoff WT, Zhu Y, Pintauer T, Jin R. Total structure determination of thiolate-protected Au_{38} nanoparticles. J Am Chem Soc 2010;132:8280–81 with permission. Copyright 2010 American Chemical Society.*

will lead to the synthesis of $Au_n(SR)_m$ clusters having core structures different from those of any of the previously reported clusters.

1.3.3 Electronic Structures

Unlike bulk Au, small $Au_n(SR)_m$ clusters have discrete electronic structures, and so multiple peak structures can be observed in the optical absorption spectra of these clusters [62,66,107]. As an example, the optical absorption spectra of a series of $Au_n(SC_{12}H_{25})_m$ clusters (Fig. 1.3B) are shown in Fig. 1.6. Here, multiple peak structures are seen across the entire visible range for the clusters up to $Au_{144}(SC_{12}H_{25})_{60}$ [66]. These semiconducting properties of $Au_n(SR)_m$ clusters are used in their application to photocatalysis [108,109] and solar cells [110,111].

These fine peak structures are not observed in the case of clusters larger than $Au_{144}(SC_{12}H_{25})_{60}$, although peaks that can be attributed to surface localized plasmon absorption can be observed at approximately 520 nm in their optical absorption spectra (Fig. 1.6). Thus, the electronic structures of $Au_n(SC_{12}H_{25})_m$ clusters tend to resemble that of the bulk material at a size of $Au_{187}(SC_{12}H_{25})_{68}$ and above [66].

1.3.4 Size-Specific Physical/Chemical Properties

$Au_n(SR)_m$ clusters exhibit several size-specific physical and chemical properties, which have been discussed in detail in the literature [112–114]. Herein, we describe only the typical properties.

Photoluminescence. Small $Au_n(SR)_m$ clusters have been shown to exhibit photoluminescence [21,23,26,62,115–133]. For example, $Au_{25}(SG)_{18}$ exhibits photoluminescence at ~700 nm (Fig. 1.7A) [122] with an associated estimated quantum yield of $\sim 1 \times 10^{-3}$. This property of $Au_n(SR)_m$ clusters can be used for sensing [134] (Fig. 1.7B) and imaging applications [135].

Redox behavior. $Au_n(SR)_m$ clusters also show redox behavior [23,43]. As an example, Fig. 1.8A shows the cyclic voltammogram of $[Au_{25}(SC_2H_4Ph)_{18}]^-$ [77]. The peaks at −1.5 V and −0.6 V are derived from the redox potentials of $[Au_{25}(SC_2H_4Ph)_{18}]^{-/2-}$ and $[Au_{25}(SC_2H_4Ph)_{18}]^{0/-}$, respectively. This redox behavior is not confined to clusters with discrete electronic structures because $Au_n(SR)_m$ clusters larger than $Au_{144}(SR)_{60}$ also exhibit redox behavior owing to the quantized double-layer charging [24,136]. This redox property of $Au_n(SR)_m$ clusters could be used in their application to single-electron transistors [137] (Fig. 1.8B).

Optical activity. Several clusters, such as $Au_{38}(SR)_{24}$ and $Au_{40}(SR)_{24}$, have optical isomers [104,138,139] with different -S(R)[-Au-S(R)-]$_x$ staple

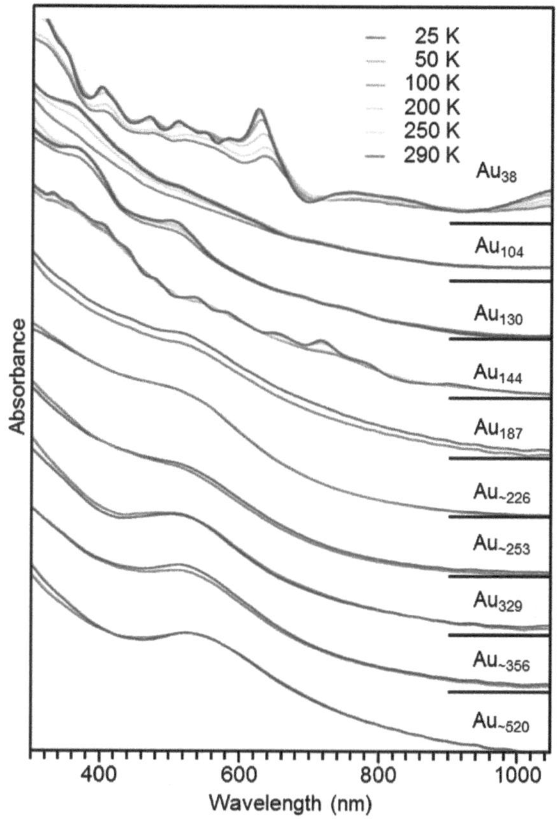

■ **FIGURE 1.6** Optical absorption spectra of the films of $Au_n(SC_{12}H_{25})_m$ clusters ($n = 38$—520) acquired at various temperatures (25—290 K). *The figure was adapted from Negishi Y, Nakazaki T, Malola S, Takano S, Niihori Y, Kurashige W, Yamazoe S, Tsukuda T, Häkkinen H. A critical size for emergence of nonbulk electronic and geometric structures in dodecanethiolate-protected Au clusters. J Am Chem Soc 2015; 137:1206—12 with permission. Copyright 2015 American Chemical Society with permission. Copyright 2015 American Chemical Society.*

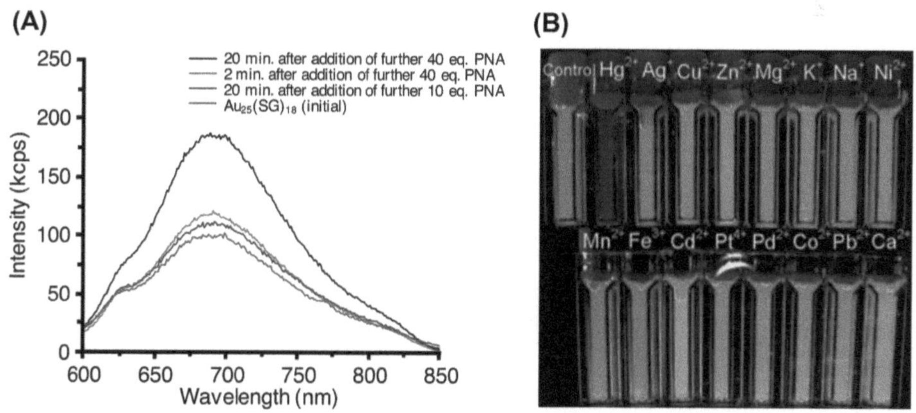

■ **FIGURE 1.7** (A) Photoluminescence spectra of $Au_{25}(SG)_{18}$ and (B) photographic images of photoluminescence of Au_{25} clusters in aqueous solution under UV light in the presence of various metal ions. *The figures were adapted from Wu Z, Jin R. On the ligand's role in the fluorescence of gold nanoclusters. Nano Lett 2010;10:2568—73 with permission. Copyright 2010 American Chemical Society; Xie J, Zheng Y, Ying JY. Highly selective and ultrasensitive detection of Hg^{2+} based on fluorescence quenching of Au nanoclusters by Hg^{2+}-Au^{+} interactions. Chem Commun 2010;46:961—63 with permission. Copyright 2010 Royal Society of Chemistry.*

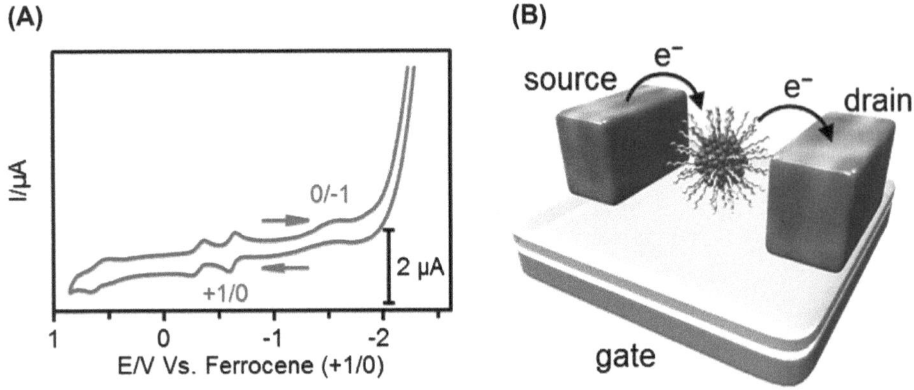

■ **FIGURE 1.8** (A) Cyclic voltammogram of $Au_{25}(SC_{12}H_{25})_{18}$ and (B) potential application of $Au_n(SR)_m$ clusters using their redox behavior. *The figure was adapted from Chaki NK, Negishi Y, Tsunoyama H, Shichibu Y, Tsukuda T. Ubiquitous 8 and 29 kDa gold: alkanethiolate cluster compounds: mass-spectrometric determination of molecular formulas and structural implications. J Am Chem Soc 2008;130:8608–10 with permission. Copyright 2008 American Chemical Society.*

($x = 1$, 2) configurations (Fig. 1.9A) and thus are optically active [39,140–142]. Fig. 1.9B shows the circular dichroism spectra of two optical isomers of $Au_{38}(SC_2H_4Ph)_{24}$ [140]. The anisotropy factor for the optical activity of this cluster is in the range of 1×10^{-3} to 4×10^{-3} and increases with the wavelength. It has also been reported that optical activity can be adopted through the use of optically active ligands [143].

Catalytic activity. Catalytic activity is another typical size-specific property of $Au_n(SR)_m$ clusters (Fig. 1.10) [25]. $Au_{25}(SR)_{18}$, for instance, catalyzes the oxidations of carbon monoxide [144,145], styrene [146,147], benzyl alcohol [148], cyclohexane [149], and sulfides [150,151]. Furthermore, $Au_{25}(SR)_{18}$

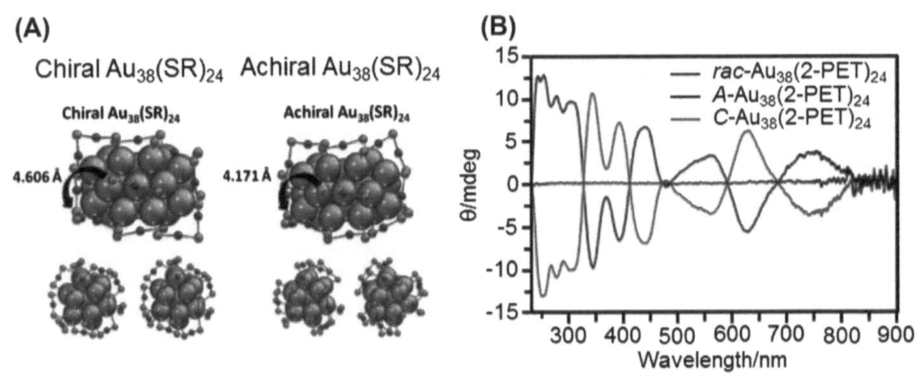

■ **FIGURE 1.9** (A) Geometrical structures and (B) circular dichroism spectra of two isomers of $Au_{38}(SC_2H_4Ph)_{24}$. *The figures were adapted from Knoppe S, Bürgi T. Chirality in thiolate-protected gold clusters. Acc Chem Res 2014;47:1318–26 with permission. Copyright 2014 American Chemical Society.*

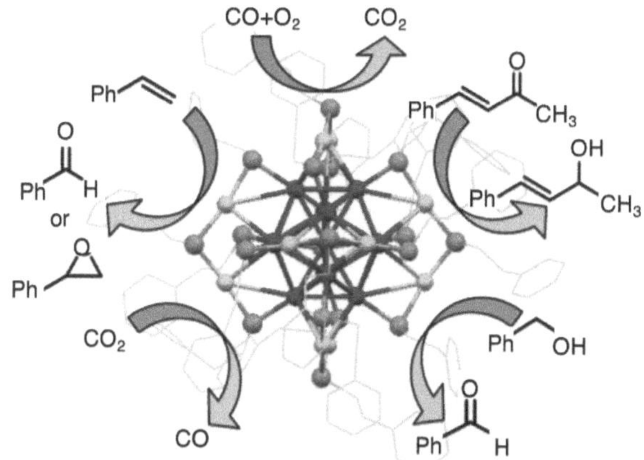

■ **FIGURE 1.10** Gold cluster-catalyzed chemical reactions. *The figure was adapted from Li G, Jin R. Atomically precise gold nanoclusters as new model catalysts. Acc Chem Res 2013;46:1749–58 with permission. Copyright 2013 American Chemical Society.*

exhibits catalytic activity for the hydrogenations of nitrophenol [152,153] and aldehydes and ketones [154–156], and also promotes C–C coupling reactions [157,158]. As noted, several $Au_n(SR)_m$ clusters have optical isomers (Fig. 1.9A) and these are expected to function as asymmetric catalysts [159].

1.4 FUNCTIONALIZATION METHODS

As described previously, $Au_n(SR)_m$ clusters show size-specific physical and chemical properties, and thus are potential candidates for new functional nanomaterials. For this reason, various methods to improve their functionalities have also been actively studied. The methods applied for this purpose can be roughly divided into the following three groups (Fig. 1.11):

1. Selection of ligand functional groups;
2. Heteroatom doping of the metal core;
3. Control of the nature of the Au-ligand bond.

Details of these three methods are presented next.

1.4.1 Selection of Ligand Functional Groups

Manipulating the functional group of the ligand is the oldest means of modulating the physical and chemical properties of $Au_n(SR)_m$ clusters (Fig. 1.11). The solubility of the cluster in a solvent or the interactions between the cluster and other materials can be controlled by choosing appropriate functional groups for the ligand. Size-specific physical and chemical

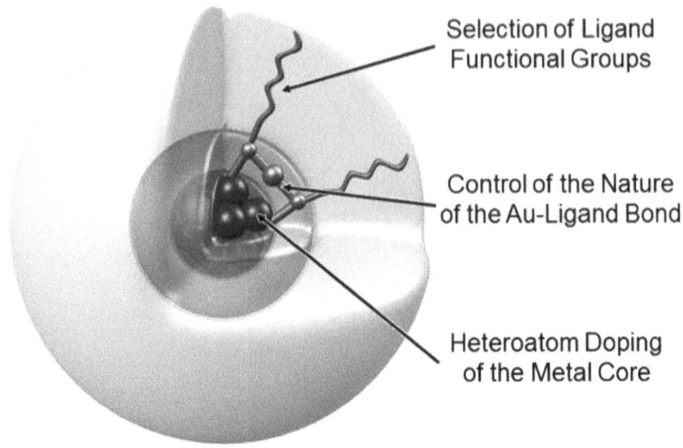

■ **FIGURE 1.11** Effective methods for controlling the physical and chemical properties of $Au_n(SR)_m$ clusters.

properties such as photoluminescence (Fig. 1.12A) [122,124], optical activity (Fig. 1.12B) [140,143], and redox behavior [160] can also be tuned through the selection of functional groups. Furthermore, new properties, such as responsivity to external stimulus [161] or molecular recognition ability [128], can be imparted to the cluster by introducing various ligand functional groups. In recent years, it has also become possible to attach specific

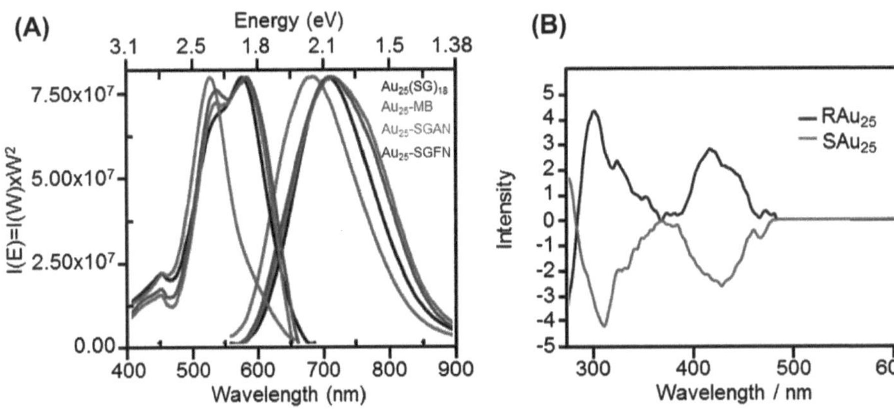

■ **FIGURE 1.12** (A) Ligand-dependent photoluminescence spectra of $Au_{25}(SR)_{18}$ and (B) circular dichroism spectra of Au_{25} clusters protected by chiral thiols (R- and S- isomers). *The figures were adapted from Zhu M, Qian H, Meng X, Jin S, Wu Z, Jin R. Chiral Au_{25} nanospheres and nanorods: synthesis and insight into the origin of chirality. Nano Lett 2011;11:3963–69 with permission. Copyright 2011 American Chemical Society; Shibu ES, Muhammed MAH, Tsukuda T, Pradeep T. Ligand exchange of $Au_{25}SG_{18}$ leading to functionalized gold clusters: spectroscopy, kinetics, and luminescence. J Phys Chem C 2008;112:12168–76 with permission. Copyright 2008 American Chemical Society.*

functional ligands while also controlling the number of such ligands [67,162–164]. It is expected that regular arrays of $Au_n(SR)_m$ clusters with specific nanoarchitectures [165] will also become possible with such techniques.

1.4.2 Heteroatom Doping of the Metal Core

The doping of heteroatoms into small metal clusters modifies the physical and chemical properties of the clusters [166–169]. Therefore, the doping of $Au_n(SR)_m$ clusters is also an effective means of imparting new properties to these clusters (Fig. 1.11). $Au_n(SR)_m$ clusters doped with Ag [170–180], Cu [180–184], Pt [185,186], Pd [178,184–193], Cd [194,195], or Hg [194–196] have been synthesized to date. Studies of these modified clusters have examined the effects of doping and provided direction for modulating the physical and chemical characteristics of the clusters with this method.

One case in which this technique has been applied involves the introduction of multiple Ag atoms into $Au_{25}(SR)_{18}$. The highest occupied molecular orbital-lowest unoccupied molecular orbital (HOMO-LUMO) gap of the cluster continuously increases (Fig. 1.13A) while its photoluminescence emission wavelength continuously shifts to longer wavelengths upon Ag doping [170]. In contrast, Cu doping reduces the HOMO-LUMO gap of the cluster (Fig. 1.13B) [181]. These results indicate that doping with Ag and Cu is an effective method for continuously modulating the electronic structure of a cluster. This technique could therefore be used to tune the HOMO-LUMO gap of a cluster for a particular purpose.

Only a single Pd or Pt atom can be doped into a $Au_{25}(SR)_{18}$ cluster, but this doping enhances the stability of the cluster against degradation in solution under severe conditions, such as high temperature or the presence of H_2O_2 (Fig. 1.13C) [185,188]. In addition, Pt doping of $Au_{25}(SC_2H_4Ph)_{18}$ enhances the catalytic activity of the cluster for the oxidation reaction of styrene [185], while Pd doping of $Au_{25}(SC_{12}H_{25})_{18}$ improves the reactivity of the cluster in ligand exchange reactions (Fig. 1.13D) [197]. It can be concluded that Pd and Pt doping is an effective means of creating $Au_{25}(SC_{12}H_{25})_{18}$ clusters with greater stability and higher reactivity.

In the future, doping is also expected to be established for elements other than Ag, Cu, Pt, Pd, Cd, and Hg. If this can be achieved, the scope of this method will be greatly expanded with the aim of imparting new functions to clusters and creating stable nanomaterials with desired functions for specific applications.

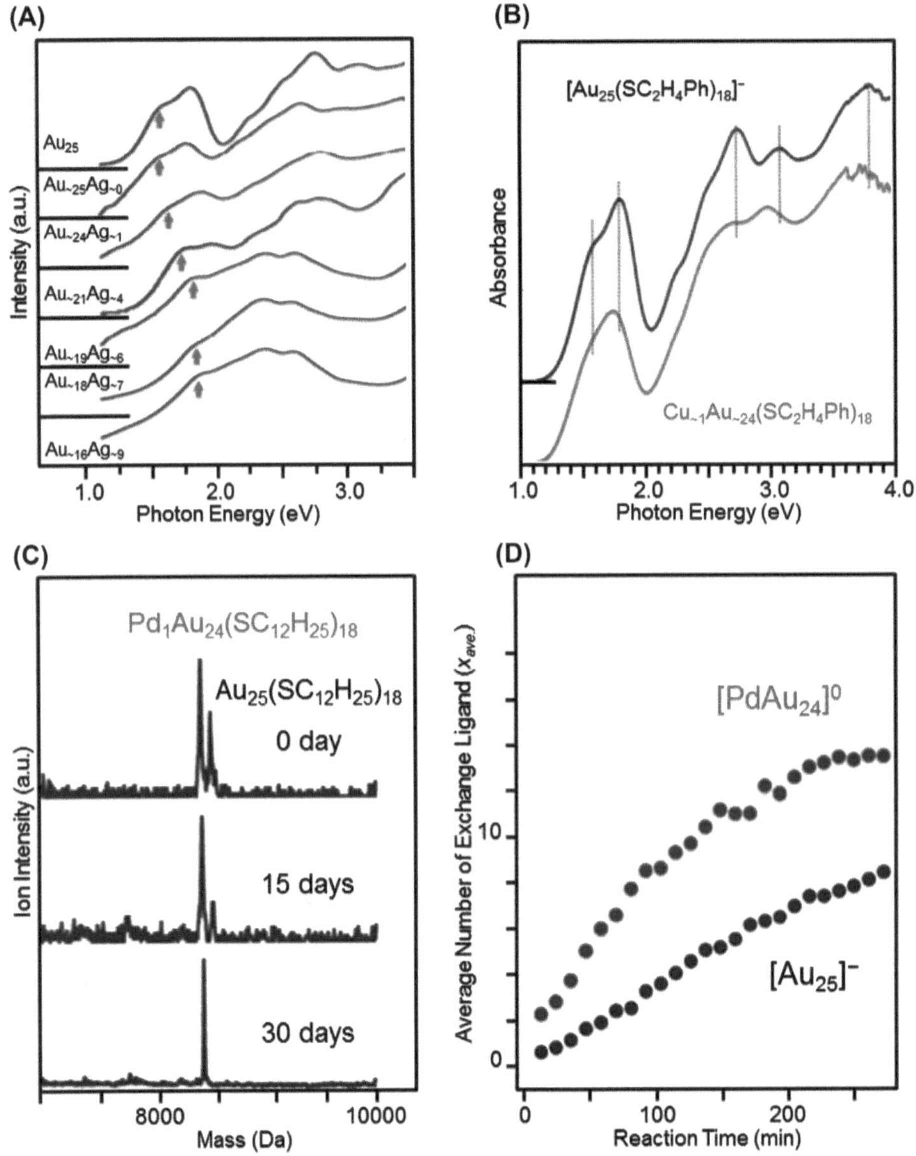

FIGURE 1.13 Effects of heteroatom doping of $Au_{25}(SR)_{18}$. (A) Optical absorption spectra of $Au_{25-n}Ag_n(SC_{12}H_{25})_{18}$ and (B) $Au_{25-n}Cu_n(SC_{12}H_{25})_{18}$. The peak in the vicinity of 1.5—1.8 eV can be attributed to a transition from the highest occupied molecular orbital (HOMO; the occupied Au 6sp orbital in the Au_{13} core) to the lowest unoccupied molecular orbital (LUMO; the unoccupied Au 6sp orbital in the Au_{13} core). (C) Comparison of stability in solution at 50°C and (D) ligand exchange reactivity between $Au_{24}Pd(SR)_{18}$ and $Au_{25}(SR)_{18}$. *The figures were adapted from Negishi Y, Iwai T, Ide M. Continuous modulation of electronic structure of stable thiolate-protected Au_{25} cluster by Ag doping. Chem Commun 2010;46:4713—15 with permission. Copyright 2010 Royal Society of Chemistry; Negishi Y, Munakata K, Ohgake W, Nobusada K. Effect of copper doping on electronic structure, geometric structure, and stability of thiolate-protected Au_{25} nanoclusters. J Phys Chem Lett 2012;3:2209—14 with permission. Copyright 2012 American Chemical Society; Negishi Y, Kurashige W, Niihori Y, Iwasa T, Nobusada K. Isolation, structure, and stability of a dodecanethiolate-protected Pd_1Au_{24} cluster. Phys Chem Chem Phys 2010;12:6219—25 with permission. Copyright 2012 Royal Society of Chemistry; Niihori Y, Kurashige W, Matsuzaki M, Negishi Y. Remarkable enhancement in ligand-exchange reactivity of thiolate-protected Au_{25} nanoclusters by Single Pd atom doping. Nanoscale 2013;5:508—12 with permission. Copyright 2013 Royal Society of Chemistry.*

1.4.3 Control of the Nature of the Au-Ligand Bond

Studies of two-dimensional self-assembled monolayers (SAMs) have shown a decrease in the charge transfer from Au to the ligand in Au−selenolate (SeR) compared with Au−SR. It was also found that changing the nature of the bonding interaction in this manner results in a stronger Au−SeR bond compared with the Au−SR bond [198,199], as well as higher conductivity in the Au−SeR bond than in the Au−SR bond [200,201]. This type of change in the bonding nature and the resulting variations in the physical and chemical properties are expected to occur even in clusters. Therefore, it can be predicted that modification of the bonding nature between the metal and the ligand will also be effective for imparting new properties (Fig. 1.11). In fact, changing the ligand in Au_{25} clusters from SR to SeR decreases the charge transfer from Au to the ligand in a manner similar to that observed in an SAM (Fig. 1.14A) [202,203]. Accordingly, the stability of the cluster against degradation in solution is also improved for Au_{25} clusters (Fig. 1.14B) [204]. These results demonstrate that the use of selenolate as a ligand can generate clusters with greater stability to degradation in solution compared with $Au_n(SR)_m$ clusters.

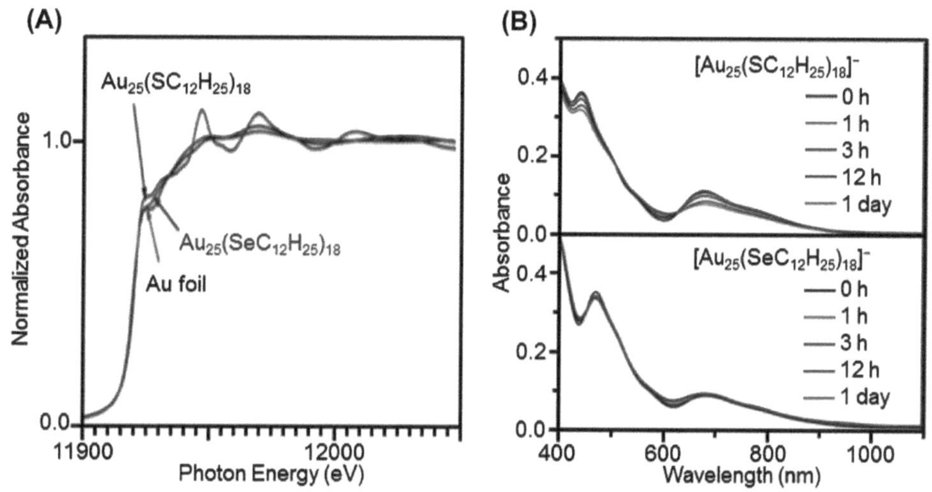

■ **FIGURE 1.14** (A) Au L_3-edge XANES spectra of $Au_{25}(SC_{12}H_{25})_{18}$, $Au_{25}(SeC_{12}H_{25})_{18}$, and Au foil. In the case of Au−S bonding, charge transfer from the Au 5d to the S 3d orbital takes place because of the difference in the electronegativities of the two elements. Therefore, the 5d electron density of $Au_n(SR)_m$ is lower than that of the Au foil, resulting in an increase in the white-line peak intensity, as seen for $Au_{25}(SC_{12}H_{25})_{18}$. The white-line peak intensity of $Au_{25}(SeC_{12}H_{25})_{18}$ is lower than that of $Au_{25}(SC_{12}H_{25})_{18}$, indicating that charge transfer is suppressed in Au−Se bonding compared with Au−S bonding. (B) Variations in the absorption spectra of $Au_{25}(SC_{12}H_{25})_{18}$ and $Au_{25}(SeC_{12}H_{25})_{18}$ toluene solutions over time at 60°C. *The figures were adapted from Kurashige W, Yamazoe S, Kanehira K, Tsukuda T, Negishi Y. Selenolate-protected Au_{38} nanoclusters: isolation and structural characterization. J Phys Chem Lett 2013;4:3181−85 with permission. Copyright 2013 American Chemical Society; Kurashige W, Yamaguchi M, Nobusada K, Negishi Y. Ligand-induced stability of gold nanoclusters: thiolate versus selenolate. J Phys Chem Lett 2012;3:2649−52 with permission. Copyright 2012 American Chemical Society.*

In addition to an improvement in stability, the use of selenolate ligands is expected to induce greater conductivity between the gold core and the ligands [200,205]. This is desirable when designing molecular electronic devices with this type of cluster. However, at present, the details of this effect have not yet been elucidated. In the future, the impact of the use of selenolate ligands on the conductivity between the gold core and the ligands is expected to be clarified by comparing the electrical conductivity between precisely synthesized $Au_n(SeR)_m$ and $Au_n(SR)_m$ clusters. It is also believed that the use of telluronate (TeR) as a ligand will generate still greater improvements in conductivity [206]. Thus, we expect that precise synthesis methods for $Au_n(TeR)_m$ clusters [207–209] will eventually be established to allow the fabrication of highly conductive stable clusters.

1.5 SUMMARY AND PERSPECTIVE

In this chapter, we focused on $Au_n(SR)_m$ clusters as an example of metal nanoclusters and described the newest techniques and knowledge regarding their precise synthesis, stability, electronic and geometrical structures, and physical and chemical properties, as well as their functionalization methods. It is evident that the study of $Au_n(SR)_m$ clusters has progressed with spectacular speed in recent years. As such, the associated synthetic techniques have also dramatically advanced, and a greater understanding of their characteristics has been obtained. It is expected that $Au_n(SR)_m$ clusters will be applied in various fields, such as sensing [134], imaging [135], cancer radiation therapy [210,211], catalysts [25,144–158], photocatalysts [108,109], solar cells [110,111], fuel cells [60], photosensitizers [151], and single electron devices [137]. If these $Au_n(SR)_m$ clusters can be regularly assembled, further new functions could be induced and their field of application might be further expanded. In the future, it is likely that intensive investigations will be conducted regarding the formation of various nanoarchitectures [165] using $Au_n(SR)_m$ clusters as structural units in addition to research regarding the $Au_n(SR)_m$ clusters themselves.

REFERENCES

[1] Corain B, Schmid G, Toshima N, editors. Metal nanoclusters in catalysis and materials science: the issue of size control. Amsterdam: Elsevier; 2007.

[2] McPartlin M, Mason R, Malatesta L. Novel cluster complexes of gold(0)-gold(I). Chem Commun 1969:334.

[3] Briant CE, Theobald BRC, White JW, Bell LK, Mingos DMP, Welch AJ. Synthesis and X-ray structural characterization of the centred icosahedral gold cluster compound $[Au_{13}(PMe_2Ph)_{10}Cl_2](PF_6)_3$; the realization of a theoretical prediction. J Chem Soc Chem Commun 1981;5:201–2.

[4] Teo BK, Shi X, Zhang H. Pure gold cluster of 1:9:9:1:9:9:1 layered structure: a novel 39-metal-atom cluster $[(Ph_3P)_{14}Au_{39}Cl_6]Cl_2$ with an interstitial gold atom in a hexagonal antiprismatic cage. J Am Chem Soc 1992;114:2743–5.

[5] Teo BK, Zhang H. Polyicosahedricity: icosahedron to icosahedron of icosahedra growth pathway for bimetallic (Au- Ag) and trimetallic (Au- Ag-M; M = Pt, Pd, Ni) supraclusters; synthetic strategies, site preference, and stereochemical principles. Coord Chem Rev 1995;143:611–36.

[6] Schmid G. Large clusters and colloids. Metals in the embryonic state. Chem Rev 1992;92:1709–27.

[7] Craighead KL, Felicissimo AMP, Krogstad DA, Nelson LTJ, Pignolet LH. Halide derivatives of palladium-gold cluster compounds. X-ray crystal and molecular structure of $(I)Pd(AuPPh_3)_7(AuI)_2$. Inorg Chim Acta 1993;212:31–9.

[8] Balasubramanian R, Guo R, Mills AJ, Murray RW. Reaction of $Au_{55}(PPh_3)_{12}Cl_6$ with thiols yields thiolate monolayer protected Au_{75} clusters. J Am Chem Soc 2005;127:8126–32.

[9] Yanagimoto Y, Negishi Y, Fujihara H, Tsukuda T. Chiroptical activity of BINAP-stabilized undecagold clusters. J Phys Chem B 2006;110:11611–4.

[10] Liu Y, Tsunoyama H, Akita T, Tsukuda T. Preparation of ~1 nm gold clusters confined within mesoporous silica and microwave-assisted catalytic application for alcohol oxidation. J Phys Chem C 2009;113:13457–61.

[11] Inomata T, Konishi K. Gold nanocluster confined within a cage: template-directed formation of a hexaporphyrin cage and its confinement capability. Chem Commun 2003:1282–3.

[12] Shichibu Y, Konishi K. HCl-induced nuclearity convergence in diphosphine-protected ultrasmall gold clusters: a novel synthetic route to "magic-number" Au_{13} Clusters. Small 2010;6:1216–20.

[13] Kamei Y, Shichibu Y, Konishi K. Generation of small gold clusters with unique geometries through cluster-to-cluster transformations: octanuclear clusters with edge-sharing gold tetrahedron motifs. Angew Chem Int Ed 2011;50:7442–5.

[14] Shichibu Y, Kamei Y, Konishi K. Unique [core+two] structure and optical property of a dodeca-ligated undecagold cluster: critical contribution of the exo gold atoms to the electronic structure. Chem Commun 2012;48:7559–61.

[15] Shichibu Y, Suzuki K, Konishi K. Facile synthesis and optical properties of magic-number Au_{13} clusters. Nanoscale 2012;4:4125–9.

[16] Shichibu Y, Zhang M, Kamei Y, Konishi K. $[Au_7]^{3+}$: a missing link in the four-electron gold cluster family. J Am Chem Soc 2014;136:12892–5.

[17] Kamei Y, Robertson N, Shichibu Y, Konishi K. Impact of skeletal isomerization of ultrasmall gold clusters on electrochemical properties: voltammetric profiles of nonspoked octanuclear clusters. J Phys Chem C 2015;119:10995–9.

[18] Kang X, Song Y, Deng H, Zhang J, Liu B, Pan C, et al. Ligand-induced change of the crystal structure and enhanced stability of the Au_{11} nanocluster. RSC Adv 2015; 5:66879–85.

[19] Kurashige W, Negishi Y. Synthesis, stability, and photoluminescence properties of $PdAu_{10}(PPh_3)_8Cl_2$ clusters. J Clust Sci 2012;23:365–74.

[20] Brust M, Walker M, Bethell D, Schiffrin DJ, Whyman R. Synthesis of thiol-derivatised gold nanoparticles in a two-phase liquid-liquid system. J Chem Soc Chem Commun 1994:801–2.

[21] Tsukuda T. Toward an atomic-level understanding of size-specific properties of protected and stabilized gold clusters. Bull Chem Soc Jpn 2012;85:151–68.

[22] Whetten RL, Shafigullin MN, Khoury JT, Schaaff TG, Vezmar I, Alvarez MM, et al. Crystal structures of molecular gold nanocrystal arrays. Acc Chem Res 1999;32:397–406.

[23] Parker JF, Fields-Zinna CA, Murray RW. The story of a monodisperse gold nanoparticle: $Au_{25}L_{18}$. Acc Chem Res 2010;43:1289–96.

[24] Murray RW. Nanoelectrochemistry: metal nanoparticles, nanoelectrodes, and nanopores. Chem Rev 2008;108:2688–720.

[25] Li G, Jin R. Atomically precise gold nanoclusters as new model catalysts. Acc Chem Res 2013;46:1749–58.

[26] Qian H, Zhu M, Wu Z, Jin R. Quantum sized gold nanoclusters with atomic precision. Acc Chem Res 2012;45:1470–9.

[27] Dass A. Nano-scaling law: geometric foundation of thiolated gold nanomolecules. Nanoscale 2012;4:2260–3.

[28] Negishi Y, Kurashige W, Niihori Y, Nobusada K. Toward the creation of stable, functionalized metal clusters. Phys Chem Chem Phys 2013;15:18736–51.

[29] Negishi Y. Toward the creation of functionalized metal nanoclusters and highly active photocatalytic materials using thiolate-protected magic gold clusters. Bull Chem Soc Jpn 2014;87:375–89.

[30] Luo Z, Nachammai V, Zhang B, Yan N, Leong DT, Jiang D-E, et al. Toward understanding the growth mechanism: tracing all stable intermediate species from reduction of Au(I)–thiolate complexes to evolution of Au_{25} nanoclusters. J Am Chem Soc 2014;136:10577–80.

[31] Häkkinen H. The gold–sulfur interface at the nanoscale. Nat Chem 2012;4: 443–55.

[32] Fernando A, Weerawardene KLDM, Karimova NV, Aikens CM. Quantum mechanical studies of large metal, metal oxide, and metal chalcogenide nanoparticles and clusters. Chem Rev 2015;115:6112–216.

[33] Sánchez-Castillo A, Noguez C, Garzón IL. On the origin of the optical activity displayed by chiral-ligand-protected metallic nanoclusters. J Am Chem Soc 2010;132:1504–5.

[34] Pei Y, Zeng XC. Investigating the structural evolution of thiolate protected gold clusters from first-principles. Nanoscale 2012;4:4054–72.

[35] Zhang P. X-ray spectroscopy of gold-thiolate nanoclusters. J Phys Chem C 2014; 118:25291–9.

[36] Jiang D-E, Kühn M, Tang Q, Weigend F. Superatomic orbitals under spin-orbit coupling. J Phys Chem Lett 2014;5:3286–9.

[37] Jadzinsky PD, Calero G, Ackerson CJ, Bushnell DA, Kornberg RD. Structure of a thiol monolayer-protected gold nanoparticle at 1.1 Å resolution. Science 2007;318: 430–3.

[38] Knoppe S, Wong OA, Malola S, Häkkinen H, Bürgi T, Verbiest T, et al. Chiral phase transfer and enantioenrichment of thiolate-protected Au_{102} Clusters. J Am Chem Soc 2014;136:4129–32.

[39] Dolamic I, Knoppe S, Dass A, Bürgi T. First enantioseparation and circular dichroism spectra of Au_{38} clusters protected by achiral ligands. Nat Commun 2012;3:798.

[40] Udayabhaskararao T, Pradeep T. New protocols for the synthesis of stable Ag and Au nanocluster molecules. J Phys Chem Lett 2013;4:1553−64.

[41] Dainese T, Antonello S, Gascón JA, Pan F, Perera NV, Ruzzi M, et al. $Au_{25}(SEt)_{18}$, a nearly naked thiolate-protected Au_{25} cluster: structural analysis by single crystal X-ray crystallography and electron nuclear double resonance. ACS Nano 2014;8:3904−12.

[42] Li Y, Zaluzhna O, Tong YYJ. Critical role of water and the structure of inverse micelles in the brust-schiffrin synthesis of metal nanoparticles. Langmuir 2011;27:7366−70.

[43] Kwak K, Kumar SS, Pyo K, Lee D. Ionic liquid of a gold nanocluster: a versatile matrix for electrochemical biosensors. ACS Nano 2014;8:671−9.

[44] Yao H. On the electronic structures of $Au_{25}(SR)_{18}$ clusters studied by magnetic circular dichroism spectroscopy. J Phys Chem Lett 2012;3:1701−6.

[45] Kouchi H, Kawasaki H, Arakawa R. A new matrix of MALDI-TOF MS for the analysis of thiolate-protected gold clusters. Anal Methods 2012;4:3600−3.

[46] Sakai N, Ikeda T, Teranishi T, Tatsuma T. Sensitization of TiO_2 with Pt, Pd, and Au clusters protected by mercapto- and dimercaptosuccinic acid. ChemPhysChem 2011;12:2415−8.

[47] Goulet PJG, Lennox RB. New insights into brust-schiffrin metal nanoparticle synthesis. J Am Chem Soc 2010;132:9582−4.

[48] Tlahuice-Flores A, Whetten RL, Jose-Yacaman M. Ligand effects on the structure and the electronic optical properties of anionic $Au_{25}(SR)_{18}$ clusters. J Phys Chem C 2013;117:20867−75.

[49] Varnavski O, Ramakrishna G, Kim J, Lee D, Goodson T. Critical size for the observation of quantum confinement in optically excited gold clusters. J Am Chem Soc 2010;132:16−7.

[50] Bahena D, Bhattarai N, Santiago U, Tlahuice A, Ponce A, Bach SBH, et al. STEM electron diffraction and high-resolution images used in the determination of the crystal structure of the $Au_{144}(SR)_{60}$ cluster. J Phys Chem Lett 2013;4:975−81.

[51] Philip R, Chantharasupawong P, Qian H, Jin R, Thomas J. Evolution of nonlinear optical properties: from gold atomic clusters to plasmonic nanocrystals. Nano Lett 2012;12:4661−7.

[52] Krommenhoek PJ, Wang J, Hentz N, Johnston-Peck AC, Kozek KA, Kalyuzhny G, et al. Bulky adamantanethiolate and cyclohexanethiolate ligands favor smaller gold nanoparticles with altered discrete sizes. ACS Nano 2012;6:4903−11.

[53] Swanick KN, Hesari M, Workentin MS, Ding Z. Interrogating near-infrared electrogenerated chemiluminescence of $Au_{25}(SC_2H_4Ph)_{18}^{+}$ clusters. J Am Chem Soc 2012;134:15205−8.

[54] Matthiesen JE, Jose D, Sorensen CM, Klabunde KJ. Loss of hydrogen upon exposure of thiol to gold clusters at low temperature. J Am Chem Soc 2012;134:9376−9.

[55] Wang ZW, Toikkanen O, Quinn BM, Palmer RE. Real-space observation of prolate monolayer-protected Au_{38} clusters using aberration-corrected scanning transmission electron microscopy. Small 2011;7:1542−5.

[56] Yao C, Chen J, Li M-B, Liu L, Yang J, Wu Z. Adding two active silver atoms on Au_{25} nanoparticle. Nano Lett 2015;15:1281−7.

[57] Suda M, Kameyama N, Suzuki M, Kawamura N, Einaga Y. Reversible phototuning of ferromagnetism at Au-S interfaces at room temperature. Angew Chem Int Ed 2008;47:160–3.
[58] Koivisto J, Malola S, Kumara C, Dass A, Häkkinen H, Pettersson M. Experimental and theoretical determination of the optical gap of the $Au_{144}(SC_2H_4Ph)_{60}$ cluster and the $(Au/Ag)_{144}(SC_2H_4Ph)_{60}$ nanoalloys. J Phys Chem Lett 2012;3:3076–80.
[59] Zhu M, Qian H, Jin R. Thiolate-protected $Au_{24}(SC_2H_4Ph)_{20}$ nanoclusters: superatoms or not? J Phys Chem Lett 2010;1:1003–7.
[60] Chen W, Chen S. Oxygen electroreduction catalyzed by gold nanoclusters: strong core size effects. Angew Chem Int Ed 2009;48:4386–9.
[61] Wu Z, Jiang D-E, Mann AKP, Mullins DR, Qiao Z-A, Allard LF, et al. Thiolate ligands as a double-edged sword for CO oxidation on CeO_2 supported $Au_{25}(SCH_2CH_2Ph)_{18}$ nanoclusters. J Am Chem Soc 2014;136:6111–22.
[62] Negishi Y, Nobusada K, Tsukuda T. Glutathione-protected gold clusters revisited: bridging the gap between gold(I)-thiolate complexes and thiolate-protected gold nanocrystals. J Am Chem Soc 2005;127:5261–70.
[63] Negishi Y, Takasugi Y, Sato S, Yao H, Kimura K, Tsukuda T. Magic-numbered Au_n clusters protected by glutathione monolayers (n = 18, 21, 25, 28, 32, 39): isolation and spectroscopic characterization. J Am Chem Soc 2004;126:6518–9.
[64] Negishi Y, Takasugi Y, Sato S, Yao H, Kimura K, Tsukuda T. Kinetic stabilization of growing gold clusters by passivation with thiolates. J Phys Chem B 2006;110:12218–21.
[65] Negishi Y, Sakamoto C, Ohyama T, Tsukuda T. Synthesis and the origin of the stability of thiolate-protected Au_{130} and Au_{187} clusters. J Phys Chem Lett 2012;3:1624–8.
[66] Negishi Y, Nakazaki T, Malola S, Takano S, Niihori Y, Kurashige W, Yamazoe S, Tsukuda T, Häkkinen H. A critical size for emergence of nonbulk electronic and geometric structures in dodecanethiolate-protected Au clusters. J Am Chem Soc 2015;137:1206–12.
[67] Niihori Y, Uchida C, Kurashige W, Negishi Y. High-resolution separation of thiolate-protected gold clusters by reversed-phase high-performance liquid chromatography. Phys Chem Chem Phys 2016;18:4251–65.
[68] Schaaff TG, Knight G, Shafigullin MN, Borkman RF, Whetten RL. Isolation and selected properties of a 10.4 kDa gold: glutathione cluster compound. J Phys Chem B 1998;102:10643–6.
[69] Schaaff TG, Whetten RL. Giant gold-glutathione cluster compounds: intense optical activity in metal-based transitions. J Phys Chem B 2000;104:2630–41.
[70] Yao H, Miki K, Nishida N, Sasaki A, Kimura K. Large optical activity of gold nanocluster enantiomers induced by a pair of optically active penicillamines. J Am Chem Soc 2005;127:15536–43.
[71] Kimura K, Sugimoto N, Sato S, Yao H, Negishi Y, Tsukuda T. Size determination of gold clusters by polyacrylamide gel electrophoresis in a large cluster region. J Phys Chem C 2009;113:14076–82.
[72] Yu Y, Luo Z, Chevrier DM, Leong DT, Zhang P, Jiang D-E, et al. Identification of a highly luminescent $Au_{22}(SG)_{18}$ nanocluster. J Am Chem Soc 2014;136:1246–9.

[73] Whetten RL, Khoury JT, Alvarez MM, Murthy S, Vezmar I, Wang ZL, et al. Nanocrystal gold molecules. Adv Mater 1996;8:428–33.

[74] Schaaff TG, Shafigullin MN, Khoury JT, Vezmar I, Whetten RL, Cullen WG, et al. Isolation of smaller nanocrystal Au molecules: robust quantum effects in optical spectra. J Phys Chem B 1997;101:7885–91.

[75] Donkers RL, Lee D, Murray RW. Synthesis and isolation of the molecule-like cluster $Au_{38}(PhCH_2CH_2S)_{24}$. Langmuir 2004;20:1945–52.

[76] Negishi Y, Chaki NK, Shichibu Y, Whetten RL, Tsukuda T. Origin of magic stability of thiolated gold clusters: a case study on $Au_{25}(SC_6H_{13})_{18}$. J Am Chem Soc 2007;129:11322–3.

[77] Chaki NK, Negishi Y, Tsunoyama H, Shichibu Y, Tsukuda T. Ubiquitous 8 and 29 kDa gold: alkanethiolate cluster compounds: mass-spectrometric determination of molecular formulas and structural implications. J Am Chem Soc 2008;130:8608–10.

[78] Kumara C, Zuo X, Cullen DA, Dass A. Faradaurate-940: synthesis, mass spectrometry, electron microscopy, high-energy X-ray diffraction, and X-ray scattering study of $Au_{\sim 940\pm 20}(SR)_{\sim 160\pm 4}$ nanocrystals. ACS Nano 2014;8: 6431–9.

[79] Wolfe RL, Murray RW. Analytical evidence for the monolayer-protected cluster $Au_{225}[(S(CH_2)_5CH_3)]_{75}$. Anal Chem 2006;78:1167–73.

[80] Choi MMF, Douglas AD, Murray RW. Ion-pair chromatographic separation of water-soluble gold monolayer-protected clusters. Anal Chem 2006;78:2779–85.

[81] Shichibu Y, Negishi Y, Tsunoyama H, Kanehara M, Teranishi T, Tsukuda T. Extremely high stability of glutathionate-protected Au_{25} clusters against core etching. Small 2007;3:835–9.

[82] Nimmala PR, Jupally VR, Dass A. Core size conversion: route for exclusive synthesis of Au_{38} or Au_{40} nanomolecules. Langmuir 2014;30:2490–7.

[83] Dharmaratne AC, Krick T, Dass A. Nanocluster size evolution studied by mass spectrometry in room temperature $Au_{25}(SR)_{18}$ synthesis. J Am Chem Soc 2009; 131:13604–5.

[84] Jin R, Qian H, Wu Z, Zhu Y, Zhu M, Mohanty A, et al. Size focusing: a methodology for synthesizing atomically precise gold nanoclusters. J Phys Chem Lett 2010;1:2903–10.

[85] Qian H, Zhu Y, Jin R. Size-focusing synthesis, optical and electrochemical properties of monodisperse $Au_{38}(SC_2H_4Ph)_{24}$ nanoclusters. ACS Nano 2009;3: 3795–803.

[86] Shichibu Y, Negishi Y, Tsukuda T, Teranishi T. Large-scale synthesis of thiolated Au_{25} clusters via ligand exchange reactions of phosphine-stabilized Au_{11} clusters. J Am Chem Soc 2005;127:13464–5.

[87] Yu Y, Luo Z, Yu Y, Lee JY, Xie J. Observation of cluster size growth in CO-directed synthesis of $Au_{25}(SR)_{18}$ nanoclusters. ACS Nano 2012;6:7920–7.

[88] Yu Y, Chen X, Yao Q, Yu Y, Yan N, Xie J. Scalable and precise synthesis of thiolated Au_{10-12}, Au_{15}, Au_{18}, and Au_{25} nanoclusters via pH controlled CO reduction. Chem Mater 2013;25:946–52.

[89] Takano S, Yamazoe S, Koyasu K, Tsukuda T. Slow-reduction synthesis of a thiolate-protected one-dimensional gold cluster showing an intense near-infrared absorption. J Am Chem Soc 2015;137:7027–30.

[90] Nishigaki J, Tsunoyama R, Tsunoyama H, Ichikuni N, Yamazoe S, Negishi Y, et al. A new binding motif of sterically demanding thiolates on a gold cluster. J Am Chem Soc 2012;134:14295–7.

[91] Zeng C, Chen Y, Das A, Jin R. Transformation chemistry of gold nanoclusters: from one stable size to another. J Phys Chem Lett 2015;6:2976–86.

[92] Zeng C, Qian H, Li T, Li G, Rosi NL, Yoon B, et al. Total Structure and electronic properties of the gold nanocrystal $Au_{36}(SR)_{24}$. Angew Chem Int Ed 2012;51: 13114–8.

[93] Zeng C, Liu C, Pei Y, Jin R. Thiol ligand-induced transformation of $Au_{38}(SC_2H_4Ph)_{24}$ to $Au_{36}(SPh\text{-}t\text{-}Bu)_{24}$. ACS Nano 2013;7:6138–45.

[94] Zeng C, Li T, Das A, Rosi NL, Jin R. Chiral structure of thiolate-protected 28-gold-atom nanocluster determined by X-ray crystallography. J Am Chem Soc 2013;135: 10011–3.

[95] Nimmala PR, Theivendran S, Barcaro G, Sementa L, Kumara C, Jupally VR, et al. Transformation of $Au_{144}(SCH_2CH_2Ph)_{60}$ to $Au_{133}(SPh\text{-}tBu)_{52}$ nanomolecules: theoretical and experimental study. J Phys Chem Lett 2015;6:2134–9.

[96] Zeng C, Chen Y, Kirschbaum K, Appavoo K, Sfeir MY, Jin R. Structural patterns at all scales in a nonmetallic chiral $Au_{133}(SR)_{52}$ nanoparticle. Sci Adv 2015;1: e1500045.

[97] Tsukuda T, Häkkinen H. Protected metal clusters: from fundamentals to applications. Amsterdam: Elsevier; 2015.

[98] Kumara C, Dass A. $Au_{329}(SR)_{84}$ nanomolecules: compositional assignment of the 76.3 kDa plasmonic faradaurates. Anal Chem 2014;86:4227–32.

[99] Qian H, Zhu Y, Jin R. Atomically precise gold nanocrystal molecules with surface plasmon resonance. Proc Natl Acad Sci USA 2012;109:696–700.

[100] Das A, Li T, Nobusada K, Zeng C, Rosi NL, Jin R. Nonsuperatomic $[Au_{23}(SC_6H_{11})_{16}]^-$ nanocluster featuring bipyramidal Au_{15} kernel and trimeric $Au_3(SR)_4$ motif. J Am Chem Soc 2013;135:18264–7.

[101] Walter M, Akola J, Lopez-Acevedo O, Jadzinsky PD, Calero G, Ackerson CJ, et al. A unified view of ligand-protected gold clusters as superatom complexes. Proc Natl Acad Sci USA 2008;105:9157–62.

[102] Heaven MW, Dass A, White PS, Holt KM, Murray RW. Crystal structure of the gold nanoparticle $[N(C_8H_{17})_4][Au_{25}(SCH_2CH_2Ph)_{18}]$. J Am Chem Soc 2008; 130:3754–5.

[103] Zhu M, Aikens CM, Hollander FJ, Schatz GC, Jin R. Correlating the crystal structure of a thiol-protected Au_{25} cluster and optical properties. J Am Chem Soc 2008;130:5883–5.

[104] Qian H, Eckenhoff WT, Zhu Y, Pintauer T, Jin R. Total structure determination of thiolate-protected Au_{38} nanoparticles. J Am Chem Soc 2010;132:8280–1.

[105] Chen Y, Zeng C, Liu C, Kirschbaum K, Gayathri C, Gil RR, et al. Crystal structure of barrel-shaped chiral $Au_{130}(p\text{-}MBT)_{50}$ nanocluster. J Am Chem Soc 2015;137: 10076–9.

[106] Liu C, Li T, Li G, Nobusada K, Zeng C, Pang G, et al. Observation of body-centered cubic gold nanocluster. Angew Chem Int Ed 2015;54:9826–9.

[107] Jin R. Atomically precise metal nanoclusters: stable sizes and optical properties. Nanoscale 2015;7:1549–65.

[108] Stamplecoskie KG, Kamat PV. Size-dependent excited state behavior of glutathione-capped gold clusters and their light-harvesting capacity. J Am Chem Soc 2014;136:11093–9.

[109] Yu C, Li G, Kumar S, Kawasaki H, Jin R. Stable $Au_{25}(SR)_{18}/TiO_2$ composite nanostructure with enhanced visible light photocatalytic activity. J Phys Chem Lett 2013;4:2847–52.

[110] Sakai N, Tatsuma T. Photovoltaic properties of glutathione-protected gold clusters adsorbed on TiO_2 electrodes. Adv Mater 2010;22:3185–8.

[111] Nakata K, Sugawara S, Kurashige W, Negishi Y, Nagata M, Uchida S, et al. Cosensitization properties of glutathione-protected Au_{25} cluster on ruthenium dye-sensitized TiO_2 photoelectrode. Int J Photoenergy 2013;2013:456583.

[112] Chen L-Y, Wang C-W, Yuan Z, Chang H-T. Fluorescent gold nanoclusters: recent advances in sensing and imaging. Anal Chem 2015;87:216–29.

[113] Yang X, Yang M, Pang B, Vara M, Xia Y. Gold nanomaterials at work in biomedicine. Chem Rev 2015;115:10410–88.

[114] Mathew A, Pradeep T. Noble metal clusters: applications in energy, environment, and biology. Par Part Syst Charact 2014;31:1017–53.

[115] Negishi Y, Tsukuda T. One-pot preparation of subnanometer-sized gold clusters via reduction and stabilization by meso-2,3-dimercaptosuccinic acid. J Am Chem Soc 2003;125:4046–7.

[116] Bigioni TP, Whetten RL, Dag Ö. Near-infrared luminescence from small gold nanocrystals. J Phys Chem B 2000;104:6983–6.

[117] Ramakrishna G, Varnavski O, Kim J, Lee D, Goodson T. Quantum-sized gold clusters as efficient two-photon absorbers. J Am Chem Soc 2008;130:5032–3.

[118] Link S, Beeby A, FitzGerald S, El-Sayed MA, Schaaff TG, Whetten RL. Visible to infrared luminescence from a 28-atom gold cluster. J Phys Chem B 2002;106: 3410–5.

[119] Lee D, Donkers RL, Wang G, Harper AS, Murray RW. Electrochemistry and optical absorbance and luminescence of molecule-like Au_{38} nanoparticles. J Am Chem Soc 2004;126:6193–9.

[120] Wang G, Huang T, Murray RW, Menard L, Nuzzo RG. Near-IR luminescence of monolayer-protected metal clusters. J Am Chem Soc 2005;127:812–3.

[121] Wang G, Guo R, Kalyuzhny G, Choi J-P, Murray RW. NIR luminescence intensities increase linearly with proportion of polar thiolate ligands in protecting monolayers of Au_{38} and Au_{140} quantum dots. J Phys Chem B 2006;110:20282–9.

[122] Wu Z, Jin R. On the ligand's role in the fluorescence of gold nanoclusters. Nano Lett 2010;10:2568–73.

[123] Zeng C, Chen Y, Li G, Jin R. Magic size $Au_{64}(S\text{-}c\text{-}C_6H_{11})_{32}$ nanocluster protected by cyclohexanethiolate. Chem Mater 2014;26:2635–41.

[124] Shibu ES, Muhammed MAH, Tsukuda T, Pradeep T. Ligand exchange of $Au_{25}SG_{18}$ leading to functionalized gold clusters: spectroscopy, kinetics, and luminescence. J Phys Chem C 2008;112:12168–76.

[125] Miller SA, Womick JM, Parker JF, Murray RW, Moran AM. Femtosecond relaxation dynamics of $Au_{25}L_{18}^{-}$ monolayer-protected clusters. J Phys Chem C 2009;113:9440–4.

[126] Wang S, Zhu X, Cao T, Zhu M. A simple model for understanding the fluorescence behavior of Au_{25} nanoclusters. Nanoscale 2014;6:5777–81.

[127] Ghosh A, Udayabhaskararao T, Pradeep T. One-step route to luminescent $Au_{18}SG_{14}$ in the condensed phase and its closed shell molecular ions in the gas phase. J Phys Chem Lett 2012;3:1997–2002.

[128] Hassinen J, Pulkkinen P, Kalenius E, Pradeep T, Tenhu H, Häkkinen H, et al. Mixed-monolayer-protected Au_{25} clusters with bulky calix[4]arene functionalities. J Phys Chem Lett 2014;5:585–9.

[129] Stamplecoskie KG, Chen Y-S, Kamat PV. Excited-state behavior of luminescent glutathione-protected gold clusters. J Phys Chem C 2014;118:1370–6.

[130] Kauffman DR, Alfonso D, Matranga C, Qian H, Jin R. Experimental and computational investigation of Au_{25} clusters and CO_2: a unique interaction and enhanced electrocatalytic activity. J Am Chem Soc 2012;134:10237–43.

[131] Yu M, Zhou C, Liu J, Hankins JD, Zheng J. Luminescent gold nanoparticles with pH-dependent membrane adsorption. J Am Chem Soc 2011;133:11014–7.

[132] Aldeek F, Muhammed MAH, Palui G, Zhan N, Mattoussi H. Growth of highly fluorescent polyethylene glycol- and zwitterion-functionalized gold nanoclusters. ACS Nano 2013;7:2509–21.

[133] Huang C-C, Yang Z, Lee K-H, Chang H-T. Synthesis of highly fluorescent gold nanoparticles for sensing mercury(II). Angew Chem Int Ed 2007;46:6824–8.

[134] Xie J, Zheng Y, Ying JY. Highly selective and ultrasensitive detection of Hg^{2+} based on fluorescence quenching of Au nanoclusters by Hg^{2+}-Au^{+} interactions. Chem Commun 2010;46:961–3.

[135] Lin S-Y, Chen N-T, Sum S-P, Lo L-W, Yang C-S. Ligand exchanged photoluminescent gold quantum dots functionalized with leading peptides for nuclear targeting and intracellular imaging. Chem Commun 2008:4762–4.

[136] Mirkin MV, Amemiya S. Nanoelectrochemistry. New York: CRC Press; 2015.

[137] Kano S, Azuma Y, Kanehara M, Teranishi T, Majima Y. Room-temperature coulomb blockade from chemically synthesized Au nanoparticles stabilized by acid-base interaction. Appl Phys Exp 2010;3:105003.

[138] Lopez-Acevedo O, Tsunoyama H, Tsukuda T, Häkkinen H, Aikens CM. Chirality and electronic structure of the thiolate-protected Au_{38} nanocluster. J Am Chem Soc 2010;132:8210–8.

[139] Malola S, Lehtovaara L, Knoppe S, Hu K-J, Palmer RE, Bürgi T, et al. $Au_{40}(SR)_{24}$ cluster as a chiral dimer of 8-electron superatoms: structure and optical properties. J Am Chem Soc 2012;134:19560–3.

[140] Knoppe S, Bürgi T. Chirality in thiolate-protected gold clusters. Acc Chem Res 2014;47:1318–26.

[141] Varnholt B, Dolamic I, Knoppe S, Bürgi T. On the flexibility of the gold-thiolate interface: racemization of the $Au_{40}(SR)_{24}$ cluster. Nanoscale 2013;5:9568–71.

[142] Knoppe S, Dolamic I, Dass A, Bürgi T. Separation of enantiomers and CD Spectra of $Au_{40}(SCH_2CH_2Ph)_{24}$: spectroscopic evidence for intrinsic chirality. Angew Chem Int Ed 2012;51:7589–91.

[143] Zhu M, Qian H, Meng X, Jin S, Wu Z, Jin R. Chiral Au_{25} nanospheres and nanorods: synthesis and insight into the origin of chirality. Nano Lett 2011;11:3963–9.

[144] Nie X, Qian H, Ge Q, Xu H, Jin R. CO oxidation catalyzed by oxide-supported $Au_{25}(SR)_{18}$ nanoclusters and identification of perimeter sites as active centers. ACS Nano 2012;6:6014–22.

[145] Nie X, Zeng C, Ma X, Qian H, Ge Q, Xu H, et al. CeO_2-supported $Au_{38}(SR)_{24}$ nanocluster catalysts for CO oxidation: a comparison of ligand-on and -off catalysts. Nanoscale 2013;5:5912–8.

[146] Zhu Y, Qian H, Jin R. An atomic-level strategy for unraveling gold nanocatalysis from the perspective of $Au_n(SR)_m$ nanoclusters. Chem Eur J 2010;16:11455–62.

[147] Zhu Y, Qian H, Das A, Jin R. Comparison of the catalytic properties of 25-atom gold nanospheres and nanorods. Chin J Catal 2011;32:1145–55.

[148] Yoskamtorn T, Yamazoe S, Takahata R, Nishigaki J, Thivasasith A, Limtrakul J, et al. Thiolate-mediated selectivity control in aerobic alcohol oxidation by porous carbon-supported Au_{25} clusters. ACS Catal 2014;4:3696–700.

[149] Liu Y, Tsunoyama H, Akita T, Xie S, Tsukuda T. Aerobic oxidation of cyclohexane catalyzed by size-controlled Au clusters on hydroxyapatite: size effect in the sub-2 nm regime. ACS Catal 2011;1:2–6.

[150] Li G, Qian H, Jin R. Gold nanocluster-catalyzed selective oxidation of sulfide to sulfoxide. Nanoscale 2012;4:6714–7.

[151] Kawasaki H, Kumar S, Li G, Zeng C, Kauffman DR, Yoshimoto J, et al. Generation of singlet oxygen by photoexcited $Au_{25}(SR)_{18}$ clusters. Chem Mater 2014;26:2777–88.

[152] Yamamoto H, Yano H, Kouchi H, Obora Y, Arakawa R, Kawasaki H. N,N-dimethylformamide-stabilized gold nanoclusters as a catalyst for the reduction of 4-nitrophenol. Nanoscale 2012;4:4148–54.

[153] Shivhare A, Ambrose SJ, Zhang H, Purves RW, Scott RWJ. Stable and recyclable Au_{25} clusters for the reduction of 4-nitrophenol. Chem Commun 2013;49:276–8.

[154] Zhu Y, Qian H, Zhu M, Jin R. Thiolate-protected Au_n nanoclusters as catalysts for selective oxidation and hydrogenation processes. Adv Mater 2010;22:1915–20.

[155] Zhu Y, Qian H, Drake BA, Jin R. Atomically precise $Au_{25}(SR)_{18}$ nanoparticles as catalysts for selective hydrogenation of α, β-unsaturated ketones and aldehydes. Angew Chem Int Ed 2010;49:1295–8.

[156] Zhu Y, Wu Z, Gayathri C, Qian H, Gil RR, Jin R. Exploring stereoselectivity of Au_{25} nanoparticle catalyst for hydrogenation of cyclic ketone. J Catal 2010;271: 155–60.

[157] Li G, Liu C, Lei Y, Jin R. Au_{25} nanocluster-catalyzed ullmann-type homocoupling reaction of aryl iodides. Chem Commun 2012;48:12005–7.

[158] Li G, Jiang D-e, Liu C, Yu C, Jin R. Oxide-supported atomically precise gold nanocluster for catalyzing sonogashira cross-coupling. J Catal 2013;306:177–83.

[159] Noyori R. Asymmetric catalysis: science and opportunities (Nobel Lecture). Angew Chem Int Ed 2002;41:2008–22.

[160] Holm AH, Ceccato M, Donkers RL, Fabris L, Pace G, Maran F. Effect of peptide ligand dipole moments on the redox potentials of Au_{38} and Au_{140} nanoparticles. Langmuir 2006;22:10584–9.

[161] Negishi Y, Kamimura U, Ide M, Hirayama M. A photoresponsive Au_{25} nanocluster protected by azobenzene derivative thiolates. Nanoscale 2012;4:4263–8.

[162] Niihori Y, Matsuzaki M, Pradeep T, Negishi Y. Separation of precise compositions of noble metal clusters protected with mixed ligands. J Am Chem Soc 2013;135: 4946–9.

[163] Yuan X, Zhang B, Luo Z, Yao Q, Leong DT, Yan N, et al. Balancing the rate of cluster growth and etching for gram-scale synthesis of thiolate-protected Au_{25} nanoclusters with atomic precision. Angew Chem Int Ed 2014;53:4623–7.

[164] Niihori Y, Matsuzaki M, Uchida C, Negishi Y. Advanced use of high-performance liquid chromatography for synthesis of controlled metal clusters. Nanoscale 2014; 6:7889–96.

[165] Ariga K, Vinu A, Yamauchi Y, Ji Q, Hill JP. Nanoarchitectonics for mesoporous materials. Bull Chem Soc Jpn 2012;85:1–32.

[166] Akutsu M, Koyasu K, Atobe J, Hosoya N, Miyajima K, Mitsui M, et al. Experimental and theoretical characterization of aluminum-based binary superatoms of $Al_{12}X$ and their cluster salts. J Phys Chem A 2006;110:12073–6.

[167] Sankar M, Dimitratos N, Miedziak PJ, Wells PP, Kiely CJ, Hutchings GJ. Designing bimetallic catalysts for a green and sustainable future. Chem Soc Rev 2012;41:8099–139.

[168] Ferrando R, Jellinek J, Johnston RL. Nanoalloys: from theory to applications of alloy clusters and nanoparticles. Chem Rev 2008;108:845–910.

[169] Zhang H, Watanabe T, Okumura M, Haruta M, Toshima N. Catalytically highly active top gold atom on palladium nanocluster. Nat Mater 2012;11:49–52.

[170] Negishi Y, Iwai T, Ide M. Continuous modulation of electronic structure of stable thiolate-protected Au_{25} cluster by Ag doping. Chem Commun 2010;46:4713–5.

[171] Kauffman DR, Alfonso D, Matranga C, Qian H, Jin R. A quantum alloy: the ligand-protected $Au_{25-x}Ag_x(SR)_{18}$ cluster. J Phys Chem C 2013;117:7914–23.

[172] Dou X, Yuan X, Yao Q, Luo Z, Zheng K, Xie J. Facile synthesis of water-soluble $Au_{25-x}Ag_x$ nanoclusters protected by mono- and bi-thiolate ligands. Chem Commun 2014;50:7459–62.

[173] Kobayashi R, Nonoguchi Y, Sasaki A, Yao H. Chiral monolayer-protected bimetallic Au-Ag nanoclusters: alloying effect on their electronic structure and chiroptical activity. J Phys Chem C 2014;118:15506–15.

[174] Kumara C, Aikens CM, Dass A. X-ray crystal structure and theoretical analysis of $Au_{25-x}Ag_x(SCH_2CH_2Ph)_{18}^-$ alloy. J Phys Chem Lett 2014;5:461–6.

[175] Kumara C, Dass A. AuAg alloy nanomolecules with 38 metal atoms. Nanoscale 2012;4:4084–6.

[176] Kumara C, Dass A. $(AuAg)_{144}(SR)_{60}$ alloy nanomolecules. Nanoscale 2011;3: 3064–7.

[177] Udayabhaskararao T, Sun Y, Goswami N, Pal SK, Balasubramanian K, Pradeep T. Ag_7Au_6: a 13-atom alloy quantum cluster. Angew Chem Int Ed 2012;51:2155–9.

[178] Jupally VR, Dass A. Synthesis of $Au_{130}(SR)_{50}$ and $Au_{130-x}Ag_x(SR)_{50}$ nanomolecules through core size conversion of larger metal clusters. Phys Chem Chem Phys 2014;16:10473–9.

[179] Jupally VR, Dharmaratne AC, Crasto D, Huckaba AJ, Kumara C, Nimmala PR, et al. $Au_{137}(SR)_{56}$ nanomolecules: composition, optical spectroscopy, electrochemistry and electrocatalytic reduction of CO_2. Chem Commun 2014;50: 9895–8.

[180] Gottlieb E, Qian H, Jin R. Atomic-level alloying and de-alloying in doped gold nanoparticles. Chem Eur J 2013;19:4238–43.

[181] Negishi Y, Munakata K, Ohgake W, Nobusada K. Effect of copper doping on electronic structure, geometric structure, and stability of thiolate-protected Au_{25} nanoclusters. J Phys Chem Lett 2012;3:2209−14.

[182] Bhattarai N, Black DM, Boppidi S, Khanal S, Bahena D, Tlahuice-Flores A, et al. ESI-MS identification of abundant copper-gold clusters exhibiting high plasmonic character. J Phys Chem C 2015;119:10935−42.

[183] Dharmaratne AC, Dass A. $Au_{144-x}Cu_x(SC_6H_{13})_{60}$ nanomolecules: effect of Cu incorporation on composition and plasmon-like peak emergence in optical spectra. Chem Commun 2014;50:1722−4.

[184] Sharma S, Kurashige W, Nobusada K, Negishi Y. Effect of trimetallization in thiolate-protected $Au_{24-n}Cu_nPd$ clusters. Nanoscale 2015;7:10606−12.

[185] Qian H, Jiang D-E, Li G, Gayathri C, Das A, Gil RR, et al. Monoplatinum doping of gold nanoclusters and catalytic application. J Am Chem Soc 2012;134: 16159−62.

[186] Kwak K, Tang Q, Kim M, Jiang D-E, Lee D. Interconversion between superatomic 6-electron and 8-electron configurations of $M@Au_{24}(SR)_{18}$ Clusters (M = Pd, Pt). J Am Chem Soc 2015;137:10833−40.

[187] Fields-Zinna CA, Crowe MC, Dass A, Weaver JEF, Murray RW. Mass spectrometry of small bimetal monolayer-protected clusters. Langmuir 2009;25: 7704−10.

[188] Negishi Y, Kurashige W, Niihori Y, Iwasa T, Nobusada K. Isolation, structure, and stability of a dodecanethiolate-protected Pd_1Au_{24} cluster. Phys Chem Chem Phys 2010;12:6219−25.

[189] Qian H, Barry E, Zhu Y, Jin R. Doping 25-atom and 38-atom gold nanoclusters with palladium. Acta Phys Chim Si 2011;27:513−9.

[190] Negishi Y, Igarashi K, Munakata K, Ohgake W, Nobusada K. Palladium doping of magic gold cluster $Au_{38}(SC_2H_4Ph)_{24}$: formation of $Pd_2Au_{36}(SC_2H_4Ph)_{24}$ with higher stability than $Au_{38}(SC_2H_4Ph)_{24}$. Chem Commun 2012;48:660−2.

[191] Barrabés N, Zhang B, Bürgi T. Racemization of chiral $Pd_2Au_{36}(SC_2H_4Ph)_{24}$: doping increases the flexibility of the cluster surface. J Am Chem Soc 2014;136: 14361−4.

[192] Kothalawala N, Kumara C, Ferrando R, Dass A. $Au_{144-x}Pd_x(SR)_{60}$ nanomolecules. Chem Commun 2013;49:10850−2.

[193] Zhang B, Kaziz S, Li H, Wodka D, Malola S, Safonova O, et al. $Pd_2Au_{36}(SR)_{24}$ cluster: structure studies. Nanoscale 2015;7:17012−9.

[194] Wang S, Song Y, Jin S, Liu X, Zhang J, Pei Y, et al. Metal exchange method using Au_{25} nanoclusters as templates for alloy nanoclusters with atomic precision. J Am Chem Soc 2015;137:4018−21.

[195] Yao C, Lin Y-j, Yuan J, Liao L, Zhu M, Weng L-H, et al. Mono-cadmium vs mono-mercury doping of Au_{25} nanoclusters. J Am Chem Soc 2015;137:15350−3.

[196] Liao L, Zhou S, Dai Y, Liu L, Yao C, Fu C, et al. Mono-mercury doping of Au_{25} and the HOMO/LUMO energies evaluation employing differential pulse voltammetry. J Am Chem Soc 2015;137:9511−4.

[197] Niihori Y, Kurashige W, Matsuzaki M, Negishi Y. Remarkable enhancement in ligand-exchange reactivity of thiolate-protected Au_{25} nanoclusters by single Pd atom doping. Nanoscale 2013;5:508−12.

[198] de la Llave E, Scherlis DA. Selenium-based self-assembled monolayers: the nature of adsorbate-surface interactions. Langmuir 2010;26:173–8.

[199] Szelagowska-Kunstman K, Cyganik P, Schüpbach B, Terfort A. Relative stability of thiol and selenol based SAMs on Au(111) - exchange experiments. Phys Chem Chem Phys 2010;12:4400–6.

[200] Romashov LV, Ananikov VP. Self-assembled selenium monolayers: from nanotechnology to materials science and adaptive catalysis. Chem Eur J 2013; 19:17640–60.

[201] Yokota K, Taniguchi M, Kawai T. Control of the electrode-molecule interface for molecular devices. J Am Chem Soc 2007;129:5818–9.

[202] Negishi Y, Kurashige W, Kamimura U. Isolation and structural characterization of an octaneselenolate-protected Au_{25} cluster. Langmuir 2011;27:12289–92.

[203] Kurashige W, Yamazoe S, Kanehira K, Tsukuda T, Negishi Y. Selenolate-protected Au_{38} nanoclusters: isolation and structural characterization. J Phys Chem Lett 2013;4:3181–5.

[204] Kurashige W, Yamaguchi M, Nobusada K, Negishi Y. Ligand-induced stability of gold nanoclusters: thiolate versus selenolate. J Phys Chem Lett 2012;3:2649–52.

[205] Li Y, Zaluzhna O, Xu B, Gao Y, Modest JM, Tong YYJ. Mechanistic insights into the brust-schiffrin two-phase synthesis of organo-chalcogenate-protected metal nanoparticles. J Am Chem Soc 2011;133:2092–5.

[206] Li Y, Love O, Tong YYJ. Small gold nanoparticles synthesized with dialkyl ditelluride. Nano Bull 2013;2:130221.

[207] Li Y, Silverton LC, Haasch R, Tong YY. Alkanetelluroxide-protected gold nanoparticles. Langmuir 2008;24:7048–53.

[208] Kurashige W, Yamazoe S, Yamaguchi M, Nishido K, Nobusada K, Tsukuda T, et al. Au_{25} clusters containing unoxidized tellurolates in the ligand shell. J Phys Chem Lett 2014;5:2072–6.

[209] Kurashige W, Niihori Y, Sharma S, Negishi Y. Recent progress in the functionalization methods of thiolate-protected gold clusters. J Phys Chem Lett 2014;5:4134–42.

[210] Zhang X-D, Luo Z, Chen J, Shen X, Song S, Sun Y, et al. Ultrasmall $Au_{10-12}(SG)_{10-12}$ nanomolecules for high tumor specificity and cancer radiotherapy. Adv Mater 2014;26:4565–8.

[211] Zhang X-D, Chen J, Luo Z, Wu D, Shen X, Song S-S, et al. Enhanced tumor accumulation of sub-2 nm gold nanoclusters for cancer radiation therapy. Adv Healthc Mater 2014;3:133–41.

Highly Luminescent Metal Nanocluster Molecules

H.-T. Sun
Soochow University, Suzhou, People's Republic of China

CHAPTER OUTLINE

2.1 INTRODUCTION

Creation of novel material systems has always been fascinating to researchers as it provides immense opportunities to study the emergence of novel physical and chemical properties and to open up ways for practical applications. During the past several decades, research that tends to shrink the dimensions of materials has been driven by the desire to access the unique material properties and performance advantages that occur in the transition to nanometer-length scales. Significantly unusual properties different from those of bulk materials, as well as a broad range of potential applications in nanoscale devices, catalysis, nanoelectronics, data storage, molecular imaging, biosensors, and nanomedicine, have been successfully demonstrated [1–7]. Among nanomaterials, the study of metal nanoparticles has a rather long history in terms of preparation, characterization, and applications. In 1857 Michael Faraday prepared a gold colloid through the reduction of a gold salt with phosphorus in a two-phase system [8]. The brilliant ruby-red color of the gold colloid greatly intrigued Faraday and also stimulated a great deal of experimental and theoretical research on metal colloids.

Supra-materials Nanoarchitectonics. http://dx.doi.org/10.1016/B978-0-323-37829-1.00002-X

Since then, understanding the properties of metal nanoparticles and exploring their application potential have become major driving forces behind the synthesis of a large variety of metal nanomaterials. Nanoscale metals are roughly divided into three size domains according to their characteristic length scales [7]: large nanoparticles, small nanoparticles, and nanoclusters (NCs). When the particle size is comparable to, or even larger than, the wavelength of photons that interact with it, the optical responses of these large metal nanoparticles to external electromagnetic fields are simply dependent on their sizes, free-electron density, and their nearly bulk-like dielectric function relative to that of the surrounding medium; in this case, their photophysical features can be quantitatively described with the Mie theory. When the particle size is close to the electron mean free path ($\sim$50 nm for Au), the dielectric function and refractive indices become strongly size dependent. The third class of metal nanoparticles possess sizes comparable with the Fermi wavelength of an electron, which show optical, electronic, and chemical properties largely different from those in the other two size regimes. Generally, the nanoparticles in such a size regime are referred to as NCs, which consist of only several to tens of metal atoms and possess characteristic sizes smaller than 2 nm. This has become a fascinating domain of widespread interest in nanomaterials during the past 10 years. One of the principal reasons for this is the excellent optical properties inherent to these metal NCs such as high-yield photoluminescence (PL), good photostability, large Stokes shift, and high emission rates.

This chapter will focus on two kinds of representative luminescent NCs composed of noble metals (Au and Ag) and Bi and will compare their respective advantages and disadvantages. Because the preparation approaches are strongly dependent on the cluster's type (that is, one synthesis method for one kind of NC may not be suitable for others), here the preparation methods and properties of typical NCs according to their compositions are summarized. The optoelectronic properties, especially the structure and PL features, are highlighted. Finally, the key scientific challenges that researchers currently face are discussed and the perspectives on future research trends in this exciting field of science are presented.

2.2 AU, AG, AND AU-AG ALLOY NANOCLUSTERS

Brust et al. reported in 1994 the first chemical synthesis of thiolate-protected Au nanoparticles, in which a strong reducing agent, sodium borohydride ($NaBH_4$), was used to reduce the thiolate-Au^I complexes to Au nanoparticles [9]. Although the quantum yields of the thiolate-Au nanocrystals from the Brust-like syntheses are more than five orders of magnitude higher than that of bulk gold ($\sim 10^{-10}$), they rarely exceed 0.1%. Such low yields of these Au NCs have limited their practical application. In general, the reduction of metal ions in aqueous solution results in large nanoparticles rather than small NCs due to the tendency of NCs to aggregate. In addition, the nature of the ligands used for capping the particle surface can remarkably affect their emission properties. As a result, choosing suitable agents capable of stabilizing clusters from aggregating and enhancing their fluorescence is of vital importance for obtaining small, highly fluorescent metal NCs. In 2003 Zheng et al. reported a novel route to water-soluble, monodisperse, blue-emitting Au_8 nanodots that, when encapsulated in and stabilized by biocompatible poly(amidoamine) dendrimers, exhibit a PL quantum yield of 41%, a more than 100-fold improvement over other reported gold NCs [10]. This work greatly enriches the methods for the synthesis of noble NCs with very high PL yields and various approaches have been developed for the synthesis of Au or Ag NCs after this breakthrough.

There are generally two main synthetic methods when using thiolate ligands as etching or protecting agents. The first is the etching method, in which large nanoparticles or small NC precursors are etched by excess ligands to produce ultrasmall luminescent NCs. Thiolate ligands are generally used as the etchants due to their unique and strong interaction with Au/Ag atoms. Large nanoparticles with a core size >3 nm have been used as precursors for the synthesis of luminescent NCs. For instance, Pradeep et al. synthesized fluorescent Au NCs from mercaptosuccinic acid (MSA)-protected Au nanoparticles by etching with excess glutathione [11]. Similarly, highly luminescent Ag NCs were prepared by etching large Ag nanoparticles through an interfacial etching method [12]. Two possible mechanisms were proposed to explain this etching-based strategy (Fig. 2.1) [4,13]. Owing to strong interaction between thiolate and noble atoms, noble atoms may be removed from the nanoparticle surfaces by excess ligands (such as glutathione) in the form of Au(I)-thiolate complexes, which then undergo strong aurophilic interactions and form luminescent Au clusters. Alternatively, ligands may etch the surface gold atoms of the nanoparticles, leading to the formation of luminescent particles with decreased size, and the removed atoms will precipitate as Au(I)-thiolate complexes after etching. By etching method, highly luminescent Au_8, Au_{23}, and Au_{25} NCs have been successfully synthesized [13]. The direct benefit resulting from NCs with different metal cores is the tunable PL emission owing to the inherently different electronic structures of NCs (Fig. 2.2). Interestingly, the etching-based approach has also been used to synthesize thiol-protected luminescent Ag NCs. A crude mixture of luminescent Ag_7 and Ag_8 clusters was finally obtained by this approach [12].

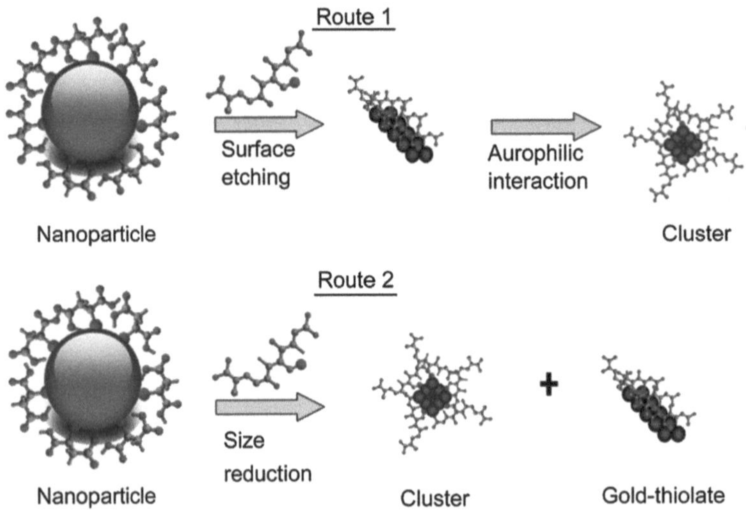

■ **FIGURE 2.1** Schematic illustration of two possible routes for the formation of luminescent Au NCs via etching of preformed MSA-protected Au nanoparticles. *Reprinted with permission from Muhammed MH, Ramesh S, Sinha S, Pal S, Pradeep T. Two distinct fluorescent quantum clusters of gold starting from metallic nanoparticles by pH-dependent ligand etching. Nano Res 2008;1:333–40. © 2008 Springer.*

The second approach is based on the insufficient reduction of the thiolate-Au^I complexes during the synthesis of thiolate-Au NCs [14,15]. Xie and coworkers developed a simple one-pot synthetic method for highly luminescent thiolate-Au NCs that uses a common thiol, glutathione (GSH), as

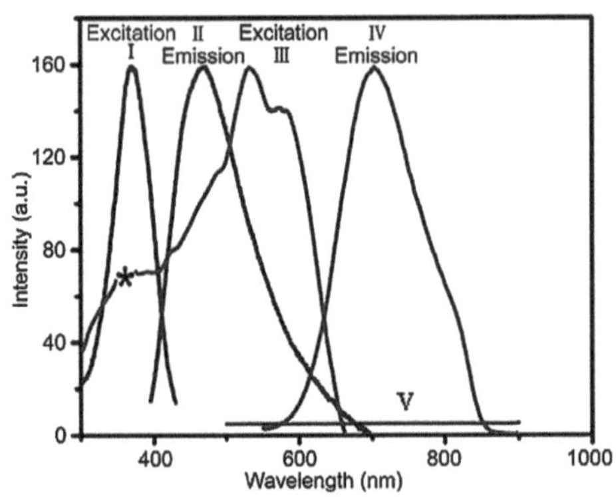

■ **FIGURE 2.2** Comparison of the PL profiles of the clusters with Au@MSA. Traces I and II are the excitation and emission spectra of Au_8, respectively. Traces III and IV are the excitation and emission spectra of Au_{25}, respectively, and trace V is the emission spectrum of Au@MSA. *Reprinted with permission from Muhammed MH, Ramesh S, Sinha S, Pal S, Pradeep T. Two distinct fluorescent quantum clusters of gold starting from metallic nanoparticles by pH-dependent ligand etching. Nano Res 2008;1: 333–40. Copyright at Springer (2008).*

both the reducing and protecting agent [14]. Under the mild reduction environment created by GSH and a relatively low thiol-to-Au ratio (1.5:1), a unique Au^0@Au^I-thiolate core−shell nanostructure was formed through the controlled aggregation of thiolate-Au^I complexes on the in situ−generated Au^0 cores (Fig. 2.3). The resulting Au NCs have a quantum yield of ~15%, and their emission stems from the aggregation-induced emission of thiolate-Au^I complexes condensed on the NC surface.

In addition to the thiolate route to luminescent noble NCs, biological and biomimetic syntheses of functional Au or Ag NCs have been extensively investigated in the nanoscience community. Xie and coworkers developed a simple one-pot synthetic route to produce highly luminescent Au NCs by using a common commercially available protein, bovine serum albumin (BSA), as both the reducing and protecting agent (Fig. 2.4). This synthetic protocol is generic and has been adapted to other proteins and other noble metals (e.g., Ag) [16−20], thereby leading to the formation of a unique family of luminescent protein-Au/Ag NCs. It is also noteworthy that other classes of ligands such as dendrimers, DNA, peptides, and even polymers were employed for the synthesis of noble luminescent NCs.

One of the most attractive features lies in the absorption and luminescence from these NCs. In general, the optical properties of metal nanoparticles are dominated by the collective oscillation of electrons at their surfaces, known as the surface plasmon resonance, that are in resonance with the incident electromagnetic radiation. As the size reduces to the subnanometer scale, metal clusters are too small to support plasmons, and a transition from metal

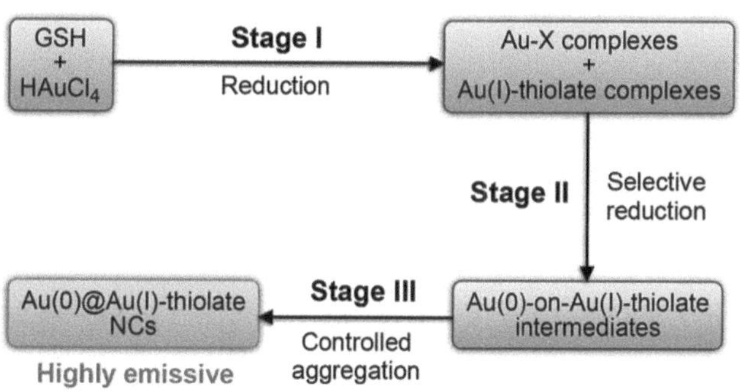

■ **FIGURE 2.3** Schematic illustration of the synthesis of highly luminescent Au^0@Au^I-thiolate NCs by the insufficient reduction method. Here, X in the AuI-X complexes denotes any nonthiolate ligand in the reaction mixture. *Reprinted with permission from Luo Z, Yuan X, Yu Y, Zhang Q, Leong DT, Lee JY, et al. From aggregation-induced emission of Au(I)−thiolate complexes to ultrabright Au(0)@Au(I)−thiolate core−shell nanoclusters. J Am Chem Soc 2012;134:16662−70. Copyright at American Chemical Society (2012).*

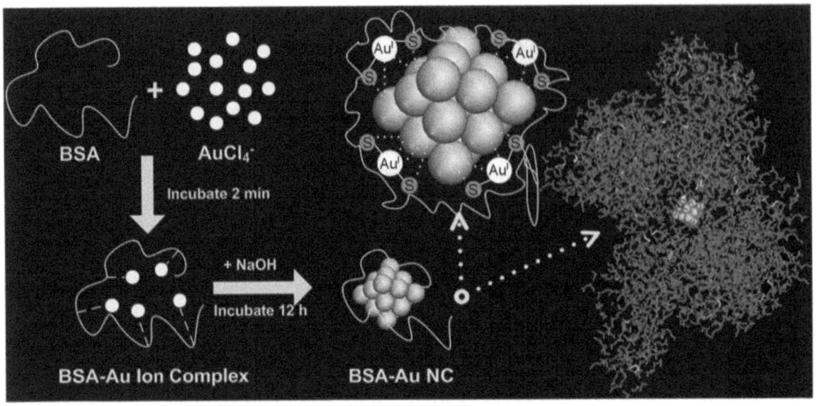

■ **FIGURE 2.4** Schematic illustration of the synthesis of highly luminescent BSA-Au NCs. Upon the introduction of Au^{III} ions to the aqueous BSA solution, the BSA molecules sequester the Au ions. The reducing capability of BSA is then activated by adjusting the reaction pH to ~12, thereby leading to the formation of BSA-Au NCs. *Reprinted with permission from Xie JP, Zheng YG, Ying JY. Protein-directed synthesis of highly fluorescent gold nanoclusters. J Am Chem Soc 2009; 131:888–889. Copyright at American Chemical Society (2009).*

to molecule occurs. Instead of forming continuous densities of states and behaving as conductors, the subnanometer-sized metal NCs exhibit discrete energy levels possibly owing to the splitting of the conduction bands. Atomically precise metal NCs are expected to possess characteristic absorption features and can be distinguished from each other by their absorption profiles. It was found that the absorption peaks of glutathione-protected Au NCs due to intraband transition tend to be blueshifted with decreasing core size. As the size decreases, the spacing between the discrete states in each band increases, thus leading to a blueshift in the absorption peaks with decreasing cluster size. Absorption spectra of a few-atom Ag NCs also exhibit discrete electronic transitions instead of collective plasmon excitations. The PL characteristics are strongly influenced by metal core sizes, ligands used, as well as the synthesis parameters such as reaction time and PH. In most cases, Au NCs are more stable than Ag counterparts, although both could display visible and near-infrared (NIR) emissions [3].

In addition to NCs composed of Au or Ag, in 2012 Pradeep et al. reported a novel alloy NC consisting of both Au and Ag, a 13-atom Ag_7Au_6 quantum cluster. This type of NC was synthesized by a galvanic exchange reaction, starting from $Ag_{7,8}$ precursor [21]. Subsequently, Pradeep et al. found that Au-Ag alloy NCs with different core sizes can be prepared by using BSA as templates and reducing agents [22]. Mass spectrometric analysis of the product of a 1:1 M ratio reaction mixture of Au_{QC}@BSA and Ag_{QC}@BSA suggested that the alloy clusters could be $Au_{38-x}Ag_x$@BSA. Detailed analyses by standard techniques revealed that the alloy cluster core of

~1.2 nm diameter is composed of nearly zero valent Au and Ag atoms that exhibit distinctly different steady-state and time-resolved excited-state luminescence profiles compared to the parent clusters (Fig. 2.5). Tuning of the alloy composition was achieved by varying the molar ratio of the parent species in the reaction mixture, and compositional changes were observed by mass spectrometry. Such noble alloy NCs can also be synthesized by other kinds of methods. For example, Jin et al. provided experimental and computational characterization of the ligand-protected mixed-metal $Au_{25-x}Ag_x(SC_2H_4Ph)_{18}$ cluster compared with the unsubstituted $Au_{25}(SC_2H_4Ph)_{18}$ cluster [23]. Density functional theory analysis has predicted that Ag heteroatoms will preferentially occupy sites on the surface of the cluster core. X-ray photoelectron spectroscopy revealed Au-Ag state mixing and charge redistribution within the $Au_{25-x}Ag_x$ cluster. Optical spectroscopy and nonaqueous electrochemistry indicate that Ag heteroatoms increased the cluster lowest unoccupied molecular orbita energy, introduced new features in the $Au_{25-x}Ag_x$ absorbance spectrum, and rendered some

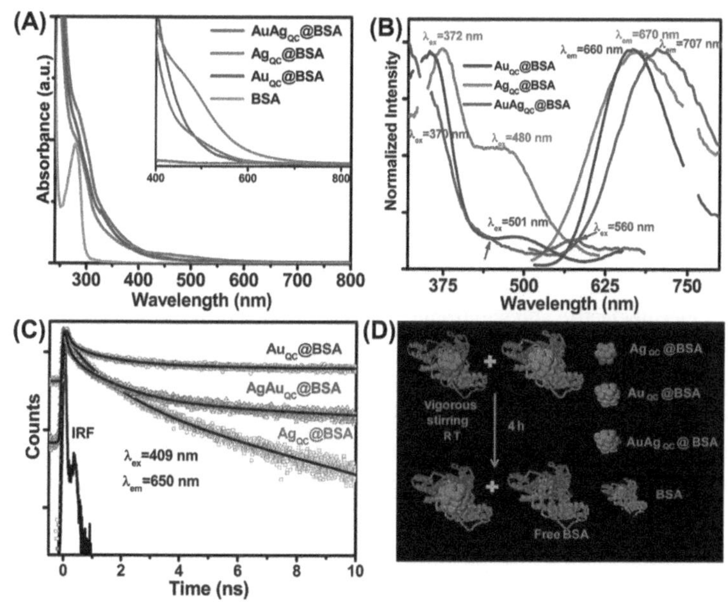

■ **FIGURE 2.5** (A) UV—Vis spectra of BSA, Au_{QC}, Ag_{QC}, and $AuAg_{QC}$@BSA. (B) PL spectra of Au_{QC}, Ag_{QC}, and $AuAg_{QC}$@BSA. Emission spectra were collected for all three clusters by exciting them at 370 nm. (C) Picosecond-resolved fluorescence transients of Au_{QC}, Ag_{QC}, $AuAg_{QC}$@BSA. (D) Schematic of the synthesis of AuAg@BSA alloy clusters through QC—QC interaction. *Reprinted with permission from Mohanty J, Xavier PL, Chaudhari K, Bootharaju MS, Goswami N, Palc SK, et al. Luminescent, bimetallic AuAg alloy quantum clusters in protein templates. Nanoscale 2012;4:4255—62. Copyright at The Royal Society of Chemistry (2012).*

optical transitions forbidden. In a subsequent work, Jin and coworkers further found that substituting silver atoms for gold in the 25-atom matrix can drastically enhance the photoluminescence [24]. The obtained Ag_xAu_{25-x} ($x = 1-13$) nanoclusters exhibit high quantum yield (40.1%), which is in striking contrast with the normally weakly luminescent Ag_xAu_{25-x} species ($x = 1-12$, quantum yield = 0.21%).

Despite great advances that have been achieved in the past decade in the synthesis of noble metal NCs, significant challenges still remain, such as the unambiguous determination of the cluster size (ie, the number of gold atoms) and atomic packing in the cluster. To determine their exact size and structure via X-ray crystallography, the ultimate approach is to grow single crystals of clusters, which unfortunately has long been a great challenge. The first major breakthrough in total structure determination of RS-Au nanoparticles was achieved by Jadzinsky et al. who attained single crystals of Au_{102}(*p*-MBA)$_{44}$ (where pMBA is *para*-mercapto benzoic acid) [25]. Later, atomic structure of the second RS-Au nanoparticles $Au_{25}(SCH_2CH_2Ph)_{18}$ was determined by several groups independently through experimental and theoretical studies [26–27]. Correlating the PL properties with the structures of NCs or nanoparticles could provide insight into the exact PL mechanisms and be helpful for the reasonable design of NCs with even excellent photophysical properties [28].

2.3 BI NANOCLUSTERS

Bi is one of the most thoroughly investigated main group elements; it exists in a broad range of functional materials such as magnets, superconductors, and thermoelectric and spintronic materials. Depending on the material systems, Bi displays two kinds of structural features. The first case is that Bi exists in functional materials in which it coordinates with other elements. In these compounds, Bi usually demonstrates diverse oxidation states. Interestingly, in some compounds such as molecular crystals containing Bi polycations and polyanions Bi can form molecule-like clusters. That is, Bi coordinates with the Bi element itself. The charges of such clusters can be balanced by other structural units such as $[AlCl_4]^-$ or [K-crypt]$^+$ (Crypt = 4,7,13,16,21,24-hexaoxa-1,10-diazabicyclo[8.8.8] hexacosane) [29–33].

Indeed, the study of such peculiar subnanometer Bi NCs dates back to half a century ago. In the 1960s Smith and coworkers studied the evolution of Bi oxidation states in molten salt systems [34,35]. Such pioneering work led to more concentrated research on the synthesis and characterization of materials containing subvalent Bi in the following decades, and much clearer pictures on the oxidation states of Bi and new synthetic approaches have been reported by many researchers [35]. These systematic studies suggest that a highly Lewis-acidic environment is a necessary prerequisite for the stabilization of Bi cationic clusters. Ruck et al. first employed the Lewis-acid ionic liquids for the synthesis of low-valent Bi compounds. Dark-red crystals of $Bi_5(AlCl_4)_3$ were isolated from the ionic liquid [36] (Fig. 2.6). Sun et al. found that Bi_5^{3+} and Bi^+ emitters stabilized by the Lewis-acidic liquid show ultrabroad PL with a lifetime of approximately 1 μs. It was also found that $Bi_5(AlCl_4)_3$ synthesized by this method exhibits extremely broad NIR PL with a full width at half maximum (FWHM) of >510 nm and an effective PL lifetime of 4.1 μs [29]. This study greatly extends the understanding of photophysical properties of materials containing subvalent Bi, owing to the establishment of structure—emitter—property relationships. In fact, the infrared emission behaviors of material systems containing Bi are rather complicated, and in many cases there is no sound and definitely convincing explanation of emission mechanisms [35]. This work, which uses the molecular crystals as model materials for the study of Bi-related emission mechanisms, represents a new and simple route to gaining deeper understanding of the photophysical behaviors of Bi-doped systems. Subsequently, it was found that high-quality $Bi_5(AlCl_4)_3$ crystal prepared by a modified approach displays a broadband emission band from 1 to 2.7 μm [32].

In addition to Bi_5^{3+}, Sun et al. experimentally and theoretically studied the photophysical properties of Bi_8^{2+} polycation stabilized by $[AlCl_4]^-$ (Fig. 2.6) [31]. As displayed in Fig. 2.7, four peaks appear at the

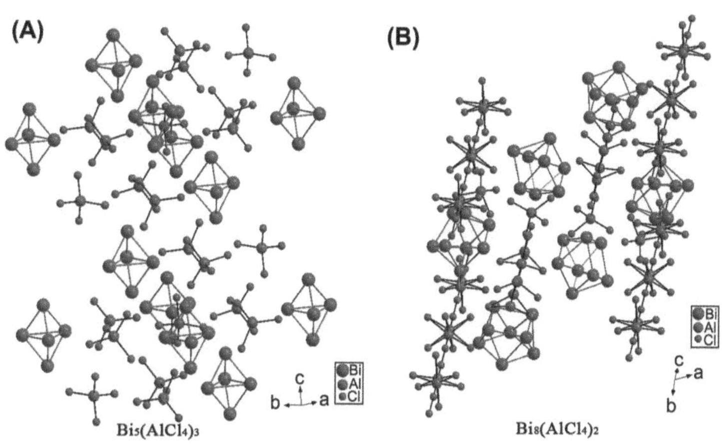

FIGURE 2.6 A general view of the structures of (A) $Bi_5(AlCl_4)_3$ and (B) $Bi_8(AlCl_4)_2$ molecular crystals.

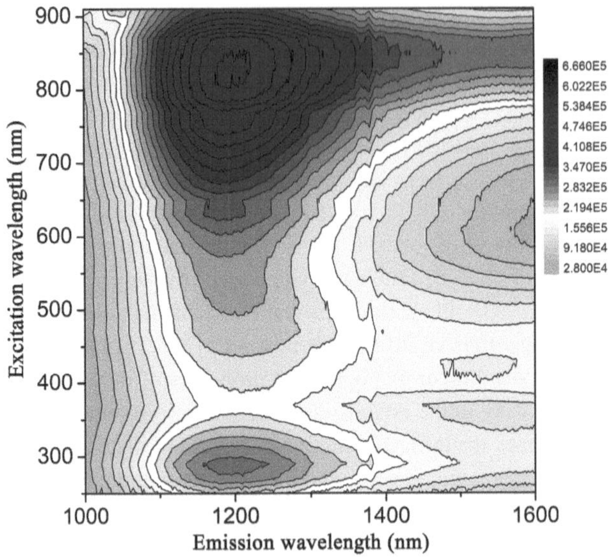

■ **FIGURE 2.7** Contour excitation-emission matrix plot of the $Bi_8(AlCl_4)_2$ sample.

excitation/emission wavelengths of 290/1180, 490/1180, 636/1187, and 830/1192 nm, among which the intensity of the fourth peak is the strongest. It is obvious that the PL peak does not display significant excitation-wavelength-dependent shift, which situates it in the range of 1180−1200 nm under different excitation wavelengths. Further detailed analyses suggest that Bi_8^{2+} polycations contribute to emission peaking at about 1180 nm, while the emission tails at longer wavelengths result from Bi_5^{3+}. This work further evidences that Bi polycations can be smart emitters, whose emission can cover important telecommunication wavelengths.

Besides molecular crystals containing Bi polycations, it is noteworthy that many compounds with Bi polyanions absorb light in the visible range of the electromagnetic spectrum since they are colored, thus giving promise for unique optical properties. In 2012 Sun and coworkers first reported that a single crystal of $(K\text{-crypt})_2Bi_2$, which contains $[Bi_2]^{2-}$ polyanions, displays ultrabroad NIR emission at around 1190 nm (Fig. 2.8) [33]. The emission spectrum under 641 nm excitation has a very broad emission range, spanning from 975 to 1400 nm. The FWHM of the spectrum is 212 nm and the emission peak is around 1190 nm. The emission spectrum by Gaussian fitting yields three components peaking at 1047, 1190, and 1331 nm with FWHMs of 99, 146 and 56 nm, respectively, whose emission energies are lower than the corresponding absorption energies at 990, 1090, and 1273 nm, respectively [33]. A qualitative explanation of the observed

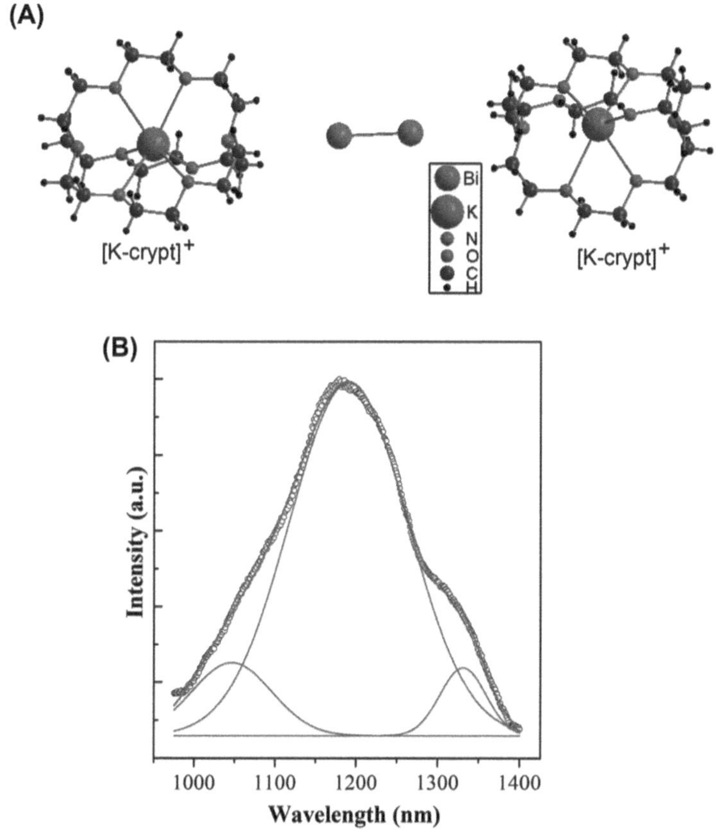

■ **FIGURE 2.8** (A) Part of the structure of (K-crypt)$_2$Bi$_2$. (B) The PL spectrum of (K-crypt)$_2$Bi$_2$ crystals under 641 nm excitation light. The *black*, *red*, and *green curves* are experimental, fitted, and three decomposed Gaussian peaks, respectively. *Reproduced with permission from Sun H, Yonezawa T, Gillett-Kunnath MM, Sakka Y, Shirahata N, Gui S, et al. Ultra-broad near-infrared photoluminescence from crystalline (K-crypt)$_2$Bi$_2$ containing $[Bi_2]^{2-}$ dimers. J Mater Chem 2012;22:20175–8. Copyright at The Royal Society of Chemistry (2012).*

photophysical behavior is proposed [33]. The $[Bi_2]^{2-}$ anion absorbs photons with energies in the NIR and visible ranges. After irradiation with high-energy photons, the electrons in the upper excited levels tend to nonradiatively relax to the first three excited levels from where the NIR emissions take place. That is, NIR PL results from the radiative electronic transitions from the first three excited levels to the ground level based on a one-photon process.

The main advantage of Bi NCs is their attractive infrared emission feature, which might lead to a wide array of applications in bioimaging, telecommunications, and lasers. However, such NCs can only be stabilized by air-sensitive units or templates that can absorb water, resulting in the air instability of the final product [29–36]. Furthermore, these NCs generally exist in crystalline matrices. It is greatly expected that monodisperse Bi NCs can be obtained for their practical applications [7]. With enough effort in these directions, it is hopeful that air-stable, monodisperse, and highly luminescent Bi NCs can be synthesized.

2.4 PERSPECTIVE

In this chapter, the recent progress in the controlled synthesis and application of luminescent metal NCs has been summarized with a particular emphasis on Au, Ag, and Bi clusters. The abundant photophysical and physiochemical properties of NCs help researchers to deepen their understanding of the rationality of synthesis approaches, as well as the inherent interaction mechanisms with other species. However, a number of challenges still remain in this exciting field of science.

Firstly, more effective synthesis protocols are greatly required for the synthesis of highly luminescent and well-dispersed clusters [7]. In particular, most of the previously mentioned metal NCs possess a relatively weak brightness in comparison to semiconducting quantum dots and many organic dyes. Furthermore, Bi NCs are stable only in suitable matrices. Thus, developing revolutionary approaches for their synthesis is important for their broader application. Secondly, more work should be done to establish relationships between the structure and emission properties [7]. It is found that in many cases the resulting NCs are polydisperse and display broad emission bands. To convincingly explain the observed luminescence from NCs, structural analyses of the obtained systems by X-ray crystallography are required, although in many cases it is difficult to culture high-quality single crystals of tiny NCs. Thirdly, the emission and excitation ranges should be optimized. Most NCs emit in the spectral range <1000 nm. Further extension into NIR or even mid-infrared will expand their application for both telecommunications and bioimaging. Furthermore, much effort is needed to build optical devices based on such NCs, which might find some commercial applications.

In summary, given their low cost, ready scalability, excellent chemical stability, colloidal stability, and photostability, it is believed that these NC materials will find broad applications in addressing important issues related to the environment, medicine, diagnosis, telecommunications, and laser science [7].

REFERENCES

[1] Zheng J, Nicovich PR, Dickson RM. Highly fluorescent noble-metal quantum dots. Annu Rev Phys Chem 2007;58:409–31.

[2] Jin RC. Quantum sized, thiolate-protected gold nanoclusters. Nanoscale 2010;2: 343–62.

[3] Choi S, Dickson RM, Yu JH. Developing luminescent silver nanodots for biological applications. Chem Soc Rev 2012;41:1867–91.

[4] Shang L, Dong SJ, Nienhaus GU. Ultra-small fluorescent metal nanoclusters: synthesis and biological applications. Nano Today 2011;6:401–18.

[5] Lu YZ, Chen W. Sub-nanometre sized metal clusters: from synthetic challenges to the unique property discoveries. Chem Soc Rev 2012;41:3594–623.

[6] Yuan X, Luo ZT, Yu Y, Yao QF, Xie JP. Luminescent noble metal nanoclusters as an emerging optical probe for sensor development. Chem Asian J 2013;8:858–71.

[7] Sun H-T, Sakka Y. Luminescent metal nanoclusters: controlled synthesis and functional applications. Sci Technol Adv Mater 2014;15:014205.

[8] Faraday M. Philos Trans R Soc Lond 1857;147:145–81.

[9] Brust M, Walker M, Bethell D, Schiffrin DJ, Whyman R. Synthesis of thiol-derivatised gold nanoparticles in a two-phase liquid–liquid system. J Chem Soc Chem Commun 1994:801–2.

[10] Zheng J, Petty JT, Dickson RM. High quantum yield blue emission from water-soluble Au_8 nanodots. J Am Chem Soc 2003;125:7780–1.

[11] Muhammed MH, Ramesh S, Sinha S, Pal S, Pradeep T. Two distinct fluorescent quantum clusters of gold starting from metallic nanoparticles by pH-dependent ligand etching. Nano Res 2008;1:333–40.

[12] Rao TUB, Pradeep T. Luminescent Ag_7 and Ag_8 clusters by interfacial synthesis. Angew Chem Int Ed 2010;49:3925–9.

[13] Muhammed M, Verma P, Pal S, Kumar R, Paul S, Omkumar R, et al. Bright, NIR-emitting Au_{23} from Au_{25}: characterization and applications including biolabeling. Chem Eur J 2009;15:10110–20.

[14] Luo Z, Yuan X, Yu Y, Zhang Q, Leong DT, Lee JY, et al. From aggregation-induced emission of Au(I)–thiolate complexes to ultrabright Au(0)@Au(I)–thiolate core–shell nanoclusters. J Am Chem Soc 2012;134:16662–70.

[15] Chakraborty I, Thumu UBR, Pradeep T. High temperature nucleation and growth of glutathione protected $\sim Ag_{75}$ clusters. Chem Commun 2012;48:6788–90.

[16] Xie JP, Zheng YG, Ying JY. Protein-directed synthesis of highly fluorescent gold nanoclusters. J Am Chem Soc 2009;131:888–9.

[17] Wen F, Dong Y, Feng L, Wang S, Zhang S, Zhang X. Horseradish peroxidase functionalized fluorescent gold nanoclusters for hydrogen peroxide sensing. Anal Chem 2011;83:1193–6.

[18] Kawasaki H, Yoshimura K, Hamaguchi K, Arakawa R. Trypsin-stabilized fluorescent gold nanocluster for sensitive and selective Hg^{2+} detection. Anal Sci 2011;27:591–6.

[19] Kawasaki H, Hamaguchi K, Osaka I, Arakawa R. ph-dependent synthesis of pepsin-mediated gold nanoclusters with blue green and red fluorescent emission. Adv Funct Mater 2011;21:3508–15.

[20] Xavier PL, Chaudhari K, Baksi A, Pradeep T. Protein-protected luminescent noble metal quantum clusters: an emerging trend in atomic cluster nanoscience. Nano Rev 2012;3:14767.

[21] Udayabhaskararao T, Sun Y, Goswami N, Pal SK, Balasubramanian K, Pradeep T. Ag_7Au_6: a 13-atom alloy quantum cluster. Angew Chem Int Ed 2012;51:2155–9.

[22] Mohanty J, Xavier PL, Chaudhari K, Bootharaju MS, Goswami N, Palc SK, et al. Luminescent, bimetallic AuAg alloy quantum clusters in protein templates. Nanoscale 2012;4:4255–62.

[23] Kauffman DR, Alfonso D, Matranga C, Qian H, Jin R. A quantum alloy: the ligand-protected $Au_{25-x}Ag_x(SR)_{18}$ cluster. J Phys Chem C 2013;117:7914–23.

[24] Wang S, Meng X, Das A, Li T, Song Y, Cao T, et al. A 200-fold quantum yield boost in the photoluminescence of silver-doped Ag_xAu_{25-x} nanoclusters: the 13th silver atom matters. Angew Chem Int Ed 2014;53:2376–80.

[25] Jadzinsky PD, Calero G, Ackerson CJ, Bushnell DA, Kornberg RD. Structure of a thiol monolayer–protected gold nanoparticle at 1.1 Å resolution. Science 2007;318:430–3.

[26] Heaven MW, Dass A, White PS, Holt KM, Murray RW. Crystal structure of the gold nanoparticle $[N(C_8H_{17})_4][Au_{25}(SCH_2CH_2Ph)_{18}]$. J Am Chem Soc 2008;130:3754–5.

[27] Zhu M, Aikens CM, Hollander FJ, Schatz GC, Jin R. Correlating the crystal structure of a thiol-protected Au_{25} cluster and optical properties. J Am Chem Soc 2008;130:5883.

[28] Yu Y, Luo Z, Chevrier DM, Leong DT, Zhang P, Jiang D, et al. Identification of a highly luminescent $Au_{22}(SG)_{18}$ nanocluster. J Am Chem Soc 2014;136:1246–9.

[29] Sun H, Sakka Y, Gao H, Miwa Y, Fujii M, Shirahata N, et al. Ultrabroad near infrared photoluminescence from $Bi_5(AlCl_4)_3$ crystal. J Mater Chem 2011;21:4060–3.

[30] Sun H, Sakka Y, Fujii M, Shirahata N, Gao H. Ultrabroad near-infrared photoluminescence from ionic liquids containing subvalent bismuth. Opt Lett 2011;36:100–2.

[31] Sun H, Sakka Y, Shirahata N, Gao H, Yonezawa T. Experimental and theoretical studies of photoluminescence from Bi_8^{2+} and Bi_5^{3+} stabilized by $[AlCl_4]^-$ in molecular crystals. J Mater Chem 2012;22:12837–41.

[32] Sun H, Xu BB, Yonezawa T, Sakka Y, Shirahata N, Fujii M, et al. Photoluminescence from $Bi_5(GaCl_4)_3$ molecular crystal. Dalton Trans 2012;41:11055–61.

[33] Sun H, Yonezawa T, Gillett-Kunnath MM, Sakka Y, Shirahata N, Gui S, et al. Ultrabroad near-infrared photoluminescence from crystalline $(K\text{-crypt})_2Bi_2$ containing $[Bi_2]^{2-}$ dimers. J Mater Chem 2012;22:20175–8.

[34] Davis H, Bjerrum N, Smith J. Ligand field theory of $p^{2,4}$ configurations and its application to the spectrum of Bi^+ in molten salt media. Inorg Chem 1967;6:1172–8.

[35] Sun H, Zhou J, Qiu J. Recent advances in bismuth activated photonic materials. Prog Mater Sci 2014;64:1–72.

[36] Ahmed E, Köhler D, Ruck M. Room-temperature synthesis of bismuth clusters in ionic liquids and crystal growth of $Bi_5(AlCl_4)_3$. Z Anorg Allg Chem 2009;635:297–300.

Biomimetic Morphology Control of Metal Oxides and Their Site-Selective Immobilization

Yoshitake Masuda
National Institute of Advanced Industrial Science and Technology (AIST), Nagoya, Japan

CHAPTER OUTLINE

3.1 INTRODUCTION

Biomimetic approaches have been attractive for chemistry, physics, biotechnology, nanotechnology, life science, mechanical engineering, aeronautical engineering, etc. Especially, they have high potential in the metal oxide nanotechnology. Metal oxides have been prepared by high temperature for a long time. They cause large energy consumption and environmental burdens. On the other hand, nano/micro structures of metal oxides are fabricated in animals or plants biomimetically. They are realized in aqueous solutions

Supra-materials Nanoarchitectonics. http://dx.doi.org/10.1016/B978-0-323-37829-1.00003-1

at ordinary temperature and atmospheric pressure. Nature gives us a valuable guideline for future materials science to save the planet. In this section, newly developed biomimetic approaches for metal oxide nanotechnology was reported [1]. Aqueous solutions processes were effectively utilized for biomimetic morphology control of metal oxides. Furthermore, site-selective immobilization of metal oxides was realized biomimetically. Surface modification with self-assembled monolayers(SAMs) or super hydrophilic surfaces showed high ability for precise control of nucleation and crystal growth. Various kinds of biomimetic science and technology were included in these processes.

3.2 LIQUID PHASE MORPHOLOGY CONTROL OF ZnO

3.2.1 Biomimetic Morphology Control of Stand-Alone ZnO Self-Assembled Film [2]

Zinc nitrate hexahydrate ($Zn(NO_3)_2 \cdot 6H_2O$) (15 mM) was dissolved in distilled water at 60°C and ethylenediamine ($H_2NCH_2CH_2NH_2$) (15 mM) was added to the solution to induce the formation of ZnO [2]. The solution was kept at 60°C using a water bath for 6 h with no stirring. The solution was then left to cool for 42 h in the bath. Polyethylene terephthalate (PET) film, glass (S-1225, Matsunami Glass Ind., Ltd.) and an Si wafer [p-type Si (100), NK Platz Co., Ltd.] were used as substrates. The solution became clouded shortly after the addition of ethylenediamine by the homogeneous nucleation and growth of ZnO particles. Ethylenediamine plays an essential role in the formation of crystalline ZnO. Zinc-ethylenediamine complex forms in the solution as shown by Eq. (3.1) [3].

$$Zn^{2+} + 3H_2NCH_2CH_2NH_2 \leftrightarrows [Zn(H_2NCH_2CH_2NH_2)_3]^{2+} \quad (3.1)$$

The chemical equilibrium in Eq. (3.1) moves to the left and the zinc–ethylenediamine complex decomposes to increase the concentration of Zn^{2+} at elevated temperature. OH^- concentration increases by the hydrolysis of ethylenediamine as shown by Eq. (3.2).

$$H_2NCH_2CH_2NH_2 + 2H_2O \leftrightarrows H_3NCH_2CH_2NH_3{}^{2+} + 2OH^- \quad (3.2)$$

ZnO and $Zn(OH)_2$ are thus formed in the aqueous solution as shown by Eq. (3.3).

$$Zn^{2+} + 2OH^- \rightarrow Zn(OH)_2 \rightarrow ZnO + H_2O \quad (3.3)$$

ZnO particles were gradually deposited to cover the bottom of the vessel, and the solution became light white after 1 h and clear after 6 h.

The supersaturation degree of the solution was high at the initial stage of the reaction for the first 1 h and decreased as the color of the solution changed. White films were formed at the air–liquid interface and they grew to large films. The films had sufficiently high strength to be obtained as stand-alone films. Additionally, a film was scooped to paste onto a desired substrate such as PET film, Si wafer, glass plate or paper, and the pasted ZnO film was then dried to bond it to the substrate. Both sides of the film can be pasted on substrate. The film physically adhered to the substrate. The film maintained its adhesion during immersion in lightly ultrasonicated water, however, it can be easily peeled off again by strong ultrasonication. The film can be handled easily from substrate to other substrate. It also can be attached strongly to substrate by annealing or addition of chemical regents such as silane coupling agent to form chemical bonds between the film and the substrate. The film grew to a thickness of about 5 μm after 48 h, i.e., 60°C for 6 h, and was left to cool for 42 h. The air side of the stand-alone film had a smooth surface over a wide area due to the flat air–liquid interface (Fig. 3.1A1), whereas the liquid side of the film had a rough surface (Fig. 3.1B1). The films consisted of ZnO nanosheets that were clearly observed from the liquid side (Fig. 3.1B2) and the fracture edge—on profile of the film (Fig. 3.1C1 and C2). The nanosheets had a thickness of 5–10 nm and were 1–5 μm in size. They mainly grew forward to the bottom of the solution, i.e., perpendicular to the air–liquid interface, such that the sheets stood perpendicular to the air–liquid interface. Thus, the liquid side of the film had many ultrafine spaces surrounded by nanosheet and had a high specific surface area. The air side of the film, on the other hand, had a flat surface that followed the flat shape of the air–liquid interface. The air–liquid interface was thus effectively utilized to form the flat surface of the film. This flatness would contribute to the strong adhesion strength to substrates for pasting of the film. The air-side surface prepared for 48 h had holes of 100–500 nm in diameter (Fig. 3.1A2), and were hexagonal, rounded hexagonal, or round in shape. The air-side surface prepared for 6 h, in contrast, had no holes on the surface. The air-side surface was well crystallized to form a dense surface, and ZnO crystals would partially grow to a hexagonal shape because of the hexagonal crystal structure. Well-crystallized ZnO hexagons were then etched to form holes on the surface by decrease in pH. The growth face of the film would be liquid side. ZnO nanosheets would grow to form a large ZnO film by Zn ion supply from the aqueous solution. Further investigation of the formation mechanism would contribute to the development of crystallography in the solution system and the creation of novel ZnO fine structures.

The film showed a very strong 0002 X-ray diffraction peak of hexagonal ZnO at $2\theta = 34.04°$ and weak 0004 diffraction peak at $2\theta = 72.16°$ with

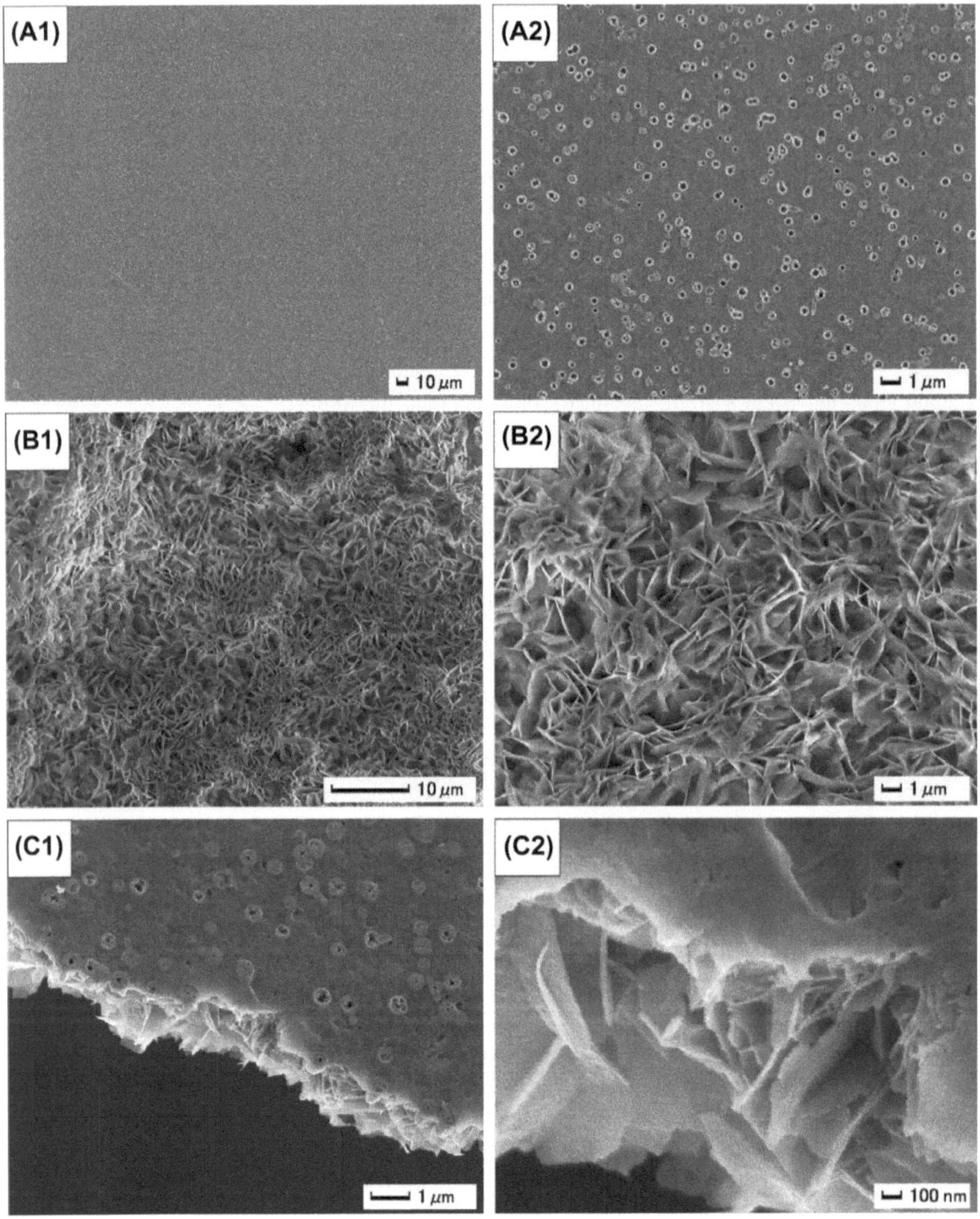

■ **FIGURE 3.1** SEM micrographs of high *c*-axis oriented stand-alone ZnO self-assembled film. (A1) Air-side surface of ZnO film. (A2) Magnified area of (A1). (B1) Liquid-side surface of ZnO film. (B2) Magnified area of (B2). (C1) Fracture cross-section of ZnO film from air side. (C2) Magnified area of (C1). *Reprinted with permission from Masuda Y, Kato K. Cryst Growth Des 2008;8(1):275. Copyright at American Chemical Society (2008).*

no other diffractions of ZnO (Fig. 3.2). (0002) planes and (0004) planes were perpendicular to the *c*-axis, and the diffraction peak only from (0002) and (0004) planes indicates high *c*-axis orientation of ZnO film. The inset figure shows that the crystal structure of hexagonal ZnO stands on a substrate to make the *c*-axis perpendicular to the substrate. Crystallite size parallel to (0002) planes was estimated from the half-maximum full width of the 0002 peak to 43 nm. This is similar to the threshold limit value of our XRD equipment and thus the crystallite size parallel to (0002) planes is estimated to be greater than or equal to 43 nm. Diffraction peaks from a silicon substrate were observed at $2\theta = 68.9°$ and $2\theta = 32.43°$. Weak diffractions at $2\theta = 12.5°$, 24.0°, 27.6°, 30.5°, 32.4°, and 57.6° were assigned to coprecipitated zinc carbonate hydroxide ($Zn_5(CO_3)_2(OH)_6$, JCPDS No. 19-1458).

Stand-alone ZnO film was further evaluated by TEM and electron diffraction. The film was crushed to sheets and dispersed in an acetone. The sheets at the air–liquid interface were skimmed by a cupper grid with a carbon

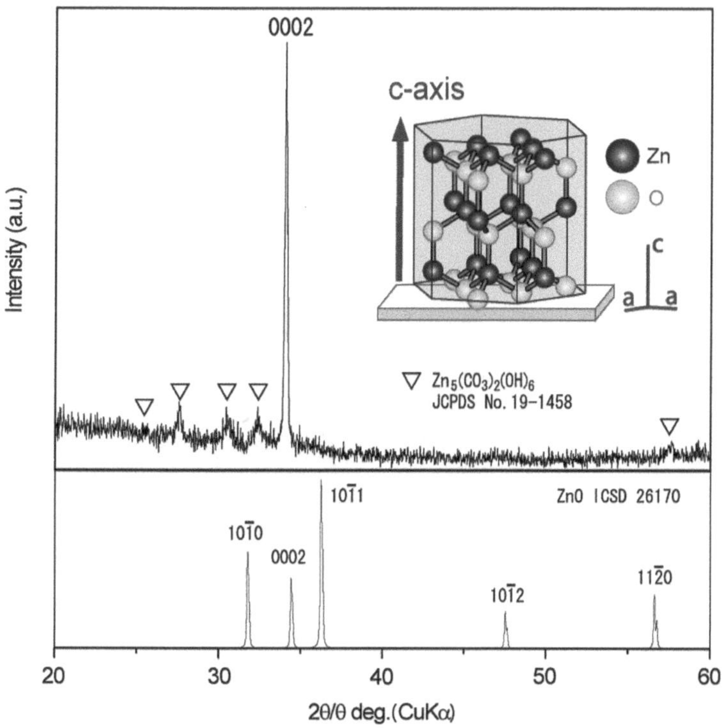

■ **FIGURE 3.2** XRD diffraction pattern of high *c*-axis oriented stand-alone ZnO self-assembled film. *Reprinted with permission from Masuda Y, Kato K. Cryst Growth Des 2008;8(1):275. Copyright at American Chemical Society (2008).*

supporting film. The sheets were shown to have uniform thickness (Fig. 3.3A). They were dense polycrystalline films constructed of ZnO nanoparticles (Fig. 3.3B). Lattice image was clearly observed to show high crystallinity of the particles. The film was shown to be single phase of ZnO by electron diffraction pattern. These observations were consistent with XRD and SEM evaluations. The film pasted on a silicon wafer was annealed at 500°C for 1 h in air to evaluate the details of the films. ZnO film maintained its structure during the annealing (Fig. 3.4). The air side of the film showed a smooth surface (Fig. 3.4A1) and the liquid side showed a relief structure having a high specific surface area (Fig. 3.4B1 and B2). The air side showed the film consisted of dense packing of small ZnO nanosheets and the size of sheets increased toward the liquid-side surface (Fig. 3.4A2). ZnO sheets would grow from the air side to the liquid side, i.e., the sheets would nucleate at the liquid–air interface and grow down toward the bottom of the solution by the supply of Zn ions from the solution. Annealed film showed X-ray diffractions of ZnO and Si substrate with no additional phases. As-deposited ZnO nanosheets were shown to be crystalline ZnO because the sheets maintained their fine structure during the annealing without any phase transition.

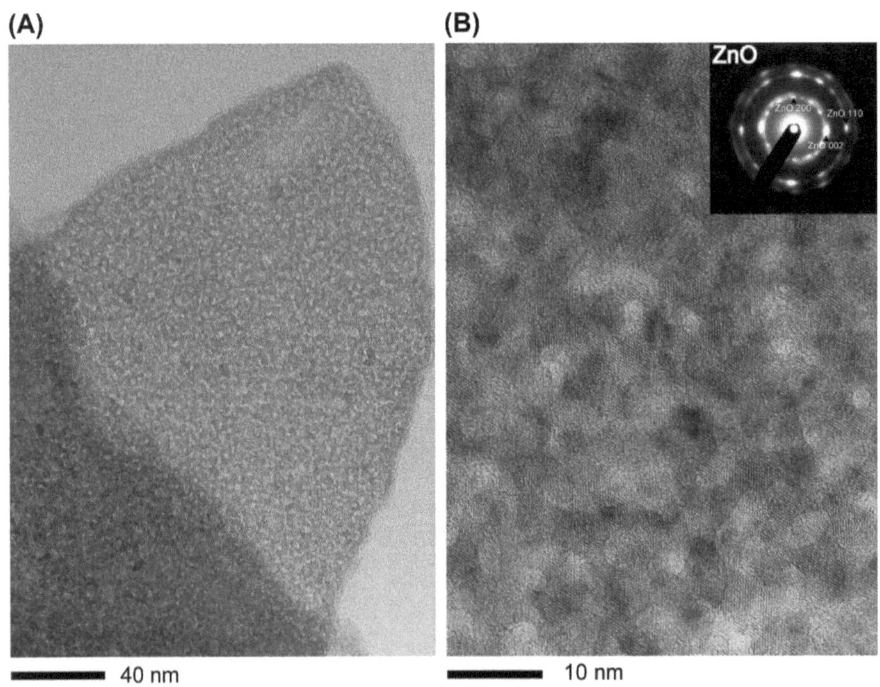

■ **FIGURE 3.3** (A) TEM micrograph of ZnO nanosheets. (B) Magnified area of (A). (Insertion) Electron diffraction pattern of ZnO. *Reprinted with permission from Masuda Y, Kato K. Cryst Growth Des 2008;8(1):275. Copyright at American Chemical Society (2008).*

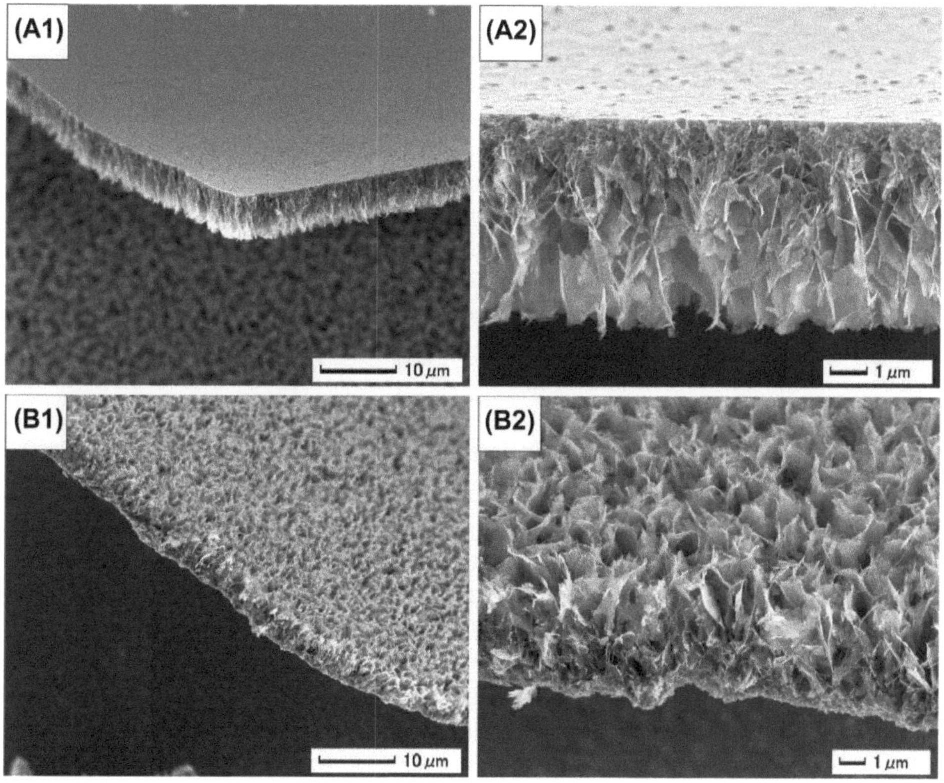

■ **FIGURE 3.4** SEM micrographs of high *c*-axis oriented stand-alone ZnO self-assembled film annealed at 500°C for 1 h in air. (A1) Fracture edge—on profile of ZnO film from air side. (A2) Cross-section profile of ZnO film from air side. (B1) Fracture edge—on profile of ZnO film from liquid side. (B2) Cross-section profile of ZnO film from liquid side. *Reprinted with permission from Masuda Y, Kato K. Cryst Growth Des 2008;8(1):275. Copyright at American Chemical Society (2008).*

High *c*-axis orientation was also maintained during the annealing, showing a very strong 0002 diffraction peak. The solution was further kept at 25°C for 1 month to evaluate the details of the deposition mechanism. The film prepared at the air–liquid interface for 1 month was not hexagonal ZnO. The film showed strong X-ray diffractions of zinc carbonate hydroxide single phase. ZnO would be dissolved by decrease in pH. ZnO would be crystallized at the initial reaction stage for the first 48 h. ZnO was then gradually etched and dissolved by nitric acid and zinc carbonate hydroxide was crystallized using Zn ions that were supplied by the dissolution of crystalline ZnO.

In summary, high *c*-axis oriented stand-alone ZnO self-assembled film was successfully fabricated using a simple solution process. The film consisted of ZnO nanosheets was crystallized at air–liquid interface. The nanosheets grew perpendicular to the film. The film had high *c*-axis orientation and showed a strong 0002 diffraction peak and weak 0004 peak. The air side of the film had a flat surface, whereas the liquid side had a rough surface having many ultrafine spaces surrounded by ZnO nanosheets. The rough surface of the liquid side can be utilized for sensors or dye-sensitized solar cells. The ZnO film was also pasted on a desired substrate such as PET films, Si substrate, or glass plates. The surface of low heat-resistant flexible polymer film was modified with high *c*-axis oriented crystalline ZnO film without heat treatment. This low-cost, low-temperature technique can be used for a wide range of applications including sensors, solar cells, electrical devices, and optical devices using the various properties of high *c*-axis oriented crystalline ZnO.

3.2.2 Biomimetic Morphology Control of Nanocrystal Assembled TiO_2 Particles [4]

Ammonium hexafluorotitanate (12.372 g) and boric acid (11.1852 g) were separately dissolved in deionized water (600 mL) at 50°C [4]. Boric acid solution was added to ammonium hexafluorotitanate solution at a concentration of 0.15 and 0.05 M, respectively. The solution was kept at 50°C for 30 min using a water bath with no stirring, after which it was centrifuged at 4000 rpm for 10 min (Model 8920, Kubota Corp.). Preparation for centrifugation, centrifugation at 4000 rpm, deceleration from 4000 to 0 rpm, and preparation for removal of supernatant solution took 4, 10, 10, and 8 min, respectively. The particles contacted with the solution, the temperature of which was gradually lowered for 32 min after maintaining 50°C for 30 min. Precipitates were dried at 60°C for 12 h after removal of supernatant solution. The solution became clouded about 10 min after mixing ammonium hexafluorotitanate solution and boric acid solution. The particles were homogeneously nucleated in the solution, turning the solution white. X-ray diffraction peaks for the particles were observed at $2\theta = 25.1°$, 37.9°, 47.6°, 54.2°, 62.4°, 69.3°, 75.1°, 82.5°, and 94.0° after evaluation of N_2 adsorption. They were assigned to the 101, 004, 200, 105 + 211, 204, 116 + 220, 215, 303 + 224 + 312, and 305 + 321 diffraction peaks of anatase TiO_2 (JCPSD No. 21-1272, ICSD No. 9852) (Fig. 3.5). The 004 diffraction intensity of randomly oriented particles is usually 0.2 times the 101 diffraction intensity as shown in JCPDS data (No. 21-1272). However, the 004 diffraction intensity of the particles deposited in our process was 0.36 times the 101 diffraction intensity. Additionally, the integral intensity of the 004 diffraction was 0.18 times the 101 diffraction intensity, indicating the *c*-axis orientation of the particles. Particles were not oriented on the glass holder for XRD measurement. Therefore, TiO_2 crystals would be

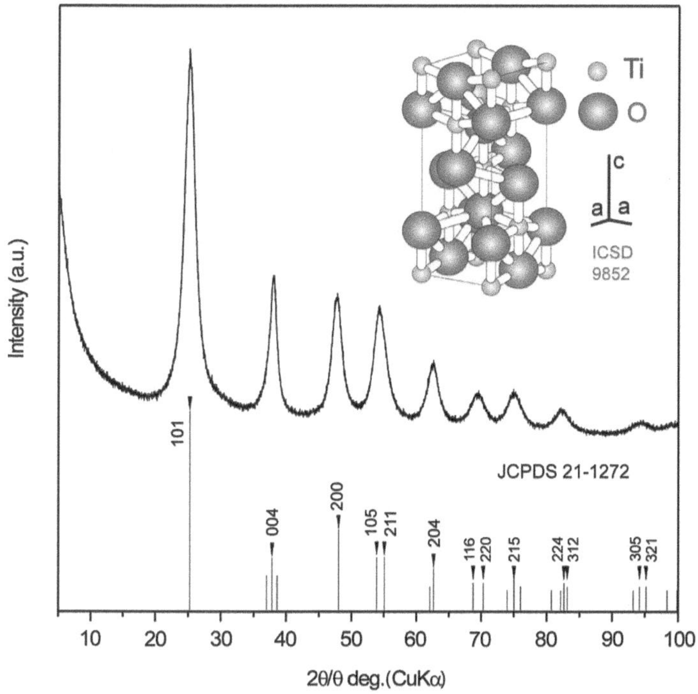

■ **FIGURE 3.5** XRD diffraction pattern of anatase TiO_2 particles. *Reprinted with permission from Masuda Y, Kato K. Cryst Growth Des 2008;8(9):3213. Copyright at American Chemical Society (2008).*

an anisotropic shape in which the crystals were elongated along the *c*-axis. The crystals would have a large number of stacks of c planes such as (001) planes compared to stacks of (101) planes. The diffraction intensity from the (004) planes would be enhanced compared to that from the (101) planes. Crystallite size perpendicular to the (101) or (004) planes was estimated from the full-width half-maximum of the 101 or 004 peak to be 3.9 nm or 6.3 nm, respectively. Elongation of crystals in the *c*-axis direction was also suggested by the difference in crystallite size.

The particles were shown to be assemblies of nano TiO_2 crystals (Fig. 3.6A). Particle diameter was estimated to be 100–200 nm. Relief structures had formed on the surfaces and open pores had formed inside because the particles were porous assemblies of nanocrystals. Nanocrystals were shown to have acicular shapes (Fig. 3.6B). They were about 5–10 nm in length. The longer direction of acicular TiO_2 is indicated by the black arrow. The inserted FFT image shows the 101 and 004 diffractions of anatase TiO_2. Nanocrystals are assigned to the single phase of anatase TiO_2. It is notable that the diffraction from the (101) planes has a ring shape due to random

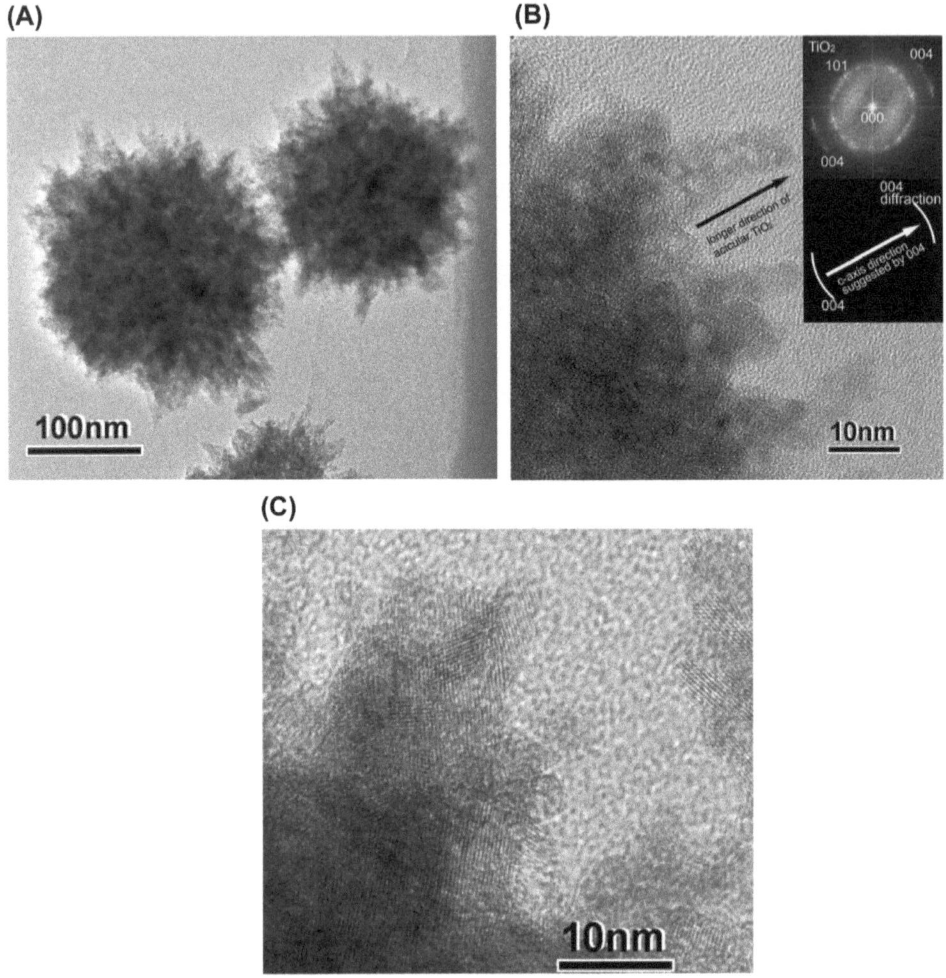

■ **FIGURE 3.6** (A) TEM micrograph of anatase TiO_2 particles. (B) Magnified area of (A) showing morphology of acicular crystals. Insertion in (B) FFT image of (B) anatase TiO_2. (C) Magnified area of (A) showing lattice images of anatase TiO_2. *Reprinted with permission from Masuda Y, Kato K. Cryst Growth Des 2008; 8(9):3213. Copyright at American Chemical Society (2008).*

orientation but that from the (004) planes was observed only in the upper-right region and lower-left region in the FFT image. Anisotropic 004 diffractions indicated the direction of the *c*-axis, which was perpendicular to the (004) planes, as shown by the white arrow. It was roughly parallel to the longer direction of acicular TiO_2. These results suggest that acicular TiO_2 grew along the *c*-axis to enhance the diffraction intensity from the (004) planes. Crystal growth of anatase TiO_2 along the *c*-axis was previously observed in TiO_2 films [5]. Anisotropic crystal growth is one of the features of liquid phase crystal deposition. Acicular nanocrystals showed lattice images of anatase TiO_2 (Fig. 3.6C). They were constructed of anatase TiO_2 crystals without amorphous or additional phases. Anatase crystals were not covered with amorphous or additional phases even at the tips. Bare anatase crystal with nanosized structure is important to achieve high performance for catalysts and devices. Crystallization of TiO_2 was effectively utilized to form assemblies of acicular nanocrystals in the process. Open pores and surface relief structures were successfully formed on the particles. The dried particles were dispersed in water to evaluate zeta potential and particle size distribution after evaluation of N_2 adsorption. The particles had positive zeta potential of 30.2 mV at pH 3.1, which decreased to 5.0, −0.6, −11.3, and −36.3 mV at pH 5.0, 7.0, 9.0, and 11.1, respectively (Fig. 3.7). The isoelectric point was estimated to be pH 6.7, slightly higher than that of anatase TiO_2 (pH 2.7−6.0) [6]. Zeta potential is very sensitive to the particle surface conditions, ions adsorbed on the particle surfaces, and the kind and concentration of ions in the solution. The variations in zeta potential were likely caused by the difference in the surface conditions of TiO_2 particles, affected by the interaction between particles and ions in the solution. Mean particle size was estimated to be ~550 nm in diameter with a standard deviation (STD) of 220 nm at pH 3.1 (Fig. 3.8A). This was larger than that observed by TEM. Slight aggregation occurred at pH 3 because the particles were

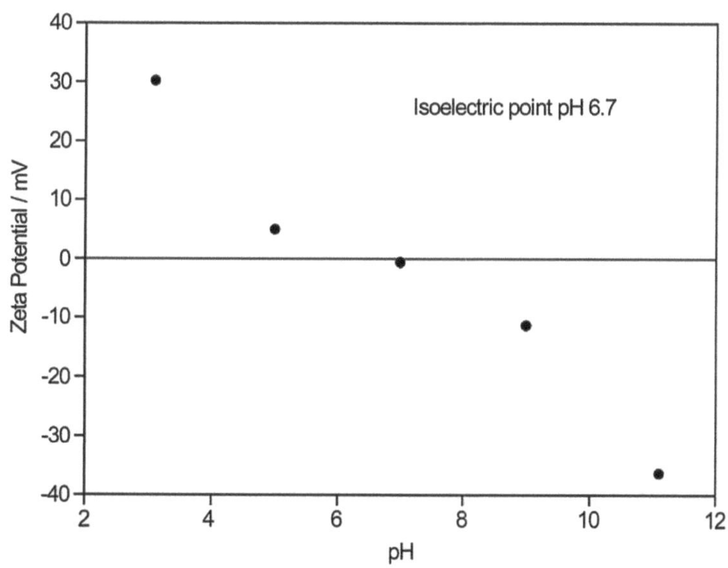

■ **FIGURE 3.7** Zeta potential of anatase TiO_2 particles as a function of pH. *Reprinted with permission from Masuda Y, Kato K. Cryst Growth Des 2008;8(9):3213. Copyright at American Chemical Society (2008).*

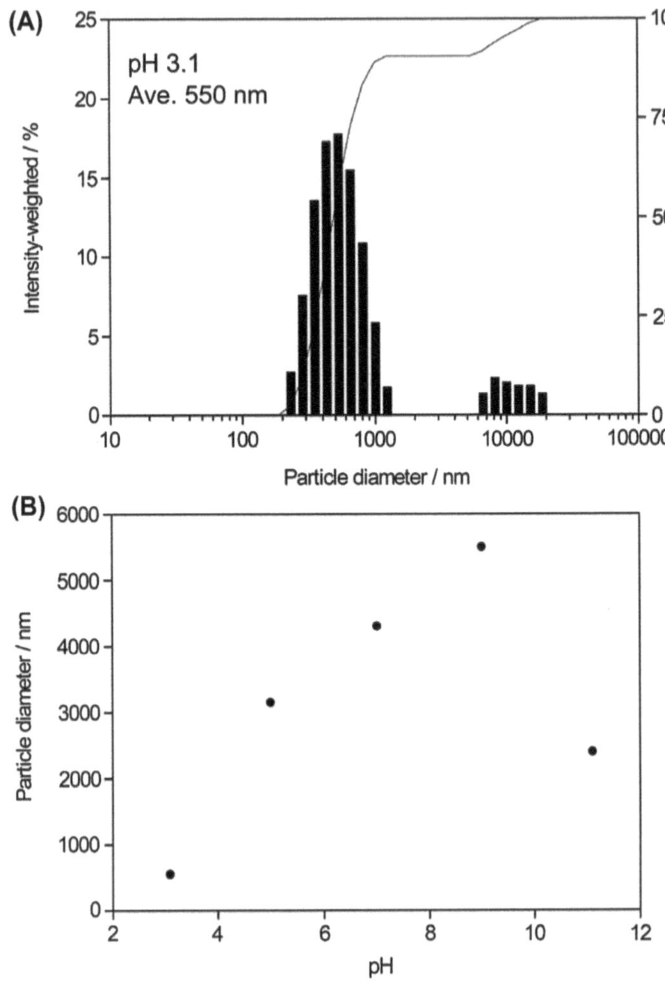

■ **FIGURE 3.8** (A) Particle size distribution of anatase TiO_2 particles at pH 3.1. (B) Particle size of anatase TiO_2 particles as a function of pH. *Reprinted with permission from Masuda Y, Kato K. Cryst Growth Des 2008;8(9):3213. Copyright at American Chemical Society (2008).*

dried completely prior to measurement. Particle size increased with pH and showed a maximum of near the isoelectric point (550 nm at pH 3.1, 3150 nm at pH 5, 4300 nm at pH 7, 5500 nm at pH 9, or 2400 nm at pH 11.1) (Fig. 3.8B). Strong aggregation resulted from the lack of repulsion force between particles near the isoelectric point. The particles were generated in the solution at pH 3.8 in this study. It would be suitable to obtain repulsion force between particles for crystallization without strong aggregation.

TiO_2 particles exhibited N_2 adsorption–desorption isotherms of Type IV (Fig. 3.9A). The desorption isotherm differed from adsorption isotherm in the relative pressure (P/P_0) range from 0.4 to 0.7, showing mesopores in the particles. BET surface area of the particles was estimated to be 270 m^2/g (Fig. 3.9B). This is higher than that of TiO_2 nanoparticles such as Aeroxide P25 (BET 50 m^2/g, 21 nm in diameter, anatase 80% + rutile 20%, Degussa), Aeroxide P90 (BET 90–100 m^2/g, 14 nm in diameter, anatase 90% + rutile 10%, Degussa), MT-01 (BET 60 m^2/g, 10 nm in diameter, rutile, Tayca Corp.), and Altair TiNano (BET 50 m^2/g, 30–50 nm in diameter, Altair Nanotechnologies Inc.) [7]. A high BET surface area cannot be obtained from particles having a smooth surface even if the particle size is less than 100 nm. A high BET surface area would be realized by the unique morphology of TiO_2 particles constructed of nanocrystal assemblies. Total pore volume and average pore diameter were estimated from pores smaller than 230 nm at P/Po = 0.99–0.431 cc/g and 6.4 nm, respectively. They were estimated to be 0.212 cc/g and 3.1 nm, respectively, from pores smaller than 11 nm at P/Po = 0.80. Total pore volume was also estimated by the BJH method from pores smaller than 154 nm to be 0.428 cc/g. Average pore diameter was estimated to be 6.3 nm using BET surface area. Pore size distribution was calculated by the BJH method using adsorption isotherms (Fig. 3.9C). It showed a pore size distribution curve having a peak at ~2.8 nm, and pores larger than 10 nm TiO_2 particles would have mesopores of ~2.8 nm surrounded by nanocrystals. Pores larger than 10 nm are considered to be interparticle spaces. The pore size distribution also suggested the existence of micropores smaller than 1 nm. Pore size distribution was further calculated by the DFT/Monte-Carlo method. The model was in fair agreement with adsorption isotherms (Fig. 3.9D). Pore size distribution showed a peak at ~3.6 nm that indicated the existence of mesopores of ~3.6 nm (Fig. 3.9E). The pore size calculated by the DFT/Monte-Carlo method was slightly larger than that calculated from the BJH method because the latter method is considered to have produced an underestimation [8–10]. The pore size distribution also suggested the existence of micropores of ~1 nm, probably resulting from microspaces surrounded by nanocrystals and the uneven surface structure of nanocrystals. The particles were shown to have a large surface area as well as micropores of ~1 nm, mesopores of ~2.8–3.6 nm, and pores larger than 10 nm, by N_2 adsorption characteristics. Assembly of acicular nanocrystals resulted in unique features and high surface area. TiO_2 particles were generated in the solutions at 90°C for 1 h using an oil bath with no stirring for comparison. The solutions became clouded after the addition of boric acid solutions into ammonium hexafluorotitanate solutions. High temperature accelerated crystal growth of TiO_2. Hydrogen chloride of 0.6 ml was added into the solutions of 200 ml to

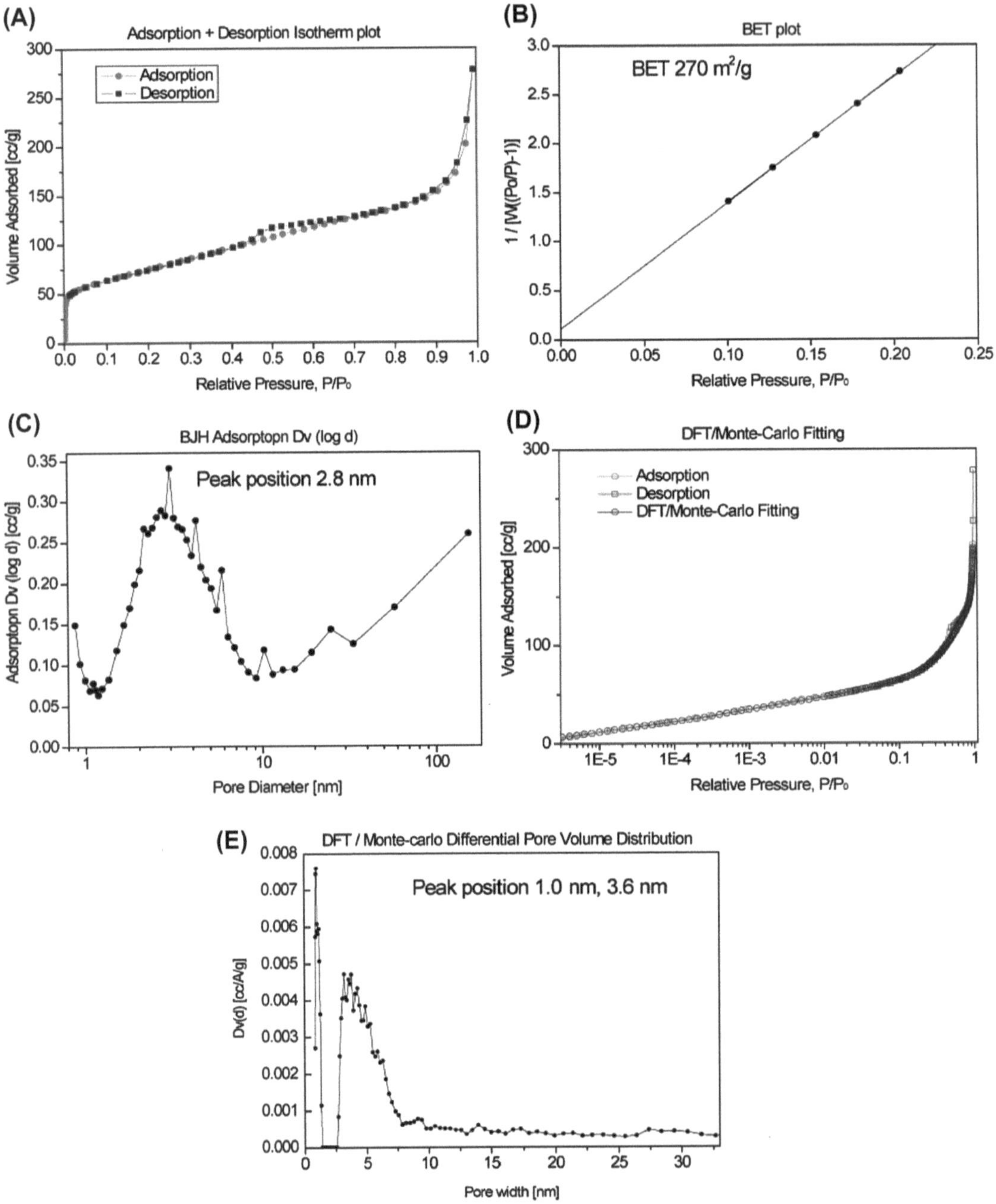

■ **FIGURE 3.9** (A) N_2 adsorption—desorption isotherm of anatase TiO_2 particles. (B) BET surface area of anatase TiO_2 particles. (C) Pore size distribution calculated from N_2 adsorption data of anatase TiO_2 particles using BJH equation. (D) N_2 adsorption—desorption isotherm and DFT/Monte-Carlo fitting curve of anatase TiO_2 particles. (E) Pore size distribution calculated from N_2 adsorption data of anatase TiO_2 particles using DFT/Monte-Carlo equation. *Reprinted with permission from Masuda Y, Kato K. Cryst Growth Des 2008;8(9):3213. Copyright at American Chemical Society (2008).*

decrease crystallization speed of TiO_2. The pH of the solutions was 2.4 1 h after mixing the solutions. BET surface area of the particles was estimated to 18 m^2/g. This is much lower than that of the particles prepared at 50°C and slightly lower than that prepared at 90°C for 8 min in our previous work (44 m^2/g) [11]. Formation of TiO_2 was accelerated at high temperature and it decreased surface area. The particles grew in the solutions to decrease surface area as function of time. Crystallization of TiO_2 was shown to be strongly affected by growth conditions such as solution temperature and growth time.

In summary, anatase TiO_2 particles, 100–200 nm in diameter, were successfully fabricated in aqueous solution. They were assemblies of nanocrystals 5–10 nm that grew anisotropically along the *c*-axis to form acicular shapes. The particles thus had nanorelief surface structures constructed of acicular crystals. They showed *c*-axis orientation due to high-intensity X-ray diffraction from the (004) crystal planes. The particles had a high BET surface area of 270 m^2/g. Total pore volume and average pore diameter were estimated from pores smaller than 230 nm at $P/Po = 0.99$–0.43 cc/g and 6.4 nm, respectively. They were also estimated from pores smaller than 11 nm at $P/Po = 0.80$–0.21 cc/g and 3.1 nm, respectively. BJH and DFT/Monte-Carlo analysis of adsorption isotherm indicated the existence of pores ~2.8 and ~3.6 nm, respectively. Additionally, the analyses suggested the existence of micropores of ~1 nm. Crystallization and self-assembly of nano TiO_2 were effectively utilized to fabricate nanocrystal assembled TiO_2 particles having high surface area and nanorelief surface structure.

3.2.3 Biomimetic Morphology Control of Acicular $BaTiO_3$ Particles [12]

Oxalic acid (252 mg) was dissolved into isopropyl alcohol (4 ml) [12]. Butyl titanate monomer (0.122 ml) was mixed with the oxalic acid solution, and the solution was then mixed with distilled water (100 ml). The pH of the solution was increased to pH = 7 by adding NaOH (1 M) and distilled water, while the volume of the solution was adjusted to 150 ml by these additions. The aqueous solution (50 ml) with barium acetate (39.3 mg) was mixed with the oxalic acid solution. The mixed solution containing barium acetate (0.77 mM), butyl titanate monomer (2 mM), and oxalic acid (10 mM) was kept at room temperature for several hours with no stirring, and the solution gradually became cloudy. Stirring causes the collision of homogeneously nucleated particles and destruction of large grown particles, and so was avoided in this process. The size of the precipitate was easily controlled

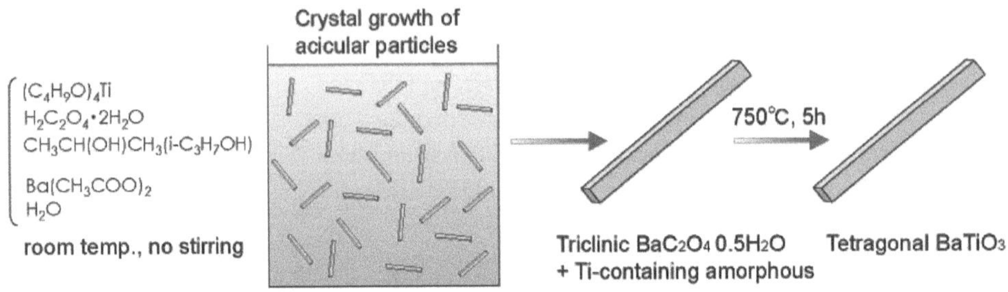

■ **FIGURE 3.10** Conceptual process for fabricating acicular $BaTiO_3$ particles. Morphology control of $BaC_2O_4 \cdot 0.5H_2O$ particles and phase transition to $BaTiO_3$. *Reprinted with permission from Masuda Y, Yamada T, Koumoto K. Cryst Growth Des 2008;8:169. Copyright at American Chemical Society (2008).*

from nanometer order to micrometer order by changing the growth period. Large particles were grown by immersion for several hours to evaluate the morphology and crystallinity in detail. Oxalate ions ($C_2O_4^{2-}$) react with barium ions (Ba^{2+}) to form barium oxalate ($BaC_2O_4 \cdot 0.5H_2O$). $BaC_2O_4 \cdot 0.5H_2O$ is dissolved in weak acetate acid provided by barium acetate ($(CH_3COO)_2Ba$), however, it can be deposited at pH 7, which is adjusted by adding NaOH. $BaC_2O_4 \cdot 0.5H_2O$ was thus successfully precipitated from the solution. Acicular particles were homogeneously nucleated and precipitated from the solution (Fig. 3.10). They were on average 23 μm (ranging from 19 to 27 μm) in width and 167 μm (ranging from 144 to 189 μm) in length, giving a high aspect ratio of 7.2. They had sharp edges and clear crystal faces, indicating high crystallinity. A gel-like solid was also coprecipitated from the solution as a second phase. XRD diffraction patterns for the mixture of acicular particles and gel-like solid showed sharp diffraction peaks of crystalline $BaC_2O_4 \cdot 0.5H_2O$ with no additional phase. Acicular particles were crystalline $BaC_2O_4 \cdot 0.5H_2O$ and the gel-like solid would be an amorphous phase. Fortunately, $BaC_2O_4 \cdot 0.5H_2O$ has a triclinic crystal structure as shown by the model calculated from structure data [13] (Fig. 3.11 XRD first step) and thus anisotropic crystal growth was allowed to proceed to produce an acicular shape. Each crystal face has a different surface energy and surface nature such as zeta potential and surface groups. Anisotropic crystal growth is induced by minimizing the total surface energy in ideal crystal growth. Additionally, site-selective adsorption of ions or molecules on specific crystal faces suppresses crystal growth perpendicular to the faces and so induces anisotropic crystal growth. These factors would cause anisotropic crystal growth of $BaC_2O_4 \cdot 0.5H_2O$ and hence allow us to control morphology and fabricate acicular $BaC_2O_4 \cdot 0.5H_2O$ particles. The positions of diffraction peaks corresponded with that of JCPDS No. 20-0134

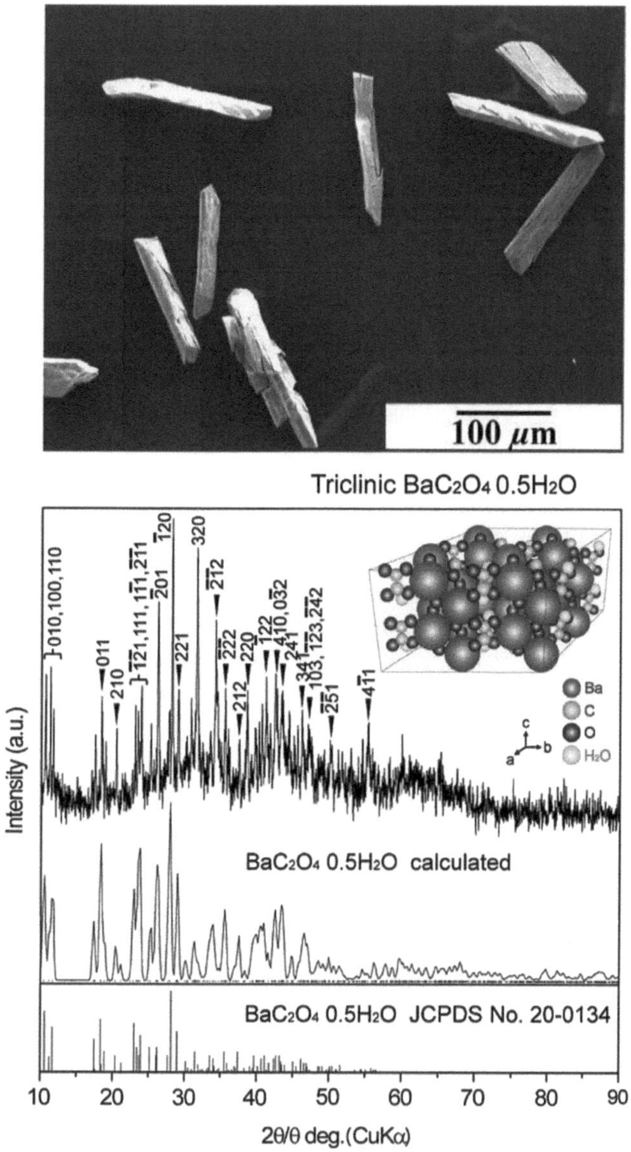

■ **FIGURE 3.11** SEM micrograph and XRD diffraction pattern of acicular $BaC_2O_4 \cdot 0.5H_2O$ particles precipitated from an aqueous solution at pH = 7. XRD diffraction measurement data (first step), XRD pattern calculated from crystal structure data16 (second step), and XRD pattern of JCPDS No. 20-134 (third step) are shown for triclinic $BaC_2O_4 \cdot 0.5H_2O$. *Reprinted with permission from Masuda Y, Yamada T, Koumoto K. Cryst Growth Des 2008;8:169. Copyright at American Chemical Society (2008).*

(Fig. 3.11 XRD third step) and that calculated from crystal structure data [13] (Fig. 3.11 XRD second step), however, several diffraction peaks, especially 320 and 201, were enhanced strongly compared to their relative intensity. The enhancement of diffraction intensity from specific crystal faces would be related to anisotropic crystal growth; a large crystal size in a specific crystal orientation increases the X-ray diffraction intensity for the crystal face perpendicular to the crystal orientation. EDX elemental analysis indicated the chemical ratio of the precipitate, which included acicular particles and gel-like solid, to be about Ba/Ti = 1 to 1.5. The chemical ratio indicated that the coprecipitated amorphous gel contained Ti ions. Additional Ba ions can be transformed into $BaCO_3$ by annealing and removed by HCl treatment in the next step. The ratio was thus controlled to slightly above Ba/Ti = 1 by adjusting the volume ratio of acicular particles and gel-like solid. Consequently, acicular particles of crystalline $BaC_2O_4 \cdot 0.5H_2O$ with Ti-containing gel-like solid were successfully fabricated in an aqueous solution process.

In comparison, isotropic particles of barium titanyl oxalate [$BaTiO(C_2O_4)_2 \cdot 4H_2O$] were precipitated at pH 2. $TiOC_2O_4$ was formed by the following reaction in which the reaction of oxalic acid ($H_2C_2O_4 \cdot 2H_2O$) with butyl titanate monomer [$(C_4H_9O)_4Ti$] and hydrolysis can take place simultaneously [14].

$$(C_4H_9O)_4Ti + H_2C_2O_4 \cdot 2H_2O \rightarrow TiOC_2O_4 + 4C_4H_9OH + H_2O \quad (3.4)$$

$TiO(C_2O_4)$ was then converted to oxalotitanic acid [$H_2TiO(C_2O_4)_2$] by the reaction:

$$TiO(C_2O_4) + H_2C_2O_4 \cdot 2H_2O \rightarrow H_2TiO(C_2O_4)_2 + 2H_2O \quad (3.5)$$

Alcoholic solution containing $H_2TiO(C_2O_4)_2$ formed by Eq. (3.5) was subjected to the following cation exchange reaction by rapidly adding an aqueous solution of barium acetate at room temperature:

$$H_2TiO(C_2O_4)_2 + Ba(CH_3COO)_2 \rightarrow BaTiO(C_2O_4)_2\downarrow + 2CH_3COOH \quad (3.6)$$

$BaTiO(C_2O_4)_2$ isotropic particles were formed by Eq. (3.6). On the other hand, neither $BaC_2O_4 \cdot 0.5H_2O$ nor $BaTiO(C_2O_4)_2$ was precipitated at pH 3 to pH 6. Gel-like solid was formed in the solution and their XRD spectra showed no diffraction peaks. The amorphous gel that precipitated at pH = 3 to 6 would be the same as the amorphous gel coprecipitated at pH 7. These comparisons show that the crystal growth and morphology control of $BaC_2O_4 \cdot 0.5H_2O$ are sensitive to the solution conditions.

The precipitate was annealed at 750°C for 5 h in air. Acicular $BaC_2O_4 \cdot 0.5H_2O$ particles were reacted with Ti-containing amorphous gel to introduce Ti ions to transform into crystalline $BaTiO_3$. X-ray diffraction of the annealed precipitate showed crystalline $BaTiO_3$ and an additional barium carbonate phase ($BaCO_3$). Excess precipitation of $BaC_2O_4 \cdot 0.5H_2O$ caused the generation of barium carbonate phase ($BaCO_3$) as expected. The annealed precipitate was further immersed in HCl solution (1 M) to dissolve barium carbonate. Acicular particles of crystalline $BaTiO_3$ were successfully fabricated with no additional phase. Particles showed acicular shape with $2.8 \times 10 \times 50$ μm and X-ray diffraction of single-phase crystalline $BaTiO_3$ (Fig. 3.12). The high aspect ratio of the particles ($17.8 = 50/2.8$) would be provided by that of $BaC_2O_4 \cdot 0.5H_2O$ particles. The particle size of acicular $BaTiO_3$ can be easily controlled by the growth period and solution concentration for $BaC_2O_4 \cdot 0.5H_2O$ precipitation, which decides the particle size of $BaC_2O_4 \cdot 0.5H_2O$. $BaTiO_3$ has a cubic crystal structure at high temperature above phase transition and has a tetragonal crystal structure at room temperature. The cubic crystal structure is completely isotropic and the tetragonal crystal structure results from stretching a cubic lattice along one of its lattice vectors. For both of the crystal structures it is difficult to control anisotropic crystal growth, however, with our newly developed process we could successfully control the morphology and fabricate acicular particles. This was achieved by controlling the morphology of triclinic $BaC_2O_4 \cdot 0.5H_2O$ to acicular shape and the phase transition to $BaTiO_3$ by introducing Ti ions from the coprecipitated amorphous phase. The novel concept can be applied to a wide variety of morphology control and crystal growth control for advanced electronic devices composed of crystalline materials.

In summary, a novel process to fabricate acicular $BaTiO_3$ particles was developed. Morphology control of crystalline $BaC_2O_4 \cdot 0.5H_2O$ to acicular shape was realized in an aqueous solution. The particles were then transformed into crystalline $BaTiO_3$ by introducing Ti ions from the coprecipitated amorphous gel phase during the annealing process. Consequently, acicular particles of tetragonal $BaTiO_3$ were produced by combining several key technologies. Morphology control in this system has high scientific value for crystal growth, and acicular particles of crystalline $BaTiO_3$ may have a great impact on ultrathin MLCC in the future.

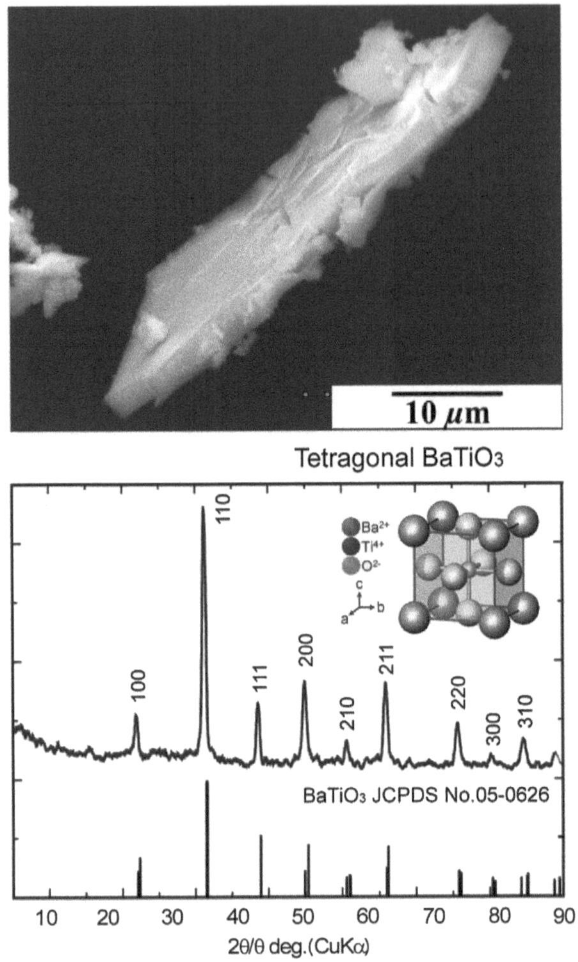

■ **FIGURE 3.12** SEM micrograph and XRD diffraction pattern of acicular $BaTiO_3$ particles after annealing at 750°C for 5 h and HCl treatment. XRD diffraction measurement data (first step) and XRD pattern of JCPDS No. 05-0626 (second step) are shown for tetragonal $BaTiO_3$. *Reprinted with permission from Masuda Y, Yamada T, Koumoto K. Cryst Growth Des 2008;8:169. Copyright at American Chemical Society (2008).*

3.3 BIOMIMETIC SITE-SELECTIVE IMMOBILIZATION OF METAL OXIDES

3.3.1 Biomimetic Site-Selective Immobilization of Anatase TiO_2 [15]

A glass substrate coated with an F-doped SnO_2 transparent conductive film (FTO, SnO_2: F, Asahi Glass Co., Ltd., 9.3–9.7 Ω/□, 26 × 50 × 1.1 mm) was blown by air to remove dust and was exposed to ultraviolet light (low-pressure mercury lamp PL16-110, air flow, 100 V, 200 W, SEN Lights Co., 14 mW/cm^2 for 184.9 nm at a distance of 10 mm from the lamp, 18 mW/cm^2 for 253.7 nm at a distance of 10 mm from the lamp) for 10 min through a photomask (Test-chart-No.1-N type, quartz substrate, 1.524 mm thickness, Toppan Printing Co., Ltd.) (Fig. 3.13) [15]. The initial SnO_2: F substrate showed a water contact angle of 96°. The UV-irradiated surface was, however, wetted completely (contact angle 0°−1°). The contact angle decreased with irradiation time (96°, 70°, 54°, 35°, 14°, 5° and 0° for 0, 0.5, 1, 2, 3, 4, and 5 min, respectively). This suggests that a small amount of adsorbed molecules on the SnO_2: F substrate was removed completely by UV irradiation. The surface of the SnO_2: F substrate would be covered by hydrophilic OH groups after irradiation. Consequently, the SnO_2: F substrate was modified to have a patterned surface with hydrophobic regions and superhydrophilic regions. Ammonium hexafluorotitanate ($[NH_4]_2TiF_6$) (2.0620 g) and boric acid (H_3BO_3) (1.8642 g) were separately dissolved in deionized water (100 mL) at 50°C. Boric acid solution was added to ammonium hexafluorotitanate solution at concentrations of 0.15 and 0.05 M, respectively. The SnO_2: F substrate having a patterned surface with hydrophobic regions and superhydrophilic regions was covered by a silicon rubber sponge sheet (Silico-sheet, SR-SG-S 5mmt RA grade, Shin-etsu Finetech Co., Ltd.) to suppress deposition of TiO_2 at the initial stage. The substrate was immersed perpendicularly in the middle of the solution (Fig. 3.13). The solution was kept at 50°C with no stirring. The silicon rubber sponge sheet was removed from the SnO_2: F substrate after 25 h, then the substrate was kept for a further 2 h at 50°C. The substrate was covered by the sheet instead of immersion of substrate at 25 h to avoid agitation of the solution. The solution became clouded in about 10 min after the mixing of ammonium hexafluorotitanate solution and boric acid solution. The particles were homogeneously nucleated in the solution and made the solution white. They then gradually

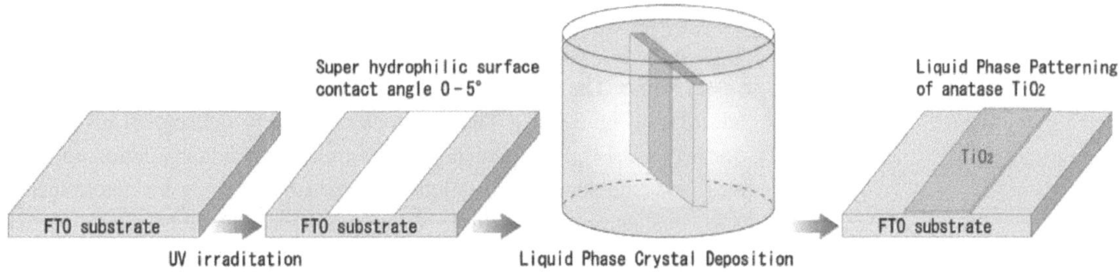

■ **FIGURE 3.13** Conceptual process for liquid phase patterning of anatase TiO_2 films using superhydrophilic surface. *Reprinted with permission from Masuda Y, Kato K. Chem Mater 2008;20(3):1057. Copyright at American Chemical Society (2008).*

precipitated and fell to the bottom of the vessel, so the solution became transparent over a period of hours. Ti ions were consumed for crystallization of TiO_2 particles, thus decreasing the ion concentration in the solution. The supersaturation degree of the solution was sufficiently low to realize slow heterogeneous nucleation without homogeneous nucleation, which forms TiO_2 particles. The silicon rubber sponge sheet was removed from the SnO_2: F substrate after 25 h, then the substrate was kept for a further 2 h at 50°C. Consequently, the patterned surface on the SnO_2: F substrate was exposed to the transparent solution including Ti ions at a low concentration for 2 h. Heterogeneous nucleation and slow crystallization of TiO_2 progressed only on the substrate. Deposition of anatase TiO_2 proceeds by the following mechanisms [5]:

$$TiF_6{}^{2-} + 2H_2O \leftrightarrows TiO_2 + 4H^+ + 6F^- \tag{3.7}$$

$$BO_3{}^{3-} + 4F^- + 6H^+ \rightarrow BF_4{}^- + 3H_2O \tag{3.8}$$

Eq. (3.7) is described in detail by the following two equations:

$$TiF_6{}^{2-} \xrightarrow{nOH^-} TiF_{6-n}(OH)_n{}^{2-} + nF^- \xrightarrow{(6-n)OH^-} Ti(OH)_6{}^{2-} + 6F^- \tag{3.9}$$

$$Ti(OH)_6{}^{2-} \rightarrow TiO_2 + 2H_2O + 2OH^- \tag{3.10}$$

Fluorinated titanium complex ions gradually change into titanium hydroxide complex ions in an aqueous solution as shown in Eq. (3.9). The increase of F^- concentration displaces Eqs. (3.7) and (3.9) to the left, however, the produced F^- can be scavenged by H_3BO_3 (BO_3^{3-}) as shown in Eq. (3.8) to displace Eqs. (3.7) and (3.9) to the right. Anatase TiO_2 formed from titanium hydroxide complex ions ($Ti(OH)_6^{2-}$) in Eq. (3.10). Liquid phase patterning was not realized in the initial solution but realized in the solution after 25 h. Solution condition was evaluated as function of time to clarify this reason. The solution was transparent immediately after the mixing of ammonium hexafluorotitanate solution and boric acid solution, became clouded after 0.5 h, and showed maximum whiteness after 1 h (Fig. 3.14A). Anatase TiO_2 particles nucleated homogeneously in the solution and grew to form large particles, which gradually precipitated and made the bottom of the vessel white. The solution became slightly white after 5 h and transparent after 25 h (Fig. 3.14A). The solutions changed to transparent by the filtrations (Fig. 3.14). Precipitated particles from the residual solution and particles from the supernatant solution trapped by filters were determined by XRD evaluation to be a single phase of anatase TiO_2. Precipitated particles from the residual solution increased as 0, 165.1, 230.9, and 424.4 mg at 0, 2, 5 and 25 h, respectively (Fig. 3.15A). The precipitation increased rapidly at

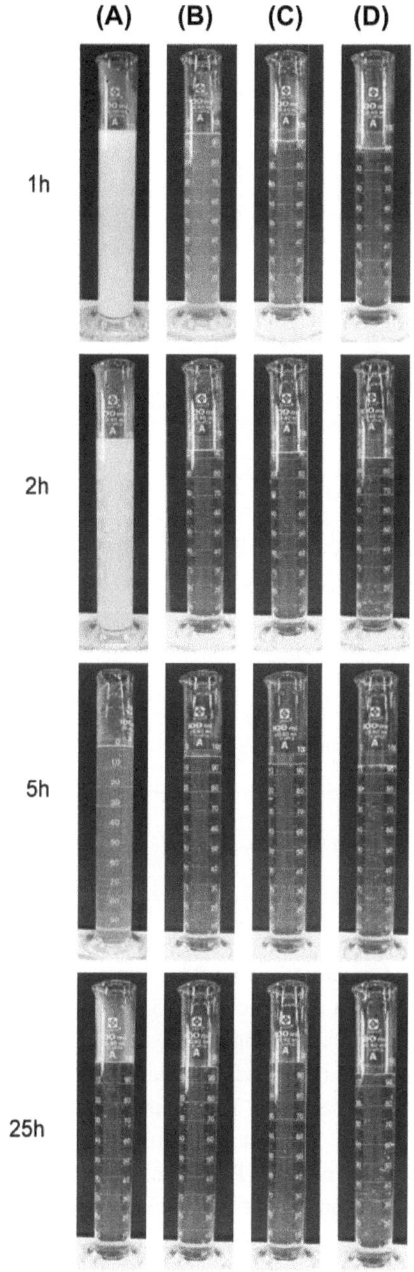

■ **FIGURE 3.14** Photographs of supernatant solutions as a function of time before and after filtration. (A) The solution before filtration contains all sizes of suspended particles. (B) The solution filtered by a filter paper (Whatman 5) contains particles $<2.5\ \mu m$. (C) The solution filtered by a glass filter paper (Whatman GF/B) contains particles $<1.0\ \mu m$. (D) The solution filtered by a filter syringe (Whatman PTFE 0.2) contains particles $<0.2\ \mu m$. *Reprinted with permission from Masuda Y, Kato K. Chem Mater 2008;20(3):1057. Copyright at American Chemical Society (2008).*

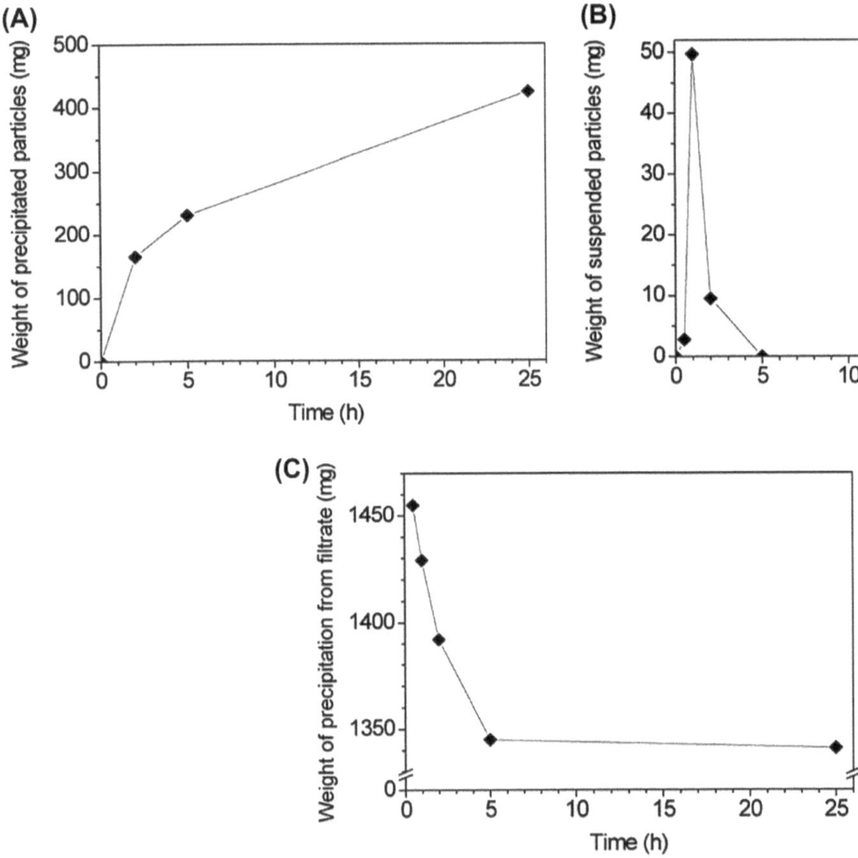

FIGURE 3.15 (A) Weight of precipitated particles >2.5 μm as a function of time. (B) Weight of suspended particles >2.5 μm in the supernatant solution as a function of time. (C) Weight of precipitation from filtrate that accumulated in all of the filters. *Reprinted with permission from Masuda Y, Kato K. Chem Mater 2008;20(3):1057. Copyright at American Chemical Society (2008).*

the initial stage and moderately after 2 h, reflecting the decrease of crystal growth rate. The weight of particles >2.5 μm in diameter was estimated to be 2.8, 49.7, 9.5, 0, and 0 mg at 0.5, 1, 2, 5, and 25 h, respectively (Fig. 3.15B). Particles were formed at the initial stage and precipitated, making the bottom of the vessel white. This is consistent with the color change of the solution (Fig. 3.14A). The precipitate from the filtrate collected by all of the filters was evaluated by XRD. The white powder contained anatase TiO_2, rutile TiO_2, and a large amount of boric acid. These were crystallized from ions in the filtrate during drying. The weight was estimated to be 1455, 1429, 1392, 1345, and 1341 mg at 0.5, 1, 2, 5, and 25 h, respectively (Fig. 3.15C). This indicated that the solution contained a high concentration of ions at the initial stage, which then decreased as a function of time. Ion concentration would decrease by the crystallization and precipitation of anatase TiO_2.

This result is consistent with the weight variation of precipitated particles (Fig. 3.15A), weight variation of suspended particles (Fig. 3.15B), and solution color change shown in the photographs (Fig. 3.14A). Liquid phase patterning was not realized in the initial clouded solution but realized in the transparent solution after 25 h. Evaluation of solution condition as function of time showed the reason for this phenomenon. TiO_2 particles formed at the initial stage around 1 h and precipitated gradually. Ions were consumed for crystallization of TiO_2 and decreased as a function of time. Heterogeneous nucleation predominantly progressed after 5 h. Consequently, TiO_2 was formed on super hydrophilic regions selectively to realize liquid phase patterning. FTO substrate was immersed in the solution for 25 h to form a thick film and ultrasonicated in water for 20 min TiO_2 film was constructed of two layers (Fig. 3.16A). Under layer with 200 nm thickness was a polycrystalline film of anatase TiO_2. Upper layer with 300 nm thickness was an assembly of acicular TiO_2 crystals that grew perpendicular to the substrate. The film was shown by electron diffraction pattern to be a single phase of anatase TiO_2 (Fig. 3.16B). Electron diffraction from the 004 plane was stronger than that of the 101, 200, 211 planes, etc. to show anisotropic crystal growth along the c-axis. Additionally, 004 diffractions were strong perpendicular to the substrate, showing that the c-axis orientation of acicular crystals was perpendicular to the substrate. The FTO layer was shown to be a single phase of SnO_2 with high crystallinity (Fig. 3.16C). Acicular TiO_2 crystals had a long shape, being ~300 nm in length and 10–100 nm in diameter (Fig. 3.16D). A lattice image of anatase TiO_2 was observed from the crystals (Fig. 3.16E). The film deposited on the substrate was evaluated by XRD analysis. Strong X-ray diffractions were observed for films deposited on FTO substrates and assigned to SnO_2 of FTO films. The 004 diffraction peak of anatase TiO_2 was not observed clearly for TiO_2 film on FTO substrates because both of the weak 004 diffraction peak of TiO_2 and the strong diffraction peak of FTO were observed at the same angle. Glass substrates with no FTO coating were immersed in the solution. Weak X-ray diffraction peaks were observed at $2\theta = 25.3°$, 37.7°, 48.0°, 53.9°, 55.1°, and 62.7° for the films deposited on glass substrates. They were assigned to 101, 004, 200, 105, 211, and 204 diffraction peaks of anatase TiO_2 (ICSD No. 9852) (Fig. 3.17). A broad diffraction peak from the glass substrate was also observed at about $2\theta = 25°$. The intensity of the 004 diffraction peak was stronger than that of the 101 diffraction peak for the film obtained by the liquid phase crystal deposition method, though the intensity of 101 was stronger than that of 004 for anatase TiO_2 powders with no orientation (ICSD No. 9852). The integral intensity or peak height of 004 was 2.6 times or 2.2 times that of 101, respectively, suggesting high c-axis orientation of anatase TiO_2 crystals. Crystallite size perpendicular to the 101 or 004

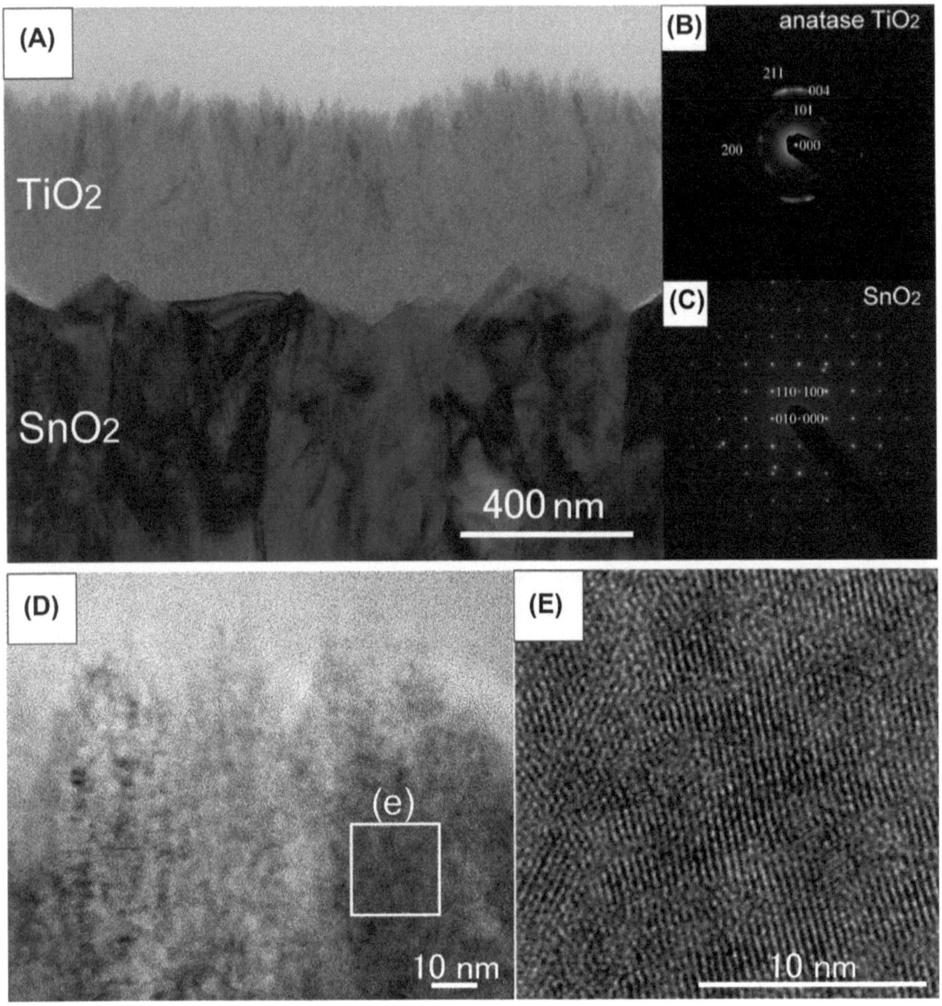

FIGURE 3.16 (A) TEM micrograph of anatase TiO_2 film on SnO_2: F substrate. (B) Electron diffraction pattern of anatase TiO_2. (C) Electron diffraction pattern of SnO_2 film. (D) Magnified area of (A) showing morphology of TiO_2 crystals. (E) Magnified area of (D) showing lattice image of TiO_2 crystals. *Reprinted with permission from Masuda Y, Kato K. Chem Mater 2008;20(3):1057. Copyright at American Chemical Society (2008).*

planes was estimated from the full-width half-maximum of the 101 or 004 peak to be 9 nm or 17 nm, respectively. Elongation of crystals in the *c*-axis direction was also suggested by the difference of crystallite size. These evaluations were consistent with high *c*-axis orientation observed by TEM and electron diffraction. Crystallite size estimated by XRD was similar to that in TiO_2 under layer rather than that of acicular crystals observed by TEM. TiO_2 thin film prepared on a glass would be constructed

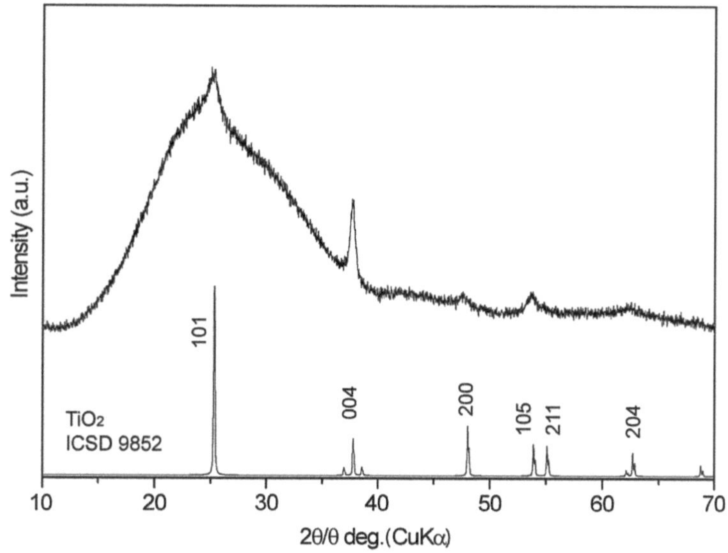

■ **FIGURE 3.17** XRD diffraction pattern of anatase TiO_2 film on a glass substrate. *Reprinted with permission from Masuda Y, Kato K. Chem Mater 2008;20(3):1057. Copyright at American Chemical Society (2008).*

not of acicular crystals but of polycrystals in under layer. After having been immersed in the solution, the substrate was rinsed with distilled water and dried in air (Fig. 3.13). The initial FTO surface appeared to be blue-green under white light due to light diffracted from the FTO layer. On the other hand, TiO_2 films deposited on the superhydrophilic surface appeared to be yellow-green. The color change would be caused by deposition of transparent TiO_2 film, which influenced the wavelength of the diffracted light. The micropattern of TiO_2 was shown by SEM evaluation to be successfully fabricated (Fig. 3.18). TiO_2 deposited on superhydrophilic regions showed black contrast, while the initial FTO regions without deposition showed white contrast in Fig. 3.18. The average line width in Fig. 3.18 is 55 μm. Line edge roughness [16], as measured by the standard deviation of the line width, is ~2.8 μm. This represents ~5% variation (i.e., 2.8/55) in the nominal line width, similar to the usual 5% variation afforded by current electronics design rules. The minimum line width of the pattern depends on the resolution of the photomask and wavelength of irradiated light (184.9 nm). It would be improved to ~1 μm by using a high-resolution photomask. The FTO layer was a particulate film having a rough surface (Fig. 3.19B1 and B2). Edged particles of 100–500 nm in diameter were observed on the surface. The micropattern of TiO_2 thin film was covered by an assembly of nanocrystals of 10–30 nm in diameter (Fig. 3.19A1

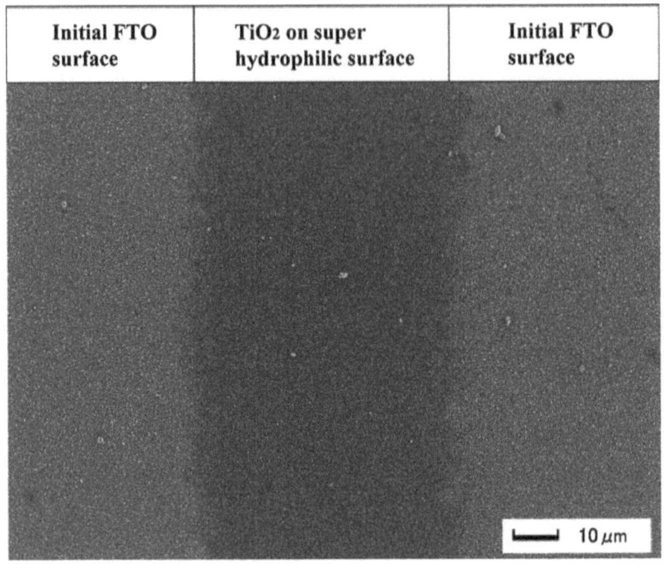

FIGURE 3.18 SEM micrograph of a micropattern of anatase TiO_2 films on SnO_2: F substrates. *Reprinted with permission from Masuda Y, Kato K. Chem Mater 2008;20(3):1057. Copyright at American Chemical Society (2008).*

and A2). The nanocrystals would be anatase TiO_2 that grew anisotropically. The TiO_2 film also had large structural relief of 100−500 nm in diameter. As the thin TiO_2 film was deposited on the edged particulate surface of the FTO layer, the surface of TiO_2 had large structural relief. The morphology of the TiO_2 layer and FTO layer was further observed by fracture cross-section profiles (Fig. 3.20). The polycrystalline FTO layer prepared on a flat glass substrate was shown to have a thickness of ~900 nm and a high roughness of 100−200 nm on the surface (Fig. 3.20A). Nano TiO_2 crystals were deposited on the superhydrophilic FTO surface (Fig. 3.20A), whereas no deposition was observed on the initial FTO surface. The superhydrophilic FTO surface was covered with an array of nano TiO_2 crystals (Fig. 3.20B and C), which had a long shape of ~150 nm in length and ~20 nm in diameter. These observations were consistent with TEM and XRD evaluations. Nano TiO_2 crystals would grow along the *c*-axis and thus enhance the 004 X-ray diffraction peak and 004 electron diffraction peak. They formed a long shape having a high aspect ratio of 7.5 (150 nm in length/20 nm in diameter) as shown in the SEM fracture cross-section profile (Fig. 3.20B and C) and TEM micrograph. The orientation of nano TiO_2 crystals with their long axis perpendicular to the FTO layer (Fig. 3.20B and C) would also enhance the 004 diffraction peak.

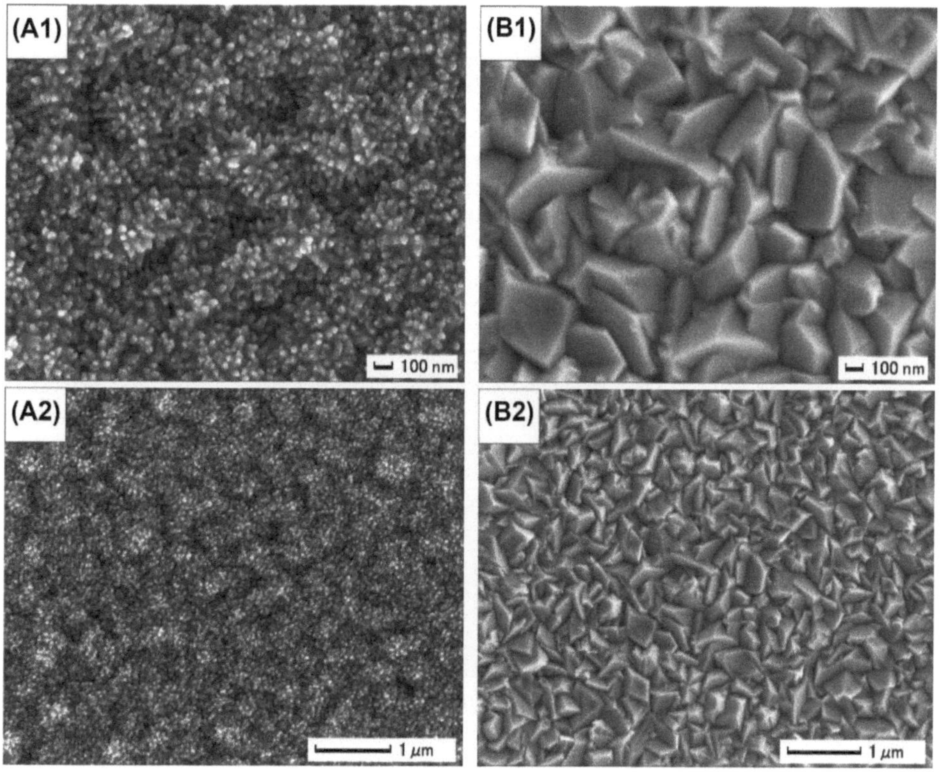

■ **FIGURE 3.19** SEM micrographs of a micropattern of anatase TiO_2 films and SnO_2: F substrates. All SEM micrographs are the magnified area of Fig. 3.18. (A1) Surface of anatase TiO_2 films deposited on superhydrophilic region. TiO_2 was formed on superhydrophilic region that was cleaned by UV irradiation before the immersion. (A2) Magnified area of (A1) showing surface morphology of anatase TiO_2 film. (B1) Surface of SnO_2: F substrate without TiO_2 deposition. TiO_2 was not formed on noncleaned region. (B2) Magnified area of (B1) showing surface morphology of SnO_2: F substrate. *Reprinted with permission from Masuda Y, Kato K. Chem Mater 2008;20(3):1057. Copyright at American Chemical Society (2008).*

In summary, a micropattern of anatase TiO_2 thin film was successfully fabricated on an SnO_2: F substrate in an aqueous solution. Crystalline anatase TiO_2 was deposited by liquid phase crystal deposition at 50°C. Nucleation and crystal growth of TiO_2 were accelerated on the superhydrophilic SnO_2: F surface, but were suppressed on the hydrophobic initial SnO_2: F surface. Consequently, liquid phase patterning of anatase TiO_2 was achieved on an SnO_2: F substrate. TiO_2 crystals were directly deposited on the SnO_2: F surface without any insulating layers that decrease the electrical conductivity between TiO_2 and the SnO_2: F substrate. The micropattern of anatase TiO_2 on the SnO_2: F substrates could be applied to electrodes of dye-sensitized solar cells or molecular sensors. Additionally, this process can be used to form a flexible micropattern of anatase TiO_2 electrodes on low-heat-resistant conductive polymer films. This process will

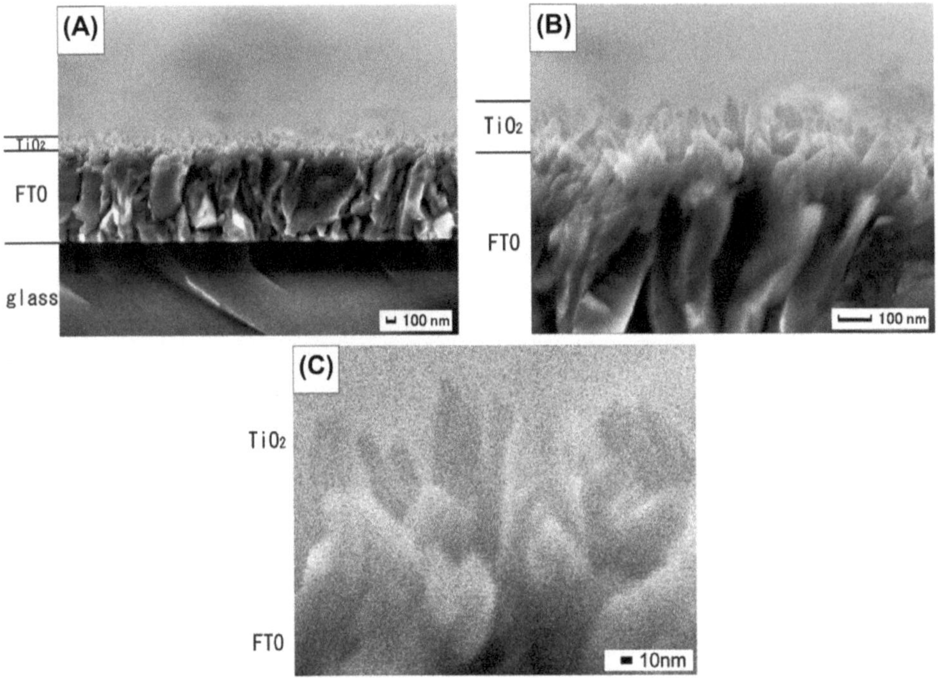

■ **FIGURE 3.20** SEM micrographs of anatase TiO_2 films on SnO_2: F substrates. (A) Fracture cross-section of TiO_2 films. (B and C) Magnified area of (A) showing morphology of nano TiO_2 crystals. *Reprinted with permission from Masuda Y, Kato K. Chem Mater 2008;20(3):1057. Copyright at American Chemical Society (2008).*

contribute to the microfabrication of TiO_2 electrodes for dye-sensitized solar cells or molecular sensors.

3.3.2 Biomimetic Site-Selective Immobilization of Eu:Y_2O_3 [17]

Si substrate [p-type (100), 1−50 Ωcm, Newwingo Co., Ltd.] was cleaned ultrasonically in acetone, ethanol, and deionized water for 5 min, respectively, in this order and was exposed to ultraviolet light and ozone gas for 10 min to remove organic contamination by using a UV/ozone cleaner (184.9 and 253.7 nm) (low-pressure mercury lamp 200 W, PL21-200, SEN Lights Co., 18 mW/cm^2, distance from lamp 30 mm, 24°C, humidity 73%, air flow 0.52 m^3/min, 100 V, 320 W) [18−21]. APTS (3-aminopropyltriethoxysilane)-SAM was prepared by immersing the Si substrate in an anhydrous toluene solution containing 1 vol% APTS for 1 h in N_2 atmosphere. The substrate was rinsed with a fresh anhydrous toluene in N_2 atmosphere. The substrate with SAM was baked at 120°C for 5 min to remove residual solvent and promote chemisorption of the

SAM. APTS-SAM was then irradiated by ultraviolet light (PL21-200) through a photomask (Test-chart-No.1-N type, quartz substrate, 1.524 mm thickness, guaranteed line width 2 μm ± 0.5 μm, Toppan Printing Co., Ltd.) for 10 min. UV irradiation modified an amino-terminated silane to a silanol forming a pattern of amino-terminated silane regions and silanol regions [18–21]. Patterned APTS-SAM having amino regions and silanol regions was used as a template for patterning of yttrium oxide. Initially deposited APTS-SAM showed water contact angles of 48°. The UV-irradiated surface of SAM was, however, wetted completely (contact angle <5°). This suggests that SAM of APTS was modified to hydrophilic OH group surfaces by UV irradiation. The patterned APTS-SAM was immersed in an aqueous solution containing $Y(NO_3)_3 \cdot 6H_2O$ (4 mM), $Eu(NO_3)_3 \cdot 6H_2O$ (0.4 mM), and NH_2CONH_2 (50 mM) at 25°C[17]. The solution was heated to 77°C gradually as shown in Fig. 3.21 since urea (NH_2CONH_2) decomposes to form ammonium ions (NH_4^+) above 70°C (Eq. (a)). The decomposition of urea at elevated temperature plays an essential role in the deposition of yttrium oxide. The aqueous solution of urea yields ammonium ions and cyanate ions (OCN^-) at temperatures above 70°C [22] (Eq. (a)). Cyanate ions react rapidly according to Eq. (b). Yttrium ions are weakly hydrolyzed [23,24] in water to $YOH(H_2O)_n^{2+}$ (Eq. (c)). The resulting release of protons (H^+) and/or

(1) Decomposition of urea (NH_2CONH_2)

$$NH_2\text{-}CO\text{-}NH_2 \rightleftarrows NH_4^+ + OCN^- \quad \text{(a)}$$

$$OCN^- + 2H^+ + H_2O \longrightarrow CO_2 + NH_4^+ \quad \text{(b)}$$

(2) Deposition of amorphous basic yttrium carbonate ($Y(OH)CO_3 \cdot XH_2O$)

$$Y(NO_3)_3 \cdot 6H_2O \longrightarrow [YOH(H_2O)_n]^{2+} + 3NO_3^- + H^+ + (5-n)H_2O \quad \text{(c)}$$

$$[YOH(H_2O)_n]^{2+} + CO_2 + H_2O \rightleftarrows Y(OH)CO_3 \cdot H_2O + 2H^+ + (n-1)H_2O \quad \text{(d)}$$

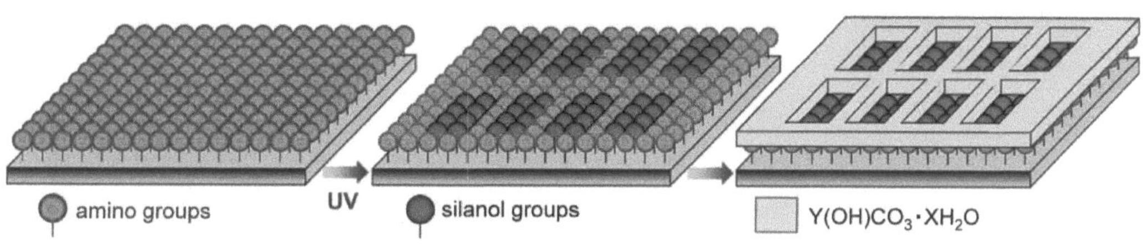

($Y(NO_3)_3 \cdot 6H_2O$ 4 mM NH_2CONH_2 50 mM
  $Eu(NO_3)_3 \cdot 6H_2O$ 0.4 mM   at 77 °C)

■ **FIGURE 3.21** Conceptual process for site-selective deposition of visible-light emitting Y_2O_3:Eu thin films using a self-assembled monolayer. *Reprinted with permission from Masuda Y, Yamagishi M, Koumoto K. Chem Mater 2007;19:1002. Copyright at American Chemical Society (2007).*

hydronium ions (H_3O^+) accelerates urea decomposition (Eq. (b)). The precipitation of the amorphous basic yttrium carbonate ($Y(OH)CO_3 \cdot xH_2O$, $x = 1$) can take place through the reaction in Eq. (d) [25,26]. The controlled release of cyanate ions by urea decomposition causes deposition of basic yttrium carbonate once the critical supersaturation in terms of reacting component is achieved. Since the decomposition of urea is quite slow, the amount needed to reach supersaturation within a given period of time must be considerably higher than the stoichiometric amount of yttrium ions, as revealed by previous studies of lanthanide compounds [27]. The temperature of the solution increased gradually and reached 77°C in about 80 min as shown in Fig. 3.22 [17]. The solution was kept at ~77°C during deposition. The pH of the solution increased from 5.2 to 5.8 in about 90 min and then gradually decreased to 5.6. Temperature and pH increased for the initial 90 min and became stable after 90 min. The average size of particles homogeneously nucleated in the solution at 100 min was about 227 nm and increased to 262 nm at 150 min, 282 nm at 180 min, 310 nm at 210 min, and 323 nm at 240 min (Fig. 3.23). Particles nucleated and grew after the solution temperature exceeded 70°C because urea decomposes above 70°C to form carbonate ions [22], which causes deposition of basic yttrium carbonate [23–26]. The particles grew rapidly at the beginning of the growth period and

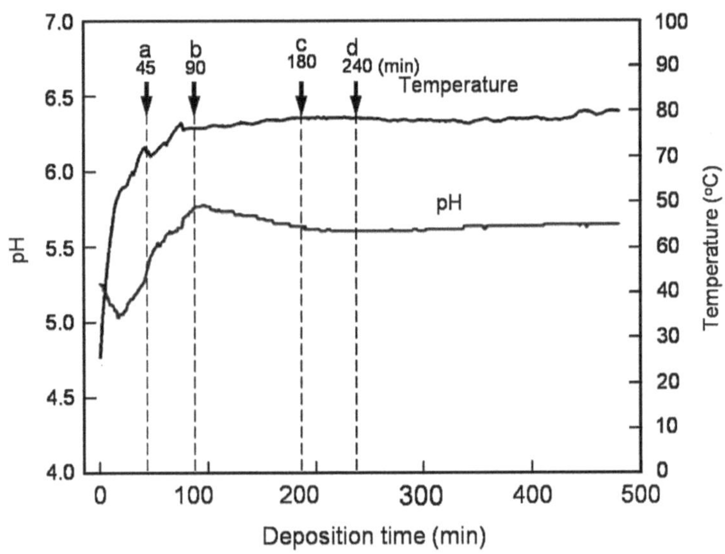

■ **FIGURE 3.22** Time variation of pH and temperature of the solution. *Reprinted with permission from Masuda Y, Yamagishi M, Koumoto K. Chem Mater 2007;19:1002. Copyright at American Chemical Society (2007).*

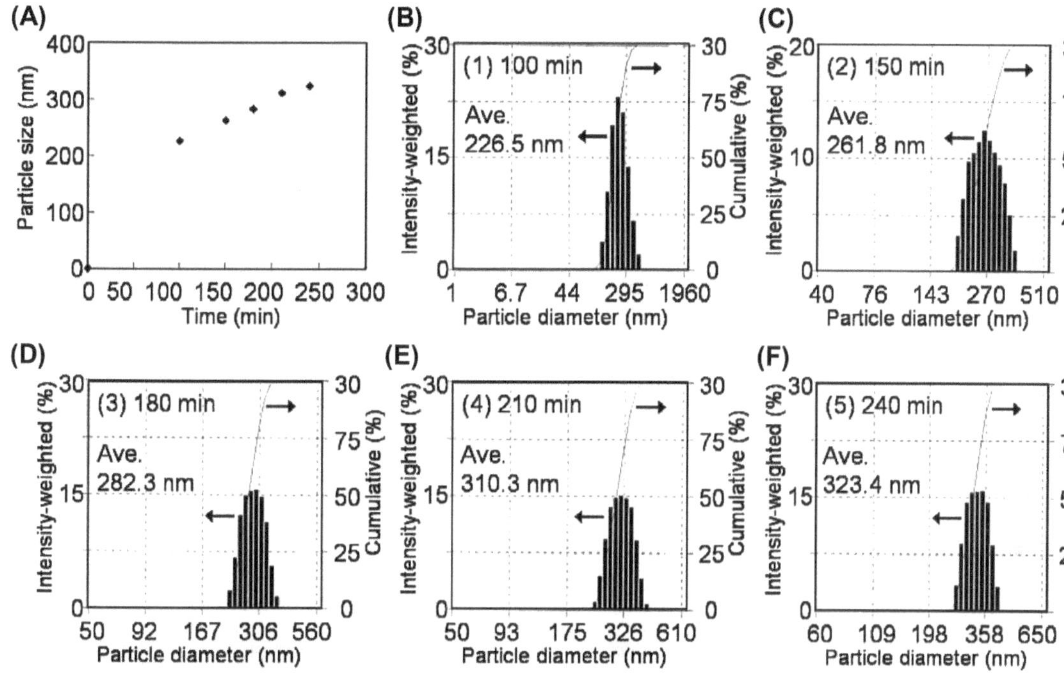

■ **FIGURE 3.23** (A) Time variation of particle size distribution. (B—F) Particle size distribution of yttrium carbonate particles at (1) 100 min, (2) 150 min, (3) 180 min, (4) 210 min, or (5) 240 min. *Reprinted with permission from Masuda Y, Yamagishi M, Koumoto K. Chem Mater 2007;19:1002. Copyright at American Chemical Society (2007).*

then their growth rate decreased exponentially (Fig. 3.23). The decrease in growth rate was caused by the decrease of supersaturation degree influenced by a decrease in solution concentration. Yttrium carbonate films were observed to deposit on amino regions of a patterned SAM after the immersion in an aqueous solution (Fig. 3.24). Deposits showed white contrast, while silanol regions without deposition showed black contrast in SEM observation. Narrow lines of depositions having 10—50 μm width were successfully fabricated in an aqueous solution. Patterned APTS-SAM showed high ability for site-selective deposition of yttrium carbonate in solution systems. Yttrium carbonate films were also deposited on the hydrophobic octadecyl surface of octadecyltrichlorosilane (OTS)-SAM having water contact angle (WCA) of 116° and as-purchased silicon wafer having WCA of about 20°—50°, which was kept in a plastic case in air. On the other hand, the films were not deposited on UV irradiated silicon wafer having WCA <5°. The super hydrophilic surface of WCA <5° suppressed film deposition, whereas the hydrophobic surface and medium surface of WCA >20°—30° accelerated film deposition possibly because of

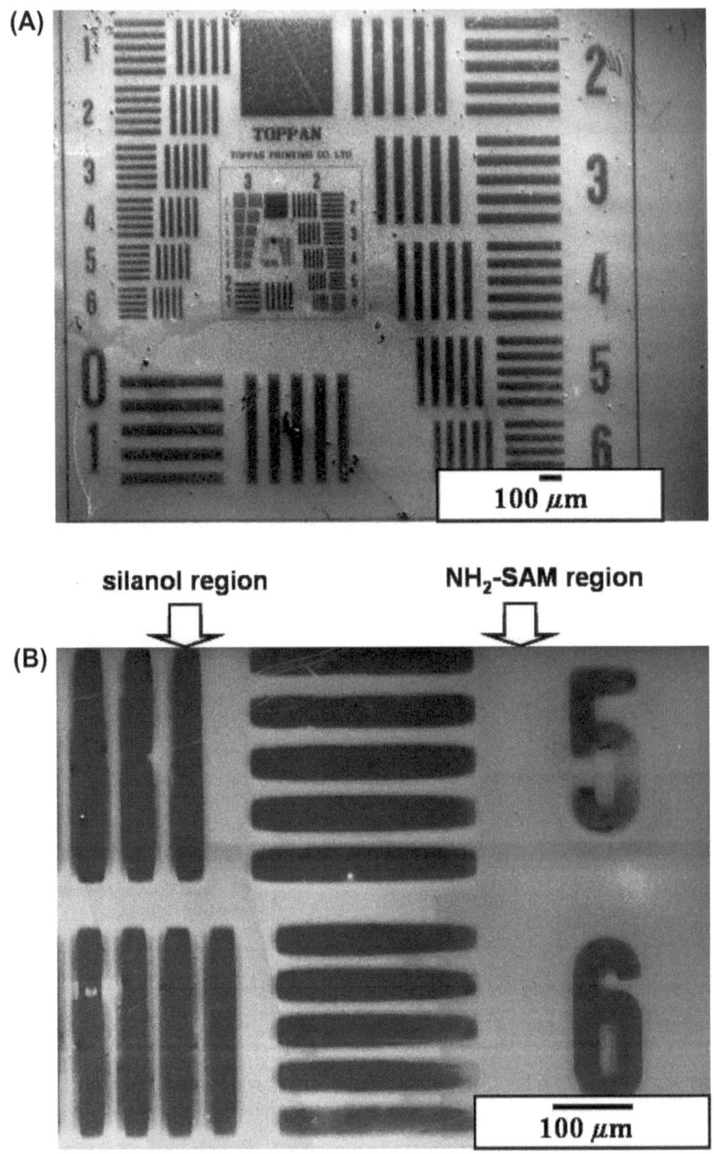

■ **FIGURE 3.24** (A) SEM micrograph of patterned Y_2O_3:Eu thin films and (B) magnified area of (A). *Reprinted with permission from Masuda Y, Yamagishi M, Koumoto K. Chem Mater 2007;19:1002. Copyright at American Chemical Society (2007).*

hydrophobic interaction between deposition and substrate surface. This is consistent with a former study [32]. Yttrium carbonate was deposited both on bare single crystal Si wafers and on Si wafers coated with sulfonate-functionalized organic self-assembled monolayers. Yttrium, europium, oxygen, and carbon were observed from as-deposited thin films on amino regions, while silicon and oxygen were detected from noncovered silanol regions by EDX (Fig. 3.25). The molecular ratio of yttrium to europium was determined to be 100: 8 (Fig. 3.26). It was close to that of

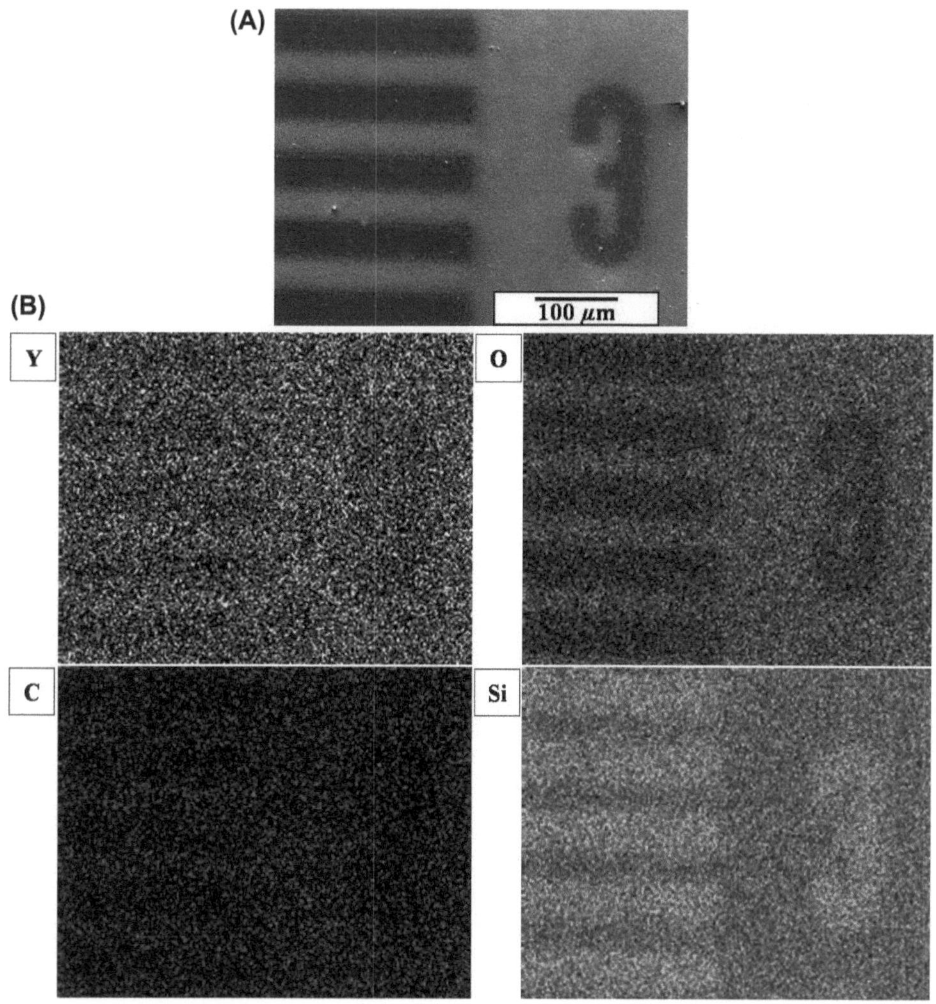

■ **FIGURE 3.25** (A) SEM micrograph of patterned Y_2O_3:Eu thin films. (B) Characteristic X-ray images of Y, O, C, and Si for (A) Y_2O_3:Eu thin films. *Reprinted with permission from Masuda Y, Yamagishi M, Koumoto K. Chem Mater 2007;19:1002. Copyright at American Chemical Society (2007).*

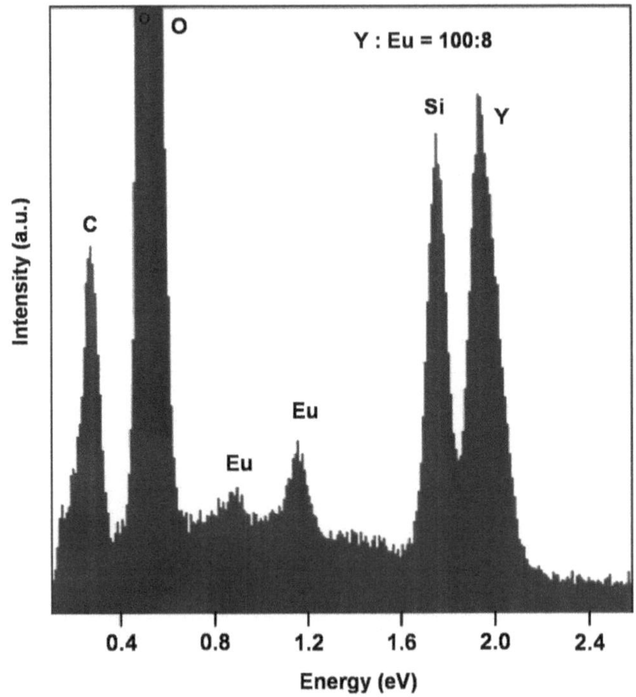

■ **FIGURE 3.26** Elemental analysis of Y_2O_3:Eu thin films deposited for 90 min. *Reprinted with permission from Masuda Y, Yamagishi M, Koumoto K. Chem Mater 2007;19:1002. Copyright at American Chemical Society (2007).*

$Y(NO_3)_3 \cdot 6H_2O$ to $Eu(NO_3)_3 \cdot 6H_2O$, i.e., 100: 10, in the solution because the chemistry of $Eu(NO_3)_3$ is similar to that of $Y(NO_3)_3$ to incorporate europium in the precipitation. The content of europium was in the range we had expected. Y_2O_3:Eu with atomic ratio Y: Eu = 100: ~8 was reported to have strong photoluminescence [28,29]. Carbon was detected from yttrium carbonate. Silicon and oxygen were detected from silicon wafer covered with a natural oxide layer (amorphous SiO_2). Amino regions were covered with thin films composed of many large particles (about 100–300 nm in diameter) and very high roughness (RMS 25.6 nm) (Fig. 3.27). Silanol regions, on the other hand, showed only nano-sized small particles (about 10–50 nm in diameter) and very low roughness (RMS 1.7 nm). The high site selectivity of deposition and the big difference in surface morphology and roughness were clearly shown by AFM observation. The thickness of the films was estimated from AFM scans across deposited and undeposited regions of the substrate. It increased with immersion time after 45 min (0 nm at 45 min, 60 nm at 70 min, and

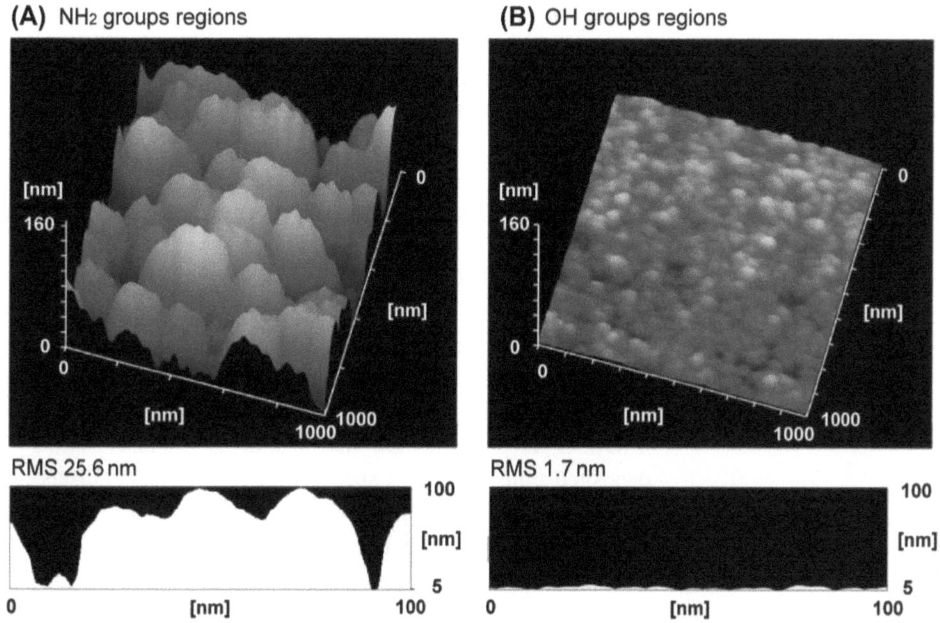

■ **FIGURE 3.27** (A) AFM images and cross-section profile of Y_2O_3:Eu thin films on NH2 groups regions. (B) AFM images and cross-section profile of Y_2O_3:Eu thin films on OH groups regions. *Reprinted with permission from Masuda Y, Yamagishi M, Koumoto K. Chem Mater 2007;19:1002. Copyright at American Chemical Society (2007).*

100 nm at 90 min (Fig. 3.27)). The average growth rate (70 nm/h = 100/90 min) was higher than that previously reported (2 nm/h = 35 nm/15 h) [26]. An amorphous yttrium basic carbonate film was deposited at 80°C from aqueous solutions of $YNO_3 \cdot 5H_2O$ and urea on Si wafers coated with sulfonate-functionalized organic self-assembled monolayers in previous studies. The thickness was then evaluated by TEM after the treatment with ultrasonication for half an hour in distilled water. The difference of growth rate was caused mainly by the difference of the substrate treatment by ultrasonication. Additionally, the thickness of our film was smaller than the particle size in the solution shown in Fig. 3.23 (227 nm at 100 min). Heterogeneous nucleation and attachment of initial particles of yttrium carbonate occurred without the attachment of aggregated large particles shown in Fig. 3.23. The yttrium carbonate was then grown on the substrate to form a film of 100 nm thickness after immersion for 90 min. The particles of about 100 nm in height were removed by ultrasonication for 30 min and the film of several nm in height remained as reported [26]. Yttrium was not detected by XPS from the substrate immersed for 45 min, however, it was clearly observed from that immersed for 90 min. This indicates that the deposition began between 45 and 90 min after immersion. The solution

temperature reached 70°C in ~45 min (Fig. 3.22) and then the solution began to decompose and release carbonate ions, causing the deposition of basic yttrium carbonate. The deposition mechanism evaluated by XPS is consistent with the change of solution temperature, decomposition temperature of urea, and chemical reaction of this system. The binding energy of Y 3d5/2 spectrum from the deposition (158.2 eV) was higher than that of metal yttrium (155.8 eV) [30]. The spectrum shifted to lower binding energy (156.7 eV) after annealing at 800°C in air for 1 h and is similar to that of Y_2O_3 (157.0 eV) [31]. The binding energies of Y 3d5/2 spectra in as-deposited films and annealed films were higher than that of metal yttrium possibly due to the chemical bonds formed between yttrium ions and oxygen ions. The chemical shift of Y 3d5/2 binding energy by annealing is consistent with crystallization of as-deposited films to crystalline Y_2O_3. C 1s spectra were detected at 289.7 and 284.6 eV from as-deposited films. The C 1s spectrum at 289.7 eV then disappeared by the annealing. C 1s at 284.6 eV was assigned to surface contamination and C 1s at 289.7 eV was detected from as-deposited yttrium carbonate. The disappearance of C 1s at 289.7 eV is consistent with the phase transition from yttrium carbonate to Y_2O_3. As-deposited film was shown to be an amorphous phase (Fig. 3.28A) by XRD measurement. The film showed no diffraction peak after annealing at 400°C for 1 h, however, it showed 222, 400, and 440 diffraction peaks of crystalline cubic Y_2O_3 [32] without any additional phase after annealing at 600°C for 1 h, and the intensities of diffraction peaks increased further by annealing at 800°C for 1 h (Fig. 3.28B). The film was shown to be a polycrystalline Y_2O_3 film constructed from randomly deposited Y_2O_3 particles without crystal–axis orientation. The crystal structure model and diffraction pattern of Y_2O_3 were calculated from the crystal structure data of ICSD #23811 as shown in Fig. 3.28. The crystallization by annealing confirmed from XRD measurement is consistent with XPS evaluation. We attempted to remove Y_2O_3 films from the silicon substrate by debonding with Scotch Tape or by ultrasonication for 5 min in water. However, the films maintained their bonds with the substrate, indicating that strong adhesion had formed between films and substrate. The thin film annealed at 800°C for 1 h, i.e., crystalline Y_2O_3:Eu thin film, was shown to be excited by 230–250 nm (center: 243 nm) and emit red light photoluminescence centered at 611 nm in the fluorescence excitation spectrum (Fig. 3.29A). Neither the as-deposited film nor the film annealed at 400°C for 1 h showed photoluminescence, on the other hand, the films annealed at 600°C or 800°C for 1 h emitted light centered at 617 nm by 250 nm in fluorescence emission spectra (Fig. 3.29B). The fluorescence intensity of the film annealed at 800°C was stronger than that of the film annealed at 600°C. Fluorescence

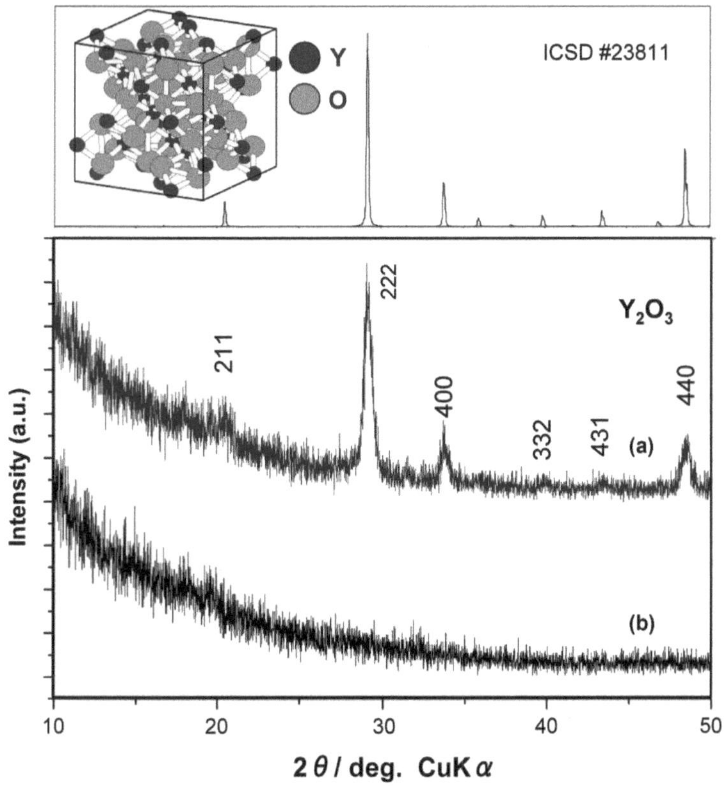

■ **FIGURE 3.28** XRD patterns of Y_2O_3:Eu thin films (A) before and (B) after annealing at 800°C for 1 h. (The upper picture) Crystal structure model and diffraction pattern of cubic Y_2O_3 calculated from crystal structure data of ICSD #23811. *Reprinted with permission from Masuda Y, Yamagishi M, Koumoto K. Chem Mater 2007;19:1002. Copyright at American Chemical Society (2007).*

intensity increased by the phase transformation from amorphous yttrium carbonate to yttrium oxide and crystal growth by the heat treatments, and is consistent with the crystallization observed by XRD [32]. The spectra are described by the well-known $^5D_0$—^7F_J line emissions (J = 0, 1, 2, ...) of the Eu^{3+} ion with the strongest emission for J = 2 at 612 nm. The thin film annealed at 800°C produced visible red light photoluminescence by excitation from Nd: YAG laser (266 nm) (Fig. 3.29, inset). The white square shows the edges of the Y_2O_3:Eu thin film and the red color shows visible red emission from the irradiated area on the substrate.

In summary, we have proposed a novel process for fabricating visible red light emitting Eu-doped Y_2O_3 and its micropattern using a self-assembled

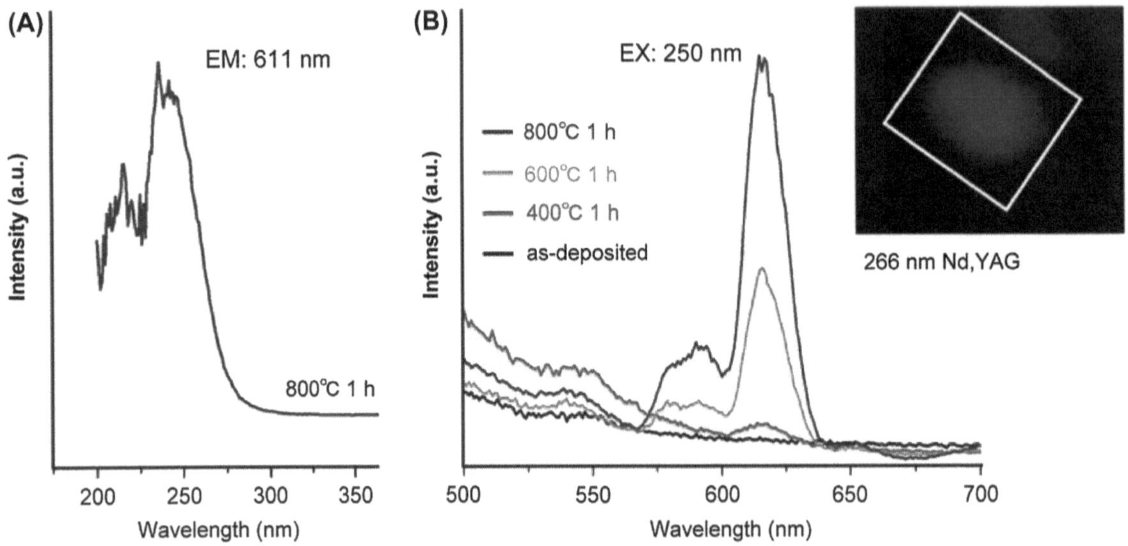

FIGURE 3.29 (A) Fluorescence excitation spectrum (emission: 611 nm) for Y_2O_3:Eu thin film after annealing at 800°C for 1 h. (B) Fluorescence emission spectra (excitation: 250 nm) for Y_2O_3:Eu thin films before and after annealing at 400°C, 600°C, or 800°C for 1 h. Inset: Photoluminescence image for Y_2O_3:Eu thin film annealed at 800°C for 1 h (excitation: 266 nm). *Reprinted with permission from Masuda Y, Yamagishi M, Koumoto K. Chem Mater 2007;19:1002. Copyright at American Chemical Society (2007).*

monolayer and an aqueous solution system. The patterned APTS-SAM having amino groups regions and silanol groups regions achieved site-selective deposition of yttrium oxide in an aqueous solution. The deposited films were crystallized by annealing at 600°C or 800°C for 1 h. Crystalline Y_2O_3:Eu produced visible red light photoluminescence centered at 611 nm by excitation from Nd: YAG laser (266 nm). This study showed the high potential of aqueous solution systems and self-assembled monolayers for the fabrication of functional metal oxide thin films and their micropatterns.

3.4 SUMMARY

Biomimetic morphology control of metal oxides and their site-selective immobilization were successfully realized biomimetically. They will contribute to development of future nanoscience and nanotechnology. Nature usually tells us a lot of valuable suggestions. We should respect nature to learn more about the way for us to go.

REFERENCES

[1] Masuda Y, http://staff.aist.go.jp/masuda-y/index.html.
[2] Masuda Y, Kato K. Cryst Growth Des 2008;8(1):275–9.
[3] Gao MD, Li MM, Yu WD. J Phys Chem B 2005;109:1155–61.
[4] Masuda Y, Kato K. Cryst Growth Des 2008;8(9):3213–8.
[5] Masuda Y, Sugiyama T, Seo WS, Koumoto K. Chem Mater 2003;15:2469–76.

[6] Furlong DN, Parfitt GD. J Colloid Interface Sci 1978;65:548–54.
[7] Wahi RK, Liu YP, Falkner JC, Colvin VL. J Colloid Interface Sci 2006;302:530–6.
[8] Kruk M, Jaroniec M. Chem Mater 2001;13:3169–83.
[9] Ravikovitch PI, Odomhnaill SC, Neimark AV, Schuth F, Unger KK. Langmuir 1995; 11:4765–72.
[10] Lastoskie C, Gubbins KE, Quirke N. J Phys Chem 1993;97:4786–96.
[11] Katagiri K, Ohno K, Masuda Y, Koumoto K. J Ceram Soc Jpn 2007;115:831–4.
[12] Masuda Y, Yamada T, Koumoto K. Cryst Growth Des 2008;8:169–71.
[13] Mutin JC, Dusausoy Y, Protas J. J Solid State Chem 1981;36:356–64.
[14] Potdar HS, Deshpande SB, Date SK. J Am Ceram Soc 1996;79:2795–7.
[15] Masuda Y, Kato K. Chem Mater 2008;20(3):1057–63.
[16] Masuda Y, Sugiyama T, Lin H, Seo WS, Koumoto K. Thin Solid Films 2001;382: 153–7.
[17] Masuda Y, Yamagishi M, Koumoto K. Chem Mater 2007;19:1002–8.
[18] Dressick WJ, Calvert JM. Jpn J Appl Phys 1993;32:5829–39.
[19] Collins RJ, Shin H, De Guire MR, Heuer AH, Sukenik CN. Appl Phys Lett 1996;69: 860–2.
[20] Masuda Y, Koumura T, Okawa T, Koumoto K. J Colloid Interface Sci 2003;263: 190–5.
[21] Masuda Y, Itoh M, Yonezawa T, Koumoto K. Langmuir 2002;18:4155–9.
[22] Shaw WHR, Bordeaux JJ. J Am Chem Soc 1955;77:4729–33.
[23] Ryabchikov DE, Ryabukhin VA. Analytical chemistry of yttrium and the lanthanide elements. Ann Arbor, MI: Humphrey Science; 1970.
[24] Baes CF, Mesmer RE. The hydrolysis of captions. New York: Wiley; 1976.
[25] Aiken B, Hsu WP, Matijevic E. J Am Ceram Soc 1988;71:845–53.
[26] Agarwal M, DeGuire MR, Heuer AH. Appl Phys Lett 1997;71:891–3.
[27] Matijevic E, Hsu WP. J Colloid Interface Sci 1987;118:506–23.
[28] Sharma PK, Jilavi MH, Nass R, Schmidt H. J Lumin 1999;82:187–93.
[29] Kwaka MG, Parkb JH, Shon SH. Solid State Commun 2004;130:199–201.
[30] Fuggle JC, Martensson N. J Electron Spectrosc Relat Phenom 1980;21:275.
[31] Wagner CD. Practical surface analysis. 2 ed., vol. 1. New York: John Wiley; 1990.
[32] Paton MG, Maslen EN. Acta Crystallogr 1967;1:1948–2023.

Molecular Surface Arrangement to Control Dynamic Dewettability

M.W. England and A. Hozumi

National Institute of Advanced Industrial Science and Technology (AIST), Nagoya, Japan

CHAPTER OUTLINE

4.1 INTRODUCTION

The adherence of liquid droplets (water or oils) to solid surfaces can cause corrosion, deterioration, degradation of appearance, and reduction of visibility, and thereby significantly damage the safety and reliability of everyday devices and industrial equipment. For this reason, the development of surface treatments that show excellent performance in liquid removal is of high importance, with extensive research on the control of adhesion and dewetting of liquid drops from solid surfaces having been reported so far. Most studies related to this field have particularly focused on the effects of surface chemical makeup [1−17], chemical structure [1,4−6,18], and topography [6,19−44] on the wetting/dewetting properties of modified surfaces. Recently, many researchers have taken inspiration from nature (most notably, from the lotus leaf surface or water strider's leg) to achieve amazing surface dewetting properties against water [45−47]. These "superhydrophobic" surfaces, which artificially mimic those found in nature, generally show

Supra-materials Nanoarchitectonics. http://dx.doi.org/10.1016/B978-0-323-37829-1.00004-3

extremely large static contact angles (θ_S, greater than 150°). Water droplets on such surfaces are unstable and can easily move, needing only low substrate tilt angles (θ_T, less than 5–10°) to roll off. On the other end of the spectrum, there are adhesive/sticky surfaces on which water drops remain pinned to the surface regardless of the magnitude of their θ_S values [25]. Rose petal surfaces are a good example; they exhibit "superhydrophobicity," but at the same time are strongly adhesive toward water (the so-called "rose petal effect") [25]. In addition, such "pinning" phenomena have also been frequently observed on conventional hydrophobic (θ_S values of ~120°) glass windows and car windshields, even after treatment with popularly used perfluoroalkylsilanes, as can be seen on rainy days. As is obvious from this example, it is clear that surface dewetting properties cannot be characterized by a single parameter, and certainly not by conventional θ_S values alone.

To accurately characterize surface dewetting properties, measurements of dynamic dewetting behavior, including the advancing (θ_A) and receding (θ_R) contact angles (CAs), CA hysteresis, and θ_T values, are required [10–17]. CA hysteresis is the mass-independent measure of the resistance to macroscopic liquid drop movement on inclined surfaces. It recognizes that in order for the gravitational force to cause a liquid drop to depin from a stationary position on a horizontal surface and begin sliding/rolling down an inclined surface, the shape of the liquid drop must first undergo a certain amount of deformation, as shown in Fig. 4.1.

Assuming that the solid–liquid contact area remains constant or experiences a relatively insignificant change, this deformation from its typical spherical dome/cap shape results in an increase in the drop's liquid–vapor interfacial area. The energy associated with this increase has been described as the activation energy barrier to liquid drop motion down an inclined surface [23,24]. From the initial horizontal position (at θ_S), the drop begins to move when its contact lines at the front and back ends are capable of

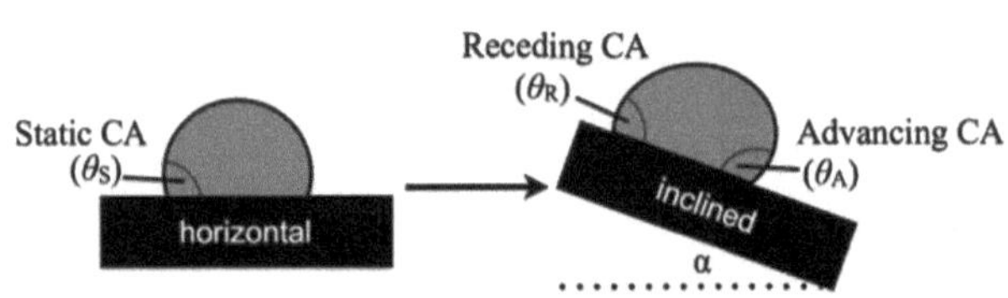

■ **FIGURE 4.1** Comparison of liquid drops on horizontal and inclined surfaces. *Reprinted from Cheng DF, Urata C, Masheder B, Hozumi A. A physical approach to specifically improve the mobility of alkane liquid drops. J Am Chem Soc 2012;134(24):10191–99 with kind permission of copyright The American Chemical Society (2012).*

advancing and receding, respectively. These advancing and receding events, which occur at θ_A and θ_R, are also the maximum and minimum possible CAs, respectively, for a liquid–surface pair. CA hysteresis can be described using θ_A and θ_R as follows [1–24].

$$\Delta\theta = \theta_A - \theta_R \tag{4.1}$$

The mobility of probe liquids on a surface can be also examined using another definition of CA hysteresis as calculated by Eq. (4.2) [11,13–17,48,49]:

$$\Delta\theta_{\cos} = \cos\theta_R - \cos\theta_A \tag{4.2}$$

The gravitational force required to initiate drop movement across a tilted solid surface dependent on $\Delta\theta_{\cos}$ is described by Eq. (4.3), where m is the mass of the drop, g is the gravitational constant, α is θ_T, k is a constant that depends on drop shape, w is the width of the drop, and γ_{LV} is the liquid–vapor surface tension [48,49]:

$$F = mg\cdot\sin\alpha = kw\gamma_{LV}(\cos\theta_R - \cos\theta_A) \tag{4.3}$$

According to Eq. (4.3), $\Delta\theta_{\cos}$ shows significantly stronger correlation with the resulting θ_T values of a liquid drop on a surface than with the conventional definition of CA hysteresis ($\Delta\theta$) [11]. Therefore, when CA hysteresis is small enough a liquid drop will only require a low θ_T value to depin from a horizontal position and begin moving down the surface, without substantial deformation of its shape. Conversely, drops can remain permanently pinned to vertically oriented surfaces, regardless of their magnitude of θ_S, if the CA hysteresis is significant.

To achieve superhydrophobic surfaces that artificially mimic lotus leaves, micro/nanostructured surfaces and subsequent surface treatments with low surface energy components (alkyl or perfluoroalkyl groups) are typically required. To create these structured surfaces a variety of approaches [26], including photolithography [19,20], electrospinning [27,28], spin/dip coating [29,30], self-assembly [31–33], polymerization [22,33,34], chemical vapor deposition (CVD) [6,35], and chemical etching [36,37] have so far been used. Moreover, superoleophobic surfaces, which demonstrate θ_S values greater than 150° for low surface tension (LST) liquids [surface tension (γ = lower than 27 dyn/cm)], such as oils and alkanes etc., have also recently emerged as a rapidly growing topic [38–40,42–44]. However, in contrast to the extensive research on superhydrophobic surfaces [6,19,20,22,26–37], there has been far less published on superoleophobic surfaces, as their fabrication is still very challenging. In addition to having extremely large CAs, they must also allow small volumes of LST liquids

(3−5 μL) to easily roll across and off the surface at small θ_T values (less than 5−10°), and in order to achieve such surfaces, their surface energies should be as low as possible. Thus, most studies related to this research field have unsurprisingly employed perfluorinated compounds (PFCs), particularly long-chain perfluoroalkylsilanes (FASs) [38−40,42,43] or perfluorinated fluids [41] to reduce interaction between the surface and the LST liquids by taking advantage of the extremely low surface energies of perfluoroalkyl groups ($CF_3(CF_2)_n$−) [50]. Unfortunately, the potential chemical and physical effects of longer-chain PFCs (in particular, $n \geq 7$) on human health and the environment have lately caused them to be viewed with suspicion [51,52]. Thus, alternative methods that do not require long-chain PFCs have been in high demand. In response to these practical and social demands, there have been increasing calls for new surfaces showing highly effective dewetting properties toward LST liquids but that do not rely on conventional surface roughening and subsequent perfluorination. In addition, from the viewpoint of practical application, alternative methods that use more ubiquitous elements are also in demand, since they have a lower cost financially and place less of a burden on the environment.

To overcome these problems, we have focused on the creation of flat/smooth surfaces showing negligible CA hysteresis toward various probe liquids [5,6,10−17]. Probe liquids, (particularly LST liquids) on these surfaces have been shown to move very easily, and roll off when only slightly tilted, regardless of the magnitude of their θ_S values. Unfortunately, there have only been a limited number of papers describing the preparation techniques and properties of flat/smooth low CA hysteresis surfaces compared to roughened superhydrophobic surfaces. McCarthy's group previously reported several such surfaces that showed low CA hysteresis toward water and *n*-hexadecane, and were prepared using ultrathin layers of low-molecular-weight poly(dimethylsiloxane) (PDMS) [7,8], specifically featured monolayers [4,18] or by controlling packing density through shortened reaction times [1]. Such surfaces typically possess a similar, "liquid-like" physical nature [1,8,23], and subsequently produce excellent dynamic dewetting behavior, allowing LST liquid (e.g., alkane) droplets to easily dewet from the surface at low θ_T values, independent of their CA magnitude. Liquid-like surface properties typically arise when flexible, surface-tethered functional groups have a high degree of freedom to move, which allows probe droplets in contact with them to experience very low energy barriers between metastable states [1,8,23]. This results in the formation of low CA hysteresis surfaces. However, the relationship between molecular structure or packing density of surface-tethered functional groups and the resulting dynamic dewetting properties has not been clearly identified. In addition, the fabrication processes required to prepare surfaces with negligible CA hysteresis are not yet well established, particularly for substrates other than Si [5,6,12].

This chapter will give an introduction to our approach toward realizing excellent dynamic dewetting behavior through molecular surface arrangement, not only against water but also LST liquids, by appropriate selection of precursor molecules (using molecular structures and molecular weights (MWs)) or controlling the packing densities of surface-tethered molecules to regulate CA hysteresis.

4.2 LOW CONTACT ANGLE HYSTERESIS SURFACES

4.2.1 Monolayers

Among the various surface modification techniques reported so far, attachment of functional organosilanes (general formula R_nSiX_{4-n}; $R =$ functional groups; $n = 1, 2, 3$; X = usually Cl, H, OH, or OR) to solid surfaces from a liquid or vapor phase, is the most common [53]. In particular, CVD is a facile and widely reproducible method for fabricating well-ordered organosilane monolayers [54–56]. It is well known that monofunctional silanes (R_3-SiX) react with hydroxyl (OH) groups on Si or metal (M) surfaces to give covalent Si(M)-O-Si bonds [1,2,4,5,18], while trifunctional silanes (R-SiX_3) react strongly with their neighbors and, to a lesser extent, the Si or metal surfaces [2,57,58]. Under certain conditions, trifunctional silanes with long-chain functional groups will form closely packed [59], highly ordered self-assembled monolayers (SAMs), and a wide variety of SAMs terminated with various functional groups have been employed to control surface properties. Among them, SAMs terminated with perfluoroalkyl groups, such as *1H, 1H, 2H, 2H*-perfluorodecyltrimethoxysilane ($CF_3(CF_2)_7CH_2CH_2Si(OCH_3)_3$; FAS17), for example, are of particular interest because they are chemically inert and exhibit both hydrophobic and oleophobic behavior [5,54]. Thus, perfluoroalkylsilane-derived SAMs are expected to reduce contact friction, adhesion, and wear, and to improve corrosion resistance [60–63]. However, in contrast to the formation of SAMs or monolayers on Si surfaces, preparations on metal surfaces, particularly on aluminum (Al) and titanium (Ti) and their oxides, have rarely been reported. In addition, no examples of metal surfaces that exhibit low CA hysteresis have been reported, in spite of their clear practical advantages. So, in accordance with previous reports of low CA hysteresis Si surfaces, we began by demonstrating surface modification of Al using a vapor of monofunctional perfluoroalkylsilane to investigate the effects of precursor molecular structures on the resulting dynamic dewettability [5]. We specifically focused on a branch-featured monofunctional silane, i.e., bis((tridecafluoro-1,1,2,2,-tetrahydrooctyl)-dimethylsiloxy)methylsilane ($CF_3(CF_2)_5CH_2CH_2Si(CH_3)_2O)_2SiCH_3H$:$R^FSi(Me)_2O)_2SiMeH$), as these bulky $R^FSi(Me)_2O)_2SiMeH$ molecules were expected to form random, covalent attachments with respect to the oxidized Al surface (Fig. 4.2) [5].

Both XPS and AFM results indicated that the CVD method produced a monolayer of ($R^FSi(Me)_2O)_2SiMeH$ molecules, with a thickness of 1.1 nm, on the oxidized Al surfaces without any discernible change in surface morphology [5]. Because of the umbrella effect of this monolayer, probe liquids were hardly able to penetrate through to interact with the residual surface aluminols (Al−OH). And, due to their random configuration and spacing, the functional groups of the ($R^FSi(Me)_2O)_2SiMeH$-derived monolayer could rotate freely, causing the surface to show liquid-like properties, as shown in Fig. 4.3. This minimized the activation barriers between contact line metastable states, resulting in the formation of a surface with virtually no CA hysteresis, giving θ_A/θ_R values for water and n-hexadecane of 110/109° [images (a) and (b) in Fig. 4.4] and 52°/50°, respectively. Droplets moved very easily on this surface and could roll off of even slightly tilted surfaces independent of the magnitude of their CAs. As a control, a conventional perfluoroalkylsilane SAM was also prepared using FAS17 vapor. In spite of the fact that the FAS17 layer's thickness was nearly identical (1.2 nm), its surface dewetting properties were completely different from those of the ($R^FSi(Me)_2O)_2SiMeH$-derived monolayer. The FAS17 layer's θ_A/θ_R values for water and n-hexadecane showed considerably greater CA hysteresis {121°/106° [images (c) and (d) in Fig. 4.4] and 76°/71°, respectively} because of the lower flexibility of its perfluoroalkyl groups. Moreover, due to the negligible CA hysteresis displayed by the ($R^FSi(Me)_2O)_2SiMeH$-treated Al surface, all samples coated with it retained

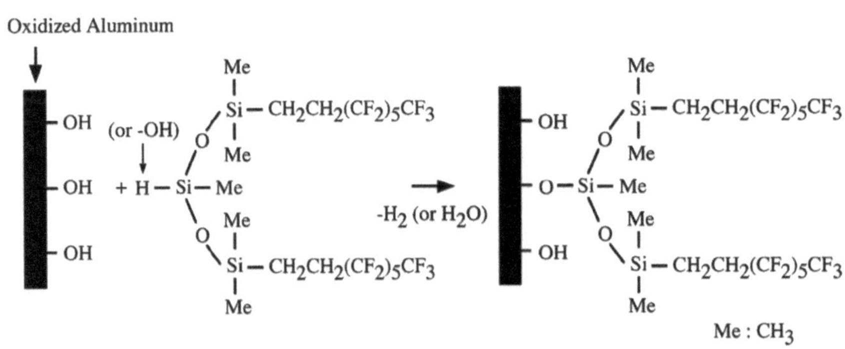

■ **FIGURE 4.2** Available reaction mechanism for an oxidized Al surface with a branch-featured monofunctional perfluoroalkylsilane (($R^FSi(Me)_2O)_2SiMeH$). *Reprinted from Hozumi A, McCarthy TJ. Ultralyophobic oxidized aluminum surfaces exhibiting negligible contact angle hysteresis. Langmuir 2010;26(4):2567–73 with kind permission of copyright The American Chemical Society (2010).*

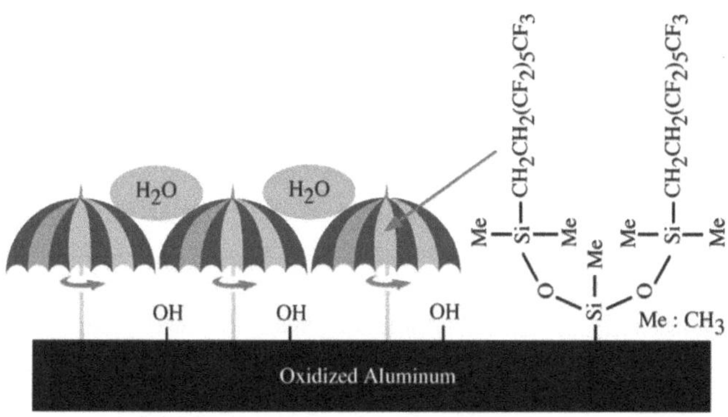

■ **FIGURE 4.3** Schematic representation of an oxidized Al surface covered with a $(R^{F}Si(Me)_2O)_2SiMeH$-derived monolayer. *Reprinted from Hozumi A, McCarthy TJ. Ultralyophobic oxidized aluminum surfaces exhibiting negligible contact angle hysteresis. Langmuir 2010;26(4):2567—73 with kind permission of copyright The American Chemical Society (2010).*

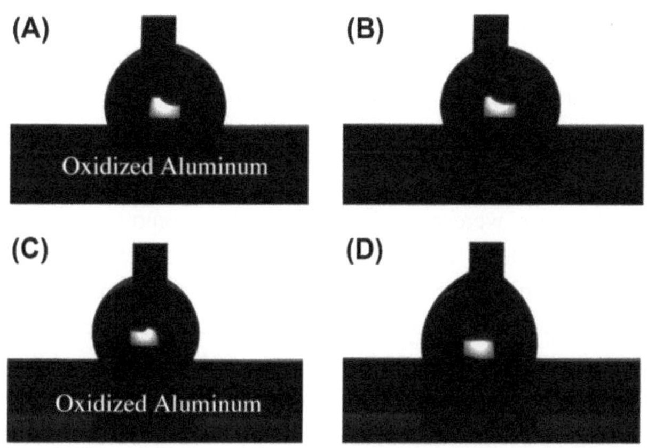

■ **FIGURE 4.4** Photographs of water droplet profiles during dynamic water CA measurements on monolayer-covered Al surfaces: (A) advancing and (B) receding events on the $(R^{F}Si(Me)_2O)_2SiMeH$-treated Al surface and (C) advancing and (D) receding events on the FAS17-SAM-covered Al surface. *Reprinted from Hozumi A, McCarthy TJ. Ultralyophobic oxidized aluminum surfaces exhibiting negligible contact angle hysteresis. Langmuir 2010;26(4), 2567—73 with kind permission of copyright The American Chemical Society (2010).*

their metallic luster, even after 1000 h of salt spray testing (JIS Z 2371). Salt-water droplets, which formed on this surface, moved very smoothly without pinning and were completely removed before they could cause corrosion. On the other hand, corrosion was observed on areas of the FAS17-SAM-covered Al surfaces.

Without the use of any perfluoroalkylsilanes, excellent dynamic dewettability against LST liquids, such as alkanes, was achieved, despite the magnitude of CAs being fairly low. In our follow-up study, we continued to focus on molecular structure by investigating the surface dewetting properties of ring-structured 1,3,5,7-tetramethylcyclotetrasiloxane (D_4^H) vapor-deposited metal substrates (on oxidized Al and Ti) [6]. Under appropriate conditions, CVD treatment was found to be highly effective in tuning dynamic wettability, resulting in several low CA hysteresis surfaces. This facile method not only produced monomeric layers but also particulate films, which were prepared by altering the experimental conditions. This resulted in marked differences in the final dewetting properties, with the reactivity of the D_4^H molecules found to be less dependent on the nature of the oxidized metal surface than on the reaction temperature.

By regulating the CVD temperatures, hydrophilic oxidized Al and Ti surfaces became low CA hysteresis surfaces. For example, when CVD was performed at 80°C, a hydrophobic D_4^H monolayer with a thicknesses of 0.5 nm was formed on both oxidized Al and Ti surfaces, and showed negligible CA hysteresis for water ($\theta_A/\theta_R = 102-104°/99-102°$ and $\Delta\theta = 2-3°$) and *n*-hexadecane ($\theta_A/\theta_R = 31-33°/28-31°$ and $\Delta\theta = 2-3°$) without any marked change in surface morphology.

Conversely, in the case of 180°C CVD treatment, fine particles with diameters of 0.5−4 μm were deposited due to the promotion of hydrolysis/condensation reactions in the gas phase at this elevated temperature (Fig. 4.5). These nano/microstructural morphologies greatly enhanced the surfaces' hydrophobicity, resulting in the formation of superhydrophobic surfaces ($\theta_A/\theta_R = 163°/160-161°$) with low CA hysteresis ($\Delta\theta = 2-3°$).

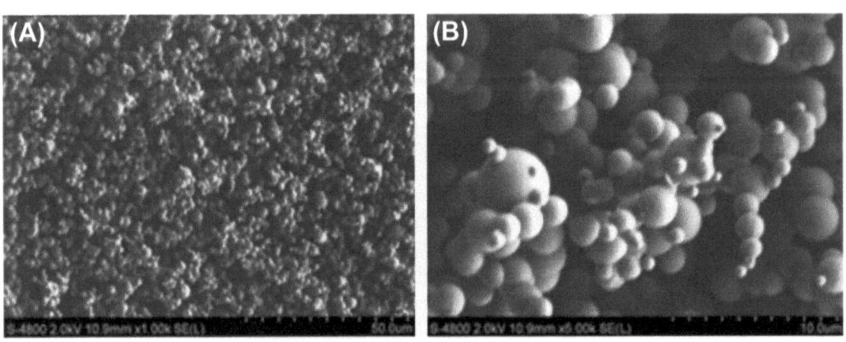

■ **FIGURE 4.5** FE-SEM images of an oxidized Al surface modified by CVD of D_4^H at 180°C for 24 h: (A) top view, (B) magnification of image (A). *Reprinted from Hozumi A, Cheng DF, Yagihashi M. Hydrophobic/superhydrophobic oxidized metal surfaces showing negligible contact angle hysteresis. J Colloid Interface Sci 2011;353(2):582−87 with kind permission of copyright Elsevier (2011).*

The reactivity of D_4^H molecules with the native oxide layers on metal surfaces was found to be excellent, and the use of D_4^H proved to be a useful strategy for tuning dynamic surface wettability, while also providing a basis for designing additional low CA hysteresis metal surfaces.

4.2.2 Polymer Brushes

Recently, we also developed another simple and effective method for controlling surface dewettability toward various LST liquids. This approach is based on the rheological and physical properties of smooth polymer brushes of linear and low MW PDMS [8–11]. The distinguishing aspect of this method is that it does not utilize the chemical properties of PDMS but rather its physical properties to control surface dewettability. In contrast to research on the dewetting behavior of monolayer-covered surfaces, investigations into the dewettability of smooth polymer brushes has been very limited, with only static situations so far reported.

In this case, vinyl-terminated PDMS (MW = 6000–117,000) [10,11] was grafted to the Si-H moieties of a D_4^H-derived monomeric layer (described in the previous section) on Si surfaces [6], through Pt-catalyzed hydrosilylation [64]. Due to the formation of the 0.5 nm-thick D_4^H-monomeric layer, the hydrophilic Si surface exhibited low CA hysteresis for both water ($\theta_A/\theta_R = 106°/105°$ and $\Delta\theta = 1°$) and *n*-hexadecane ($\theta_A/\theta_R = 33°/31°$ and $\Delta\theta = 2°$) [6], and without the use of Pt catalyst, PDMS was not grafted to the surface (as confirmed by CA analysis and ellipsometry). For example, grafting PDMS (MW = 6000) to the D_4^H layer's Si-H surface moieties produced extremely smooth, defect-free, 3.0-nm-thick brush films with a root mean square roughness (R_{rms}) of only 0.374 nm. Thus, the dynamic dewettability of the PDMS surface was not influenced by surface morphology but was determined only by physicochemical properties. As a result of PDMS brush formation, our surface exhibited excellent low CA hysteresis values for water and diiodomethane ($\Delta\theta = 4.6°$ and $4.8°$, respectively). Although our surfaces also showed statically oleophilic properties ($\theta_S = 33°$), which indicated strong interactions between the LST liquids and the sample surface, both CA hysteresis and θ_T values were very small ($\Delta\theta = 2.0°$ and $2.1°$; $\theta_T = 5.0°$ and $3.8°$ for 3 μL droplets of *n*-hexadecane and *n*-dodecane, respectively). By employing low-MW PDMS polymer brush films, we successfully controlled dynamic dewetting behavior toward a variety of probe liquids. We also found that liquid drop movement of these probe liquids was critically dependent on polymer chain mobility as it relates to polymer MW. Our low-MW PDMS (MW = 6000) brush surfaces displayed the lowest resistance to liquid drop movement, while their mobility became gradually worse with increasing MWs [11].

The physical nature of our polymer brush film surface also allowed for enhancement of probe liquid mobility by moderate heating of the surface. As shown in Table 4.1, at 70°C significant decreases in CA hysteresis were detected for all probe liquids, clearly demonstrating that our PDMS (MW = 6000) brush-covered surfaces exhibited thermoresponsive dynamic dewettability. As a liquid polymer, PDMS exhibits lower viscosity with increasing temperature [65]. Therefore, by increasing the temperature from 25°C to 70°C, the liquidity of the bulk molten polymer was visibly altered to more closely resemble water. Our results offered clear evidence that the physical and rheological properties of smooth PDMS brush films play key roles in determining their final dynamic behavior against various probe liquids. The principles we demonstrated here are simple and effective, and provided a new technique for further fundamental scientific studies on the dynamic dewettability of a wide range of liquids, and the industrial applications thereof. Finally, we also showed that simply heating the substrate could induce pronounced transitions in our surfaces' dynamic dewetting behavior.

Another method for creating polymer brush—covered surfaces is to use the specialized synthesis technique known as surface-initiated atom transfer radical polymerization (SI-ATRP). This technique typically requires high concentrations of monomer and catalyst, use of organic solvents, elevated reaction temperatures, and for the reaction to be rigorously purged of oxygen [66—69]. These stringent conditions had made it difficult to prepare polymer brushes outside of laboratory conditions and, due to practical difficulties associated with the reaction setup (large size of airtight reaction vessel and heating apparatus), impractical for covering large areas (~1 m^2 or greater).

Table 4.1 Variation of Dynamic CAs (θ_A/θ_R), CA Hysteresis ($\Delta\theta$), and Substrate TAs (θ_T) of Probe Liquids (3 μL) With Different Surface Tensions (γ) on the Poly(dimethylsiloxane) Polymer Brush Surface Before and After Heating

Probe Liquid	γ (dyn/cm)	T (°C)	θ_A/θ_R (°)	$\Delta\theta$ (°)	$\Delta(\Delta\theta)$ (°)	θ_T (°)
Water	72.8	25	108.8/104.2	4.6	−2.5	18.7
		70	112.8/110.7	2.1		—
CH_2I_2	50.8	25	72.7/67.9	4.8	−4.1	9.7
		70	69.2/68.5	0.7		5.0
$C_{16}H_{34}$	27.5	25	34.5/32.5	2.0	−1.7	5.0
		70	27.8/27.5	0.3		2.2
$C_{12}H_{26}$	25.4	25	21.4/19.3	2.1	−1.3	3.8
		70	16.6/15.8	0.8		1.8

To overcome these limitations, several improvements to conventional ATRP were proposed. Firstly, it was reported that adding water to ATRP reactions significantly increases polymerization rate for both hydrophilic and hydrophobic monomers [70,71], indicating that these reactions no longer needed to be heated and could be carried out at ambient temperatures (we hereafter refer to this method as water-accelerated ATRP). Secondly, activators (re)generated by electron transfer [A(R)GET] can remove the oxygen sensitivity of ATRP reactions by employing a reducing agent to regenerate oxidized catalyst, which has been extensively employed for the preparation of polymers in solution but less frequently for polymer brushes [72–74]. When using this A(R)GET technique, reaction solutions do not need to be purged of oxygen (although they normally are) and in some cases the concentration of Cu catalyst can be reduced to a few parts per million (ppm).

In our first study using this method, we combined the advantages of water-accelerated ATRP and A(R)GET, with a particularly facile monomer 2-(dimethylamino)ethyl methacrylate (DMAEMA), to prepare stimulus (pH)-responsive superoleophobic surfaces from low-chemical-content reaction solutions, on Si substrates, in a particularly easy and low-cost manner [75,76]. Using these techniques, we could prepare extremely thick poly[2-(dimethylamino)ethyl methacrylate] polymer brushes, using low-chemical-content reaction solutions consisting of 99% v/v water. The resulting surfaces were pH-responsive oleophobic *p*DMAEMA brushes, which showed excellent underwater oil repellency (Fig. 4.6). At pH 2, they became superoleophobic, exhibiting large dynamic CAs ($\theta_A/\theta_R = 163°/161°$) and negligible CA hysteresis ($\Delta\theta = 2°$) against *n*-hexadecane droplets. These high-dynamic CAs resulted in excellent nonstick behavior, high oil drop mobility on the surface, and a low θ_T value ($\sim 3°$). Additionally, while using other methods to generate surfaces with these superior properties could be complicated and expensive, our preparation methods were facile, low cost, and applicable to large-scale real-life substrates. Notably, this system offered several additional advantages; it could (1) be performed under mild conditions (in aqueous solution at ambient temperature), without the need to rigorously purge the reaction solutions of oxygen; (2) produce extremely thick brush layers of up to ~700 nm thick (polymer brush thicknesses are typically less than 100 nm); (3) use monomer concentrations as low as 1%v/v; and (4) be vastly less expensive and require smaller quantities of toxic chemicals than conventional ATRP. Based on these techniques, we also developed an easy and inexpensive method to directly paste reaction solutions onto various large-scale real-life substrates in air, which we referred to as "paint on"-ATRP [75]. Through these practical advantages, we could

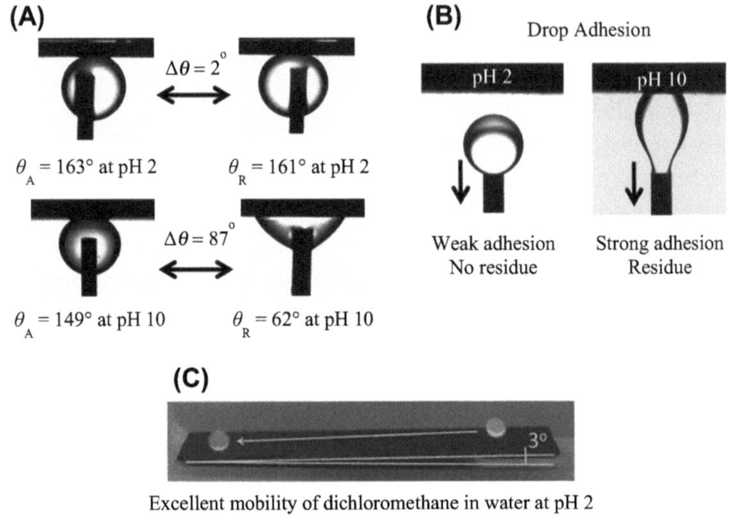

■ **FIGURE 4.6** Underwater dewetting behavior of *p*DMAEMA brush surfaces prepared by ARGET-ATRP. (A) CA hysteresis of *n*-hexadecane (C_{16}) at high and low pH. (B) Drops of dichloromethane slide along *p*DMAEMA brush surface at a small angle of incline. (C) C_{16} drop adhesion. At pH 2, C_{16} drop is easily detached from the surface and leaves no residue, whereas at pH 10, it is strongly adhered. *Reprinted from Dunderdale GJ, Urata C, Miranda DF, Hozumi A. Large-scale and environmentally friendly synthesis of pH-responsive oil-repellent polymer brush surfaces under ambient conditions. ACS Appl Mater Interfaces 2014;6(15):11864–68 with kind permission of copyright The American Chemical Society (2014).*

significantly simplify polymer brush preparation and simultaneously reduce the preparation cost by almost three orders of magnitude. In real-life applications this synthesis could also reduce environmental impact by reducing the toxic chemicals required by 99.9%.

Changes in the physical and chemical properties of our *p*DMAEMA brushes, such as brush thickness, degree of hydration, and oleophobicity, were further investigated in terms of several triggers; pH, ionic strength, and temperature were all studied as external stimuli [76]. In this case, *n*-hexane and dichloromethane were employed as oil phases. Adjusting the water's pH resulted in the most significant changes in *p*DMAEMA brush thickness and hydration, followed by changing ionic strength, and lastly changing the temperature, which only caused small changes in both properties. Coincident with these structural changes, the underwater dynamic oleophobicity of the surface also changed. In particular, θ_R values (and therefore CA hysteresis) were greatly influenced by the addition of alkali or salt but not increasing temperature.

Contrary to our expectations, several differences between the dynamic dewetting behavior of the *p*DMAEMA brushes and the brush structures estimated by ellipsometric measurement were observed, as shown in Fig. 4.7. First, the transition in oleophobicity could not be assumed from the reported pK_b values; in our study, such transitions occurred at lower pH than

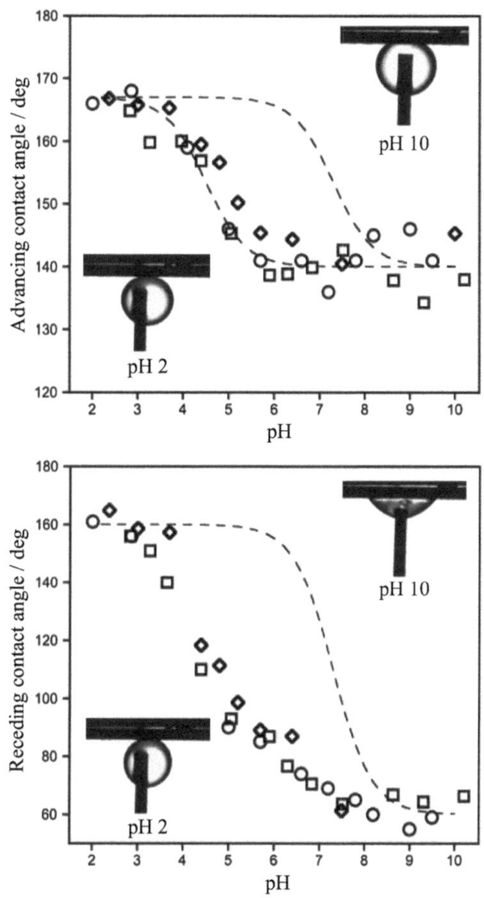

■ **FIGURE 4.7** Changes in θ_A (top figure) and θ_R (bottom figure) values of *n*-hexadecane on a *p*DMAEMA brush surface at different pH values. Insets are profiles of *n*-hexadecane droplets at pH 2 and 10 either advancing or receding. *Dotted lines* represent a change in CA consistent with a change in the fraction of ionization with a pK_b of either 7.7 (rightmost line) or 4.5 (leftmost line). Data points of squares or circles are repeats of experiments changing pH from high to low, and data points of diamonds are from low to high pH. *Reprinted from Dunderdale GJ, Urata C, Hozumi A. An underwater superoleophobic surface which can be activated/deactivated via external triggers. Langmuir 2014;30(44):13438–46 with kind permission of copyright The American Chemical Society (2014).*

expected. Second, our *p*DMAEMA brush surfaces showed stimulus-responsive oleophobicity when oil drops were gently placed on the surface, but these responses were not observed when oil drops were impacted onto the same sample. At high pH or high-ionic strength, our *p*DMAEMA brushes showed large CA hysteresis against oils, but impacting drops did not adhere under these conditions either. These types of dewetting properties are disadvantageous in practical applications because they remove the stimulus-responsive properties of the polymer brush surface.

The transition points between oleophobic and superoleophobic did not always occur at the expected values, as we observed with pH-responsive oleophobicity [76]. We also found that large changes in the degree of hydration and the chemical nature of the polymer brush resulted in larger changes in the oleophobic response. This point should be considered when designing stimulus-responsive oleophobic surfaces in the future.

4.2.3 Alkylsilane-Derived Hybrid Films

As mentioned in the previous section, the mobility of surface-tethered functional groups ("liquid-like" surface properties) is important in achieving excellent dynamic dewettability [1,8,23]. Consequently, another promising approach is to use the sol–gel method because it is a simple and widely reproducible approach that allows arbitrary control of the packing densities of surface-tethered functional groups via compositional variation of binary silane mixtures [12,13,15–17]. We developed such a system, which did not require either conventional surface roughening nor long-chain perfluoroalkylsilanes [FASs, with long perfluoroalkyl chains ($CF_3(CF_2)_n$–, where $n > 7$)] [38–40,42,43] or perfluorinated fluids [41] to reduce the interaction between the surface and liquid. We envisioned that controlling the composition of binary alkyltrialkoxysilane and tetraalkoxysilane mixtures could reduce the packing of alkyl chains on the resulting coating's surface, leading to enhancement of the surface dewetting properties by increasing the ability of the functional groups to move and become liquid-like (Scheme 4.1).

Based on this assumption, we prepared alkylsilane-derived hybrid films using *n*-decyltriethoxysilane (C_{10}) and tetramethoxysilane (TMOS) as precursors, through a conventional cohydrolysis and cocondensation reaction [12]. The typical molar ratios of precursor solutions were 1 DTES: 2–10 TMOS: 50 EtOH: 22 H_2O: 4×10^{-4} HCl. For comparative purposes we also prepared C_{10}-monolayer-covered surfaces using a conventional CVD method [54,55], with the resulting surfaces exhibiting θ_S, θ_A, and θ_R for water and *n*-hexadecane of 99°/109°/86° and 20°/21°/8°, respectively, and a thickness of 0.9 nm, as measured by ellipsometry.

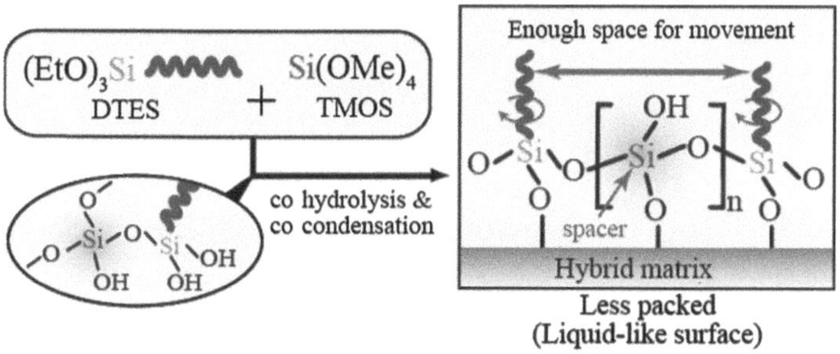

■ **SCHEME 4.1** Formation (left) and structure (right) of C_{10}-hybrid films from DTES and TMOS. *Reprinted from Urata C, Cheng DF, Masheder B, Hozumi A. Smooth, transparent and nonperfluorinated surfaces exhibiting unusual contact angle behavior toward organic liquids. RSC Adv 2012;2(26):9805–08 with kind permission of copyright The Royal Society of Chemistry (2012).*

The C_{10}-hybrid films were optically transparent, with a total thickness in the range of 0.63–1.2 μm, and showed well-ordered lamellar structures in which each layer was oriented parallel to the substrate, as confirmed by X-ray diffraction [12]. When prepared at a TMOS to DTES ratio ($R_{T/D}$) of 10, the hybrid films were fairly robust, and in simple scratch tests could not be damaged using polypropylene tweezers, but were damaged by stainless steel tweezers. They were also very smooth ($R_{rms} = 0.2–2.4$ nm), indicating that dynamic dewettability between the C_{10}-hybrid and monolayer film surfaces were not influenced by surface morphology but determined only by surface chemical and physical properties. All C_{10}-hybrid films prepared over the stated range of $R_{T/D}$ ratios were hydrophobic, and their dynamic CAs for water ($\theta_A/\theta_R = 109°/100°$, $R_{T/D} = 4$) were almost identical to those of C_{10}-monolayer films prepared by CVD [12]. However, our C_{10}-hybrid film surfaces also showed unusual dynamic CA behavior toward organic liquids. For example, the large CA hysteresis of *n*-hexadecane on a C_{10}-monolayer-covered glass surface ($\Delta\theta = 13°$) caused it to spread fully across and strongly adhere to the surface (Fig. 4.8B and C, left) when sprayed as a mist. However, because the of lower CA hysteresis (less than 2°), the same mist quickly formed droplets and rolled across and off our C_{10}-hybrid film surface without pinning at θ_T of 3.4° (Fig. 4.8b and c, right). As indicated in Table 4.2, in addition to *n*-hexadecane, excellent dynamic dewetting behavior was similarly observed for other LST organic liquids, including *n*-dodecane and *n*-decane.

The marked difference in the final dynamic oleophobicity observed between the C_{10}-monolayer- and C_{10}-hybrid-covered samples was considered to be due to differences in the mobility of their surface-tethered alkyl chains, which we attributed to the distances between the carbon chains, and which in the case of the hybrid films lowered the packing concentration. The

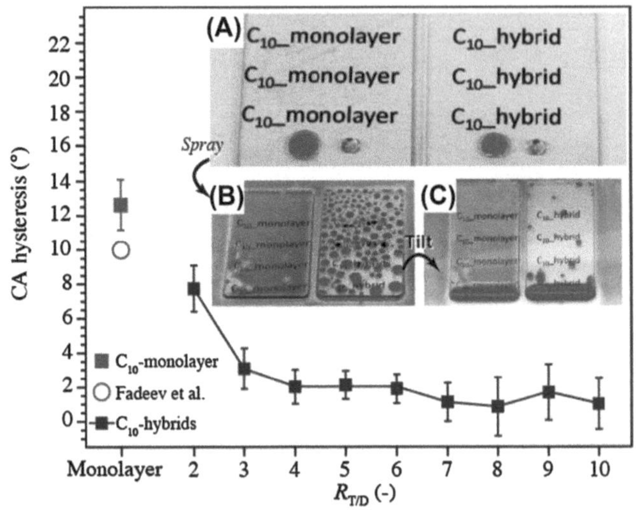

■ **FIGURE 4.8** CA hysteresis of a C_{10}-monolayer-covered glass slide (monolayer surface by Fadeev and McCarthy)[2] and C_{10}-hybrid-film-covered glass slides for *n*-hexadecane. Data represents the average value from three independent experiments ($n = 3$). (Inset images: (A) Red-colored *n*-hexadecane and blue-colored water droplets (5 μL) were placed on each sample. (B) Appearance of films following the spraying of a red-colored *n*-hexadecane mist onto them and (C) after tilting the substrates at 40°. The slides on the left of the images inset are C_{10}-monolayer-covered samples and those on the right are C_{10}-hybrid film ($R_{T/D} = 4$)-covered samples. *Reprinted from Urata C, Cheng DF, Masheder B, Hozumi A. Smooth, transparent and nonperfluorinated surfaces exhibiting unusual contact angle behavior toward organic liquids. RSC Adv 2012;2(26):9805–08 with kind permission of copyright The Royal Society of Chemistry (2012).*

Table 4.2 Variation of Dynamic CAs (θ_A/θ_R, 3 μL), CA Hysteresis ($\Delta\theta$), and Substrate TA (θ_T, 5 μL) of Several Probe Liquids on the C_{10}-Hybrid- ($R_{T/D} = 4$, Upper Row) and C_{10}-Monolayer (Lower Row)-Covered Glass Slides

Probe liquid	Surface tension (dyn/cm)	θ_A/θ_R (°/°)	$\Delta\theta$ (°)	θ_T (°)
Water	72.8	109/100	9	40
		109/86	23	90
Diiodomethane	50.8	73/71	2	6.2
		65/59	6	27
Oleic Acid	32.8	52/47	5	9.0
		42/27	15	57
Soybean Oil	31.3	52/47	5	11
		45/27	18	54

Table 4.2 Variation of Dynamic CAs (θ_A/θ_R, 3 μL), CA Hysteresis ($\Delta\theta$), and Substrate TA (θ_T, 5 μL) of Several Probe Liquids on the C_{10}-Hybrid- ($R_{T/D} = 4$, Upper Row) and C_{10}-Monolayer (Lower Row)-Covered Glass Slides *continued*

Probe liquid	Surface tension (dyn/cm)	θ_A/θ_R (°/°)	$\Delta\theta$ (°)	θ_T (°)
Toluene	28.4	37/36	1	2.5
		23/14	9	26
p-Xylene	28.3	39/37	2	3.2
		25/14	11	18
n-Hexadecane	27.5	36/34	2	3.4
		21/8	13	21
Turpentine oil	27.0	31/31	0	2.7
		16/<5	—	—
n-Dodecane	25.4	27/25	2	2.5
		7/<5	—	—
n-Decane	23.8	21/20	1	2.4
		<5/<5	—	—
Ethanol	22.1	29/25	4	8.8
		<5/<5	—	—

detailed mechanism of the unusual dynamic dewetting behavior observed in our samples will be described in the next section.

While the conjecture on reduced surface packing affecting wetting was fully supported by the CA hysteresis data shown in Fig. 4.8, the detailed mechanism of our hybrid surfaces' unusual dynamic dewetting behavior had not yet been clearly identified. We therefore conducted a follow-up study to further investigate the changes in their chemical and physical properties when in contact with probe liquids [17]. It is well known that contact with a liquid can alter the termination states of functional (alkyl) groups [77,78], and that the conformation of loosely packed alkyl chains is very sensitive to the liquid environment around them [77,79,80]. Moreover, such responses directly affect both the static and dynamic dewetting properties of a surface and probe liquid at the interface. Accordingly, we expected that changes in the conformational states of surface-tethered alkyl groups, before and after contact with the probe liquids, could influence the depinning and dewetting of liquid drops. We therefore focused particularly on changes in molecular conformation of the alkyl groups in contact with probe liquids, in terms of various physicochemical properties including surface tension, molecular weight/volume, polarity (dielectric constant), density, and viscosity.

The dynamic dewettability of our smooth, C_{10}-hybrid film surfaces was systematically investigated using static/dynamic CAs, CA hysteresis ($\Delta\theta$ and $\Delta\theta_{cos}$), and substrate TA (θ_T) measurements of 17 probe liquids [17]. As can be seen in Fig. 4.9 (A), surface tensions of the probe liquids correlated with θ_S values, while showing little or no correlation with dynamic dewetting behavior ($\Delta\theta$, $\Delta\theta_{cos}$, and θ_T in Fig. 4.9 (B–D), respectively). However, the dynamic dewettability of our hybrid films showed a strong correlation with the probe liquids' dielectric constants, which was not shown with the surface tension.

For example, for probe liquids with dielectric constants less than 30, excellent dynamic dewettability ($\Delta\theta < 3°$, $\Delta\theta_{cos} < 0.043$, and $\theta_T < 5°$) could be achieved regardless of θ_S values (15°–69°). However, for liquids with dielectric constants greater than 34, the mobility of probe liquids was significantly worse. To illustrate this point, four typical probe liquids (toluene, IPA, DMF, and water) were placed on our C_{10}-hybrid film surface, which was then inclined at a gradually increasing angle, demonstrated in Fig. 4.10.

Consistent with the dielectric constants of the probe liquids, toluene and IPA, which have low dielectric constants (2.4 and 18, respectively), rolled off at low θ_T values (1° and 2°, respectively), followed by DMF (dielectric constant = 38, $\theta_T = 15°$) and eventually, water (dielectric constant = 80, $\theta_T = 19°$). If this sequence had been dictated by surface tension, the order would have been water, DMF, toluene, and IPA, which highlighted that the dielectric constant dictated the drop mobility, irrespective of the probe liquids' surface tension. It is also noteworthy that the dynamic dewettability of our hybrid film surfaces against high surface tension liquids with low dielectric constants (such as diiodomethane, 1-bromonaphtalene, aniline, methyl cyanoacetate, and iodobenzene) was markedly superior to that of liquids with similar surface tensions and high dielectric constants (such as DMSO, DMF, and dimethyl acetamide).

In order to understand the mechanism of these unusual dynamic dewetting phenomena, the conformational states of surface-tethered alkyl groups, under dry (at the air/C_{10}-hybrid film interface) and wet (probe liquid/C_{10}-hybrid film interface) conditions, were investigated using the sum-frequency generation (SFG) technique [17]. Using this technique, it was possible to ascertain conformational data from the surface-tethered alkyl groups in the topmost regions of the films, specifically at the areas that came into contact with the probe liquids, and so directed dewetting behavior. The SFG revealed that the excellent mobility of alkane drops on the film surfaces was due to local changes in surface energy at the three-phase contact lines of the liquid droplets. Droplet mobility was found to be strongly

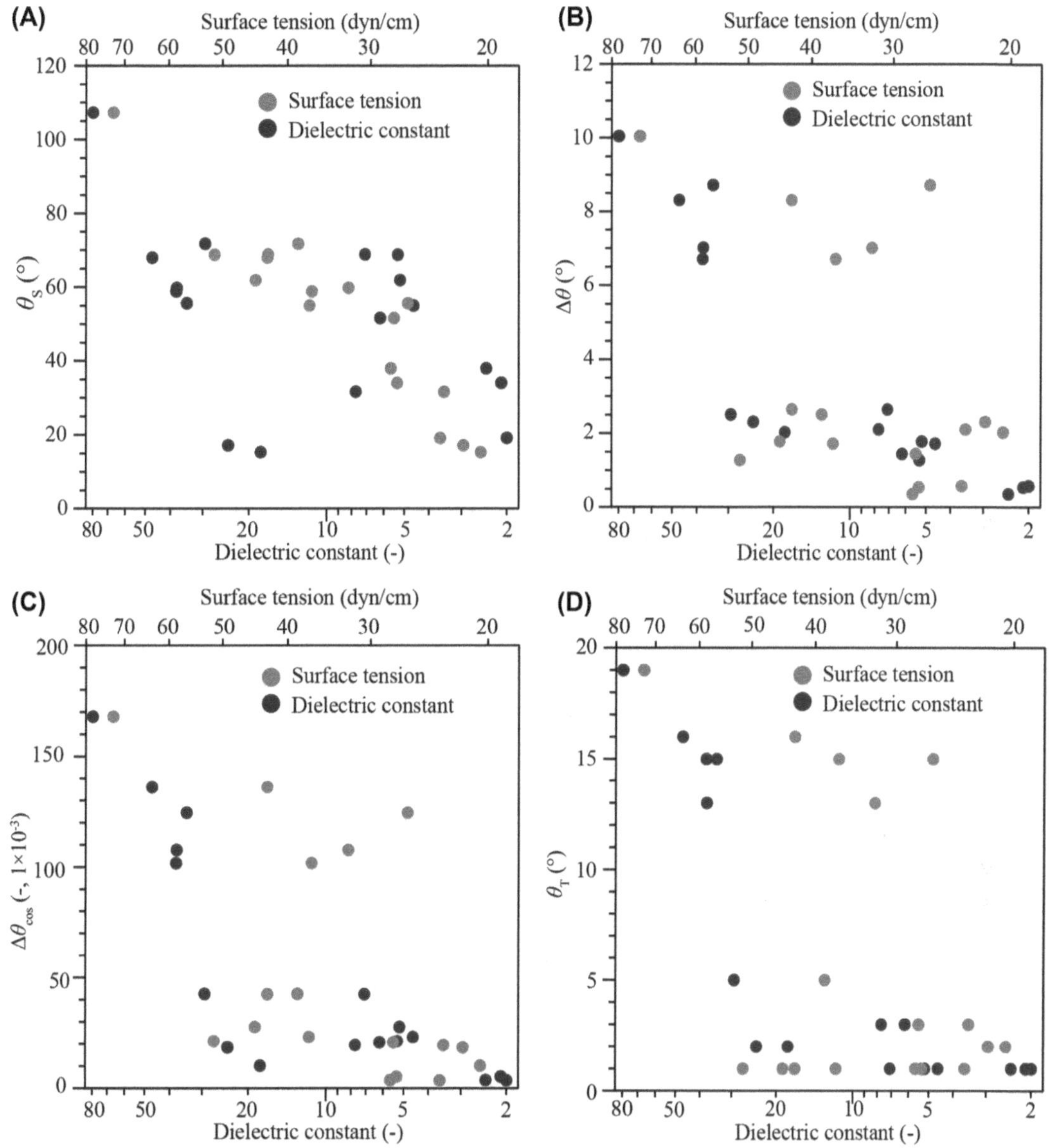

■ **FIGURE 4.9** Changes in (A) θs, (B) the CA hysteresis ($\Delta\theta$), (C) the cosine of CA hysteresis ($\Delta\theta_{cos}$), and (D) the minimum substrate TAs (θ_T) for a C_{10}-hybrid film as a function of the surface tensions and dielectric constants of 17 probe liquids. *Reprinted from Urata C, Mashder B, Cheng DF, Miranda DF, Hozumi A. Why can organic liquids move easily on smooth alkyl-terminated surfaces? Langmuir 2014;30(14):4049–55 with kind permission of copyright The American Chemical Society (2014).*

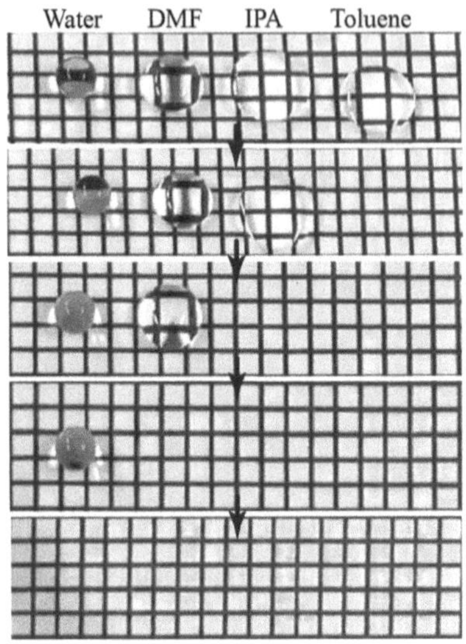

■ **FIGURE 4.10** Dynamic dewetting behavior of four probe liquids (10 μL) on a C_{10}-hybrid film surface as the surface is inclined to an increasing degree. *Reprinted from Urata C, Mashder B, Cheng DF, Miranda DF, Hozumi A. Why can organic liquids move easily on smooth alkyl-terminated surfaces? Langmuir 2014;30(14):4049—55 with kind permission of copyright The American Chemical Society (2014).*

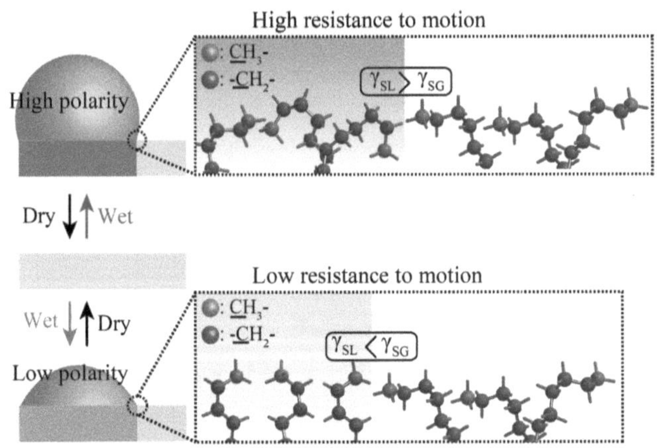

■ **SCHEME 4.2** Schematic illustration of the conformation of alkyl chains at the three-phase contact line when high- and low-dielectric-constant probe liquids are in contact with C_{10}-hybrid film interfaces (not to scale), corresponding to the influence of the conformational changes on the mobility of droplets on the C_{10}-hybrid surface. γ_{SL} and γ_{SG} indicate the surface energy of a solid at solid/liquid and solid/gas interfaces, respectively. *Reprinted from Urata C, Mashder B, Cheng DF, Miranda DF, Hozumi A. Why can organic liquids move easily on smooth alkyl-terminated surfaces? Langmuir 2014;30(14):4049—55 with kind permission of copyright The American Chemical Society (2014).*

correlated to the magnitude of the conformational change of surface-tethered alkyl chains before and after contact with the probe liquids [17]. As illustrated in Scheme 4.2, CH_3 groups were preferentially exposed at the solid/liquid interface when low-dielectric-constant probe liquids were in contact with the surface. When the samples were slightly tilted, the three-phase contact line of droplets continuously moved from low-surface-energy (solid/liquid) areas to surrounding high-surface-energy (solid/air) areas, which was induced by the rotation of surface-tethered alkyl groups, independently of the magnitude of CAs [17].

4.3 SUMMARY

This chapter details our recent work on arranging functional molecules, on topmost surfaces, using various materials, such as specific featured monolayers, polymer brushes, and alkylsilane-derived hybrid films, to control their dynamic dewetting properties. Contrary to the majority of published works on this topic, which have primarily focused on conventional static CAs (magnitude of CAs), we predominantly characterized these surfaces using dynamic CAs (advancing and receding CAs), CA hysteresis, and substrate tilt angles of various probe liquid, as these give a clearer picture of a surface's true dewetting properties than static CAs alone. Using three different surface modification techniques (CVD, A(R)GET-ATRP, and sol–gel synthesis), we have successfully created various functional surfaces, with thicknesses ranging from a few to several hundred nanometers. We have also shown that we can tune surface dewetting properties against various types of liquid types, including polar and nonpolar liquids, in air, water, and oil, simply using molecular surface arrangement, and without the use of any surface roughening techniques or perfluorinated compounds. Although such conventional processes are effective for controlling surface dewetting properties, they possess notable shortcomings, such as cost, complexity, and potential environmental/health issues. Using relatively simple chemistry and elements, we have successfully managed to circumvent these issues, by using facile and cost-effective means to create dynamic dewetting surfaces with similar or superior properties. The SFG technique also confirmed that molecular surface arrangement was a primary contributor to the unusual dynamic dewettability of our alkylsilane-derived hybrid films, although the results achieved were very limited. The universal mechanism for the excellent mobility of liquid droplets, particularly of low-surface-tension liquids, on our other low CA hysteresis surfaces is still under investigation. While simple, these systems show great potential for uses in both industrial and everyday items, and could save large sums of money while preventing unnecessary use of harmful chemicals, presenting positive outcomes for both the economy and the environment.

REFERENCES

[1] Fadeev AY, McCarthy TJ. Trialkylsilane monolayers covalently attached to silicon surfaces: wettability studies indicating that molecular topography contributes to contact angle hysteresis. Langmuir 1999;15(11):3759–66.

[2] Fadeev AY, McCarthy TJ. Self-assembly is not the only reaction possible between alkyltrichlorosilanes and surfaces: monomolecular and oligomeric covalently attached layers of dichloro- and trichloroalkylsilanes on silicon. Langmuir 2000; 16(18):7268–74.

[3] Extrand CW. Contact angles and hysteresis on surfaces with chemically heterogeneous islands. Langmuir 2003;19(9):3793–6.

[4] Fadeev AY, McCarthy TJ. Binary monolayer mixtures: modification of nanopores in silicon-supported tris(trimethylsiloxy)silyl monolayers. Langmuir 1999;15(21): 7238–43.

[5] Hozumi A, McCarthy TJ. Ultralyophobic oxidized aluminum surfaces exhibiting negligible contact angle hysteresis. Langmuir 2010;26(4):2567–73.

[6] Hozumi A, Cheng DF, Yagihashi M. Hydrophobic/superhydrophobic oxidized metal surfaces showing negligible contact angle hysteresis. J Colloid Interface Sci 2011; 353(2):582–7.

[7] Krumpfer JW, McCarthy TJ. Contact angle hysteresis: a different view and a trivial recipe for low hysteresis hydrophobic surfaces. Faraday Discuss 2010;146:103–11.

[8] Krumpfer JW, McCarthy TJ. Rediscovering silicones: "Unreactive" silicones react with inorganic surfaces. Langmuir 2011;27(18):11514–9.

[9] Wier KA, Gao L, McCarthy TJ. Two-dimensional fluidics based on differential lyophobicity and gravity. Langmuir 2006;22(11):4914–6.

[10] Cheng DF, Urata C, Yagihashi M, Hozumi A. A statically oleophilic but dynamically oleophobic smooth nonperfluorinated surface. Angew Chem Int Ed 2012;51(12): 2956–9.

[11] Cheng DF, Urata C, Masheder B, Hozumi A. A physical approach to specifically improve the mobility of alkane liquid drops. J Am Chem Soc 2012;134(24): 10191–9.

[12] Urata C, Cheng DF, Masheder B, Hozumi A. Smooth, transparent and nonperfluorinated surfaces exhibiting unusual contact angle behavior toward organic liquids. RSC Adv 2012;2(26):9805–8.

[13] Urata C, Masheder B, Cheng DF, Hozumi A. How to reduce resistance to movement of alkane liquid drops across tilted surfaces without relying on surface roughening and perfluorination. Langmuir 2012;28(51):17681–9.

[14] Cheng DF, Masheder B, Urata C, Hozumi A. Smooth perfluorinated surfaces with different chemical and physical natures: their unusual dynamic dewetting behavior toward polar and nonpolar liquids. Langmuir 2013;29(36):11322–9.

[15] Park J, Urata C, Masheder B, Cheng DF, Hozumi A. Long perfluoroalkyl chains are not required for dynamically oleophobic surfaces. Green Chem 2013;15(1):100–4.

[16] Urata C, Masheder B, Cheng DF, Hozumi A. Unusual dynamic dewetting behavior of smooth perfluorinated hybrid films: potential advantages over conventional textured and liquid-infused perfluorinated surfaces. Langmuir 2013;29(40):12472–82.

[17] Urata C, Mashder B, Cheng DF, Miranda DF, Hozumi A. Why can organic liquids move easily on smooth alkyl-terminated surfaces? Langmuir 2014;30(14):4049–55.

[18] Chen W, Fadeev AY, Hsieh MC, Öner D, Youngblood J, McCarthy TJ. Ultrahydrophobic and ultralyophobic surfaces: some comments and examples. Langmuir 1999;15(10):3395–9.

[19] Öner D, McCarthy TJ. Ultrahydrophobic surfaces. Effects of topography length scales on wettability. Langmuir 2000;16(20):7777–82.

[20] Gao L, McCarthy TJ. The 'lotus effect' explained: two reasons why two length scales of topography are important. Langmuir 2006;22(7):2966–7.

[21] Gao L, McCarthy TJ. 'Artificial lotus leaf' prepared using a 1945 patent and a commercial textile. Langmuir 2006;22(14):5998–6000.

[22] Gao L, Mccarthy TJ. A perfectly hydrophobic surface (θ_A/θ_R=180°/180°). J Am Chem Soc 2006;128(28):9052–3.

[23] Gao L, McCarthy TJ. Contact angle hysteresis explained. Langmuir 2006;22(14):6234–7.

[24] Gao L, McCarthy TJ. Wetting 101°. Langmuir 2009;25(24):14105–15.

[25] Jin M, Feng X, Feng L, Sun T, Zhai J, Li T, et al. Superhydrophobic aligned polystyrene nanotube films with high adhesive force. Adv Mater 2005;17(16):1977–81.

[26] Bhushan B, Jung YC. Natural and biomimetic artificial surfaces for superhydrophobicity, self-cleaning, low adhesion, and drag reduction. Prog Mater. Sci 2011;56(1):1–108.

[27] Acatay K, Simsek E, Ow-Yang C, Menceloglu YZ. Tunable, superhydrophobically stable polymeric surfaces by electrospinning. Angew Chem Int Ed 2004;43(39):5210–3.

[28] Han D, Steckl AJ. Superhydrophobic and oleophobic fibers by coaxial electrospinning. Langmuir 2009;25(16):9454–62.

[29] Xu L, Karunakaran RG, Guo J, Yang S. Transparent, superhydrophobic surfaces from one-step spin coating of hydrophobic nanoparticles. ACS Appl Mater Interfaces 2012;4(2):1118–25.

[30] Nguyen DD, Tai N-H, Lee S-B, Kuo W-S. Superhydrophobic and superoleophilic properties of graphene-based sponges fabricated using a facile dip coating method. Energy Environ Sci 2012;5(7):7908–12.

[31] Song X, Zhai J, Wang Y, Jiang L. Fabrication of superhydrophobic surfaces by self-assembly and their water-adhesion properties. J Phys Chem B 2005;109(9):4048–52.

[32] Han JT, Zheng Y, Cho JH, Xu X, Cho K. Stable superhydrophobic organic-inorganic hybrid films by electrostatic self-assembly. J Phys Chem B 2005;109(44):20773–8.

[33] Ke Q, Li G, Liu Y, He T, Li X-M. Formation of superhydrophobic polymerized *n*-octadecylsiloxane nanosheets. Langmuir 2010;26(5):3579–84.

[34] Yuan S, Pehkonen SO, Liang B, Ting YP, Neoh KG, Kang ET. Superhydrophobic fluoropolymer-modified copper surface via surface graft polymerisation for corrosion protection. Corros Sci 2011;53(9):2738–47.

[35] Hozumi A, Takai O. Preparation of ultra water-repellent films by microwave plasma-enhanced CVD. Thin Solid Films 1997;303(1–2):222–5.

[36] Qian B, Shen Z. Fabrication of superhydrophobic surfaces by dislocation-selective chemical etching on aluminum, copper, and zinc substrates. Langmuir 2005; 21(20):9007–9.

[37] Hozumi A, Kim B, McCarthy TJ. Hydrophobicity of perfluoroalkyl isocyanate monolayers on oxidized aluminum surfaces. Langmuir 2009;25(12):6834–40.

[38] Tuteja A, Choi W, Ma M, Mabry JM, Mazzella SA, Rutledge GC, et al. Designing superoleophobic surfaces. Science 2007;318(5856):1618–22.

[39] Tuteja A, Choi W, McKinley GH, Cohen RE, Rubner MF. Design parameters for superhydrophobicity and superoleophobicity. MRS Bull 2008;33(8):752–8.

[40] Tuteja A, Choi W, Mabry JM, McKinley GH, Cohen RE. Robust omniphobic surfaces. Proc Natl Acad Sci U.S.A 2008;105(47):18200–5.

[41] Wong T-S, Kang SH, Tang SKY, Smythe EJ, Hatton BD, Grinthal A, et al. Bioinspired self-repairing slippery surfaces with pressure-stable omniphobicity. Nature 2011;477(7365):443–7.

[42] Zhang J, Seeger S. Superoleophobic coatings with ultralow sliding angles based on silicone nanofilaments. Angew Chem Int Ed 2011;50(29):6652–6.

[43] Deng X, Mammen L, Butt H-J, Vollmer D. Candle soot as a template for a transparent robust superamphiphobic coating. Science 2012;335(6064):67–70.

[44] Liu TL, Kim CJ. Turning a surface superrepellent even to completely wetting liquids. Science 2014;346(6213):1096–100.

[45] Barthlott W, Neinhuis C. Purity of the sacred lotus, or escape from contamination in biological surfaces. Planta 1997;202(1):1–8.

[46] Neinhuis C, Barthlott W. Characterization and distribution of water-repellent, self-cleaning plant surfaces. Ann Bot 1997;79(6):667–77.

[47] Wagner P, Fürstner R, Barthlott W, Neinhuis C. Quantitative assessment to the structural basis of water repellency in natural and technical surfaces. J Exp Bot 2003;54(385):1295–303.

[48] Kawasaki K. Study of wettability of polymers by sliding of water drop. J Colloid Sci 1960;15(5):402–7.

[49] Furmidge CGL. The sliding of liquid drops on solid surfaces and a theory for spray retention. J Colloid Sci 1962;17(4):309–24.

[50] Hare EF, Shafrin EG, Zisman WA. Properties of films of adsorbed fluorinated acids. J Phys Chem 1954;58(3):236–9.

[51] Lindstrom AB, Strynar MJ, Libelo EL. Polyfluorinated compounds: past, present, and future. Environ Sci Technol 2011;45(19):7954–61.

[52] Zaggia A, Ameduri B. Recent advances on synthesis of potentially non-bioaccumulable fluorinated surfactants. Curr Opin Colloid Interface Sci 2012; 17(4):188–95.

[53] Pujari SP, Scheres L, Marcelis ATM, Zuilhof H. Covalent surface modification of oxide surfaces. Angew Chem Int Ed 2014;53(25):6322–56.

[54] Hozumi A, Ushiyama K, Sugimura H, Takai O. Fluoroalkylsilane monolayers formed by chemical vapor surface modification on hydroxylated oxide surfaces. Langmuir 1999;15(22):7600–4.

[55] Sugimura H, Hayashi K, Hozumi A, Takai O. Micropatterning of alkyl- and fluoroalkylsilane self-assembled monolayers using vacuum ultraviolet light. Langmuir 2000;16(3):885–8.

[56] Hozumi A, Yokogawa Y, Kameyama T, Sugimura H, Hayashi K, Shirayama H, et al. Amino-terminated self-assembled monolayer on a SiO_2 surface formed by chemical vapor deposition. J Vac Sci Technol A 2001;19(4):1812–6.
[57] Brzoska JB, Azouz IB, Rondelez F. Silanization of solid substrates: a step toward reproducibility. Langmuir 1994;10(11):4367–73.
[58] Allara DL, Parikh AN, Rondelez F. Evidence for a unique chain organization in long chain silane monolayers deposited on two widely different solid substrates. Langmuir 1995;11(7):2357–60.
[59] Maoz R, Sagiv J. On the formation and structure of self-assembling monolayers. I. A comparative atr-wettability study of Langmuir—Blodgett and adsorbed films on flat substrates and glass microbeads. J Colloid Interface Sci 1984;100(2):465–96.
[60] Li K, Xu S, Shi W, He M, Li H, Li S, et al. Investigating the effects of solid surfaces on ice nucleation. Langmuir 2012;28(29):10749–54.
[61] Tambe NS, Bhushan B. Nanotribological characterization of self-assembled monolayers deposited on silicon and aluminium substrates. Nanotechnology 2005; 16:1549.
[62] DeRose JA, Hoque E, Bhushan B, Mathieu HJ. Characterization of perfluorodecanoate self-assembled monolayers on aluminum and comparison of stability with phosphonate and siloxy self-assembled monolayers. Surf Sci 2008;602(7):1360–7.
[63] Hoque E, Derose JA, Houriet R, Hoffmann P, Mathieu HJ. Stable perfluorosilane self-assembled monolayers on copper oxide surfaces: Evidence of siloxy-copper bond formation. Chem Mater 2007;19(4):798–804.
[64] Stein J, Lewis LN, Gao Y, Scott RA. In situ determination of the active catalyst in hydrosilylation reactions using highly reactive Pt(0) catalyst precursors. J Am Chem Soc 1999;121(15):3693–703.
[65] Barlow AJ, Harrison G, Lamb J. Viscoelastic relaxation of polydimethylsiloxane liquids. Proc R Soc A 1964;282(1389):228–51.
[66] Barbey R, Lavanant L, Paripovic D, Schüwer N, Sugnaux C, Tugulu S, et al. Polymer brushes via surface-initiated controlled radical polymerization: synthesis, characterization, properties, and applications. Chem Rev 2009;109(11):5437–527.
[67] Edmondson S, Osborne VL, Huck WTS. Polymer brushes via surface-initiated polymerizations. Chem Soc Rev 2004;33:14–22.
[68] Xu FJ, Cai QJ, Kang ET, Neoh KG. Surface-initiated atom transfer radical polymerization from halogen-terminated Si(111) (Si-X, X = Cl, Br) surfaces for the preparation of well-defined polymer-Si hybrids. Langmuir 2005;21(8):3221–5.
[69] Lattuada M, Hatton TA. Functionalization of monodisperse magnetic nanoparticles. Langmuir 2007;23(4):2158–68.
[70] Jones DM, Huck WTS. Controlled surface-initiated polymerizations in aqueous media. Adv Mater 2001;13(16):1256–9.
[71] Kizhakkedathu JN, Norris-Jones R, Brooks DE. Synthesis of well-defined environmentally responsive polymer brushes by aqueous ATRP. Macromolecules 2004;37(3):734–43.
[72] Simakova A, Averick SE, Konkolewicz D, Matyjaszewski K. Aqueous ARGET ATRP. Macromolecules 2012;45(16):6371–9.
[73] Dong H, Matyjaszewski K. ARGET ATRP of 2-(dimethylamino) ethyl methacrylate as an intrinsic reducing agent. Macromolecules 2008;41(19):6868–70.

[74] Matyjaszewski K, Hongchen D, Jakubowski W, Pietrasik J, Kusumo A. Grafting from surfaces for "everyone": ARGET ATRP in the presence of air. Langmuir 2007;23(8):4528–31.

[75] Dunderdale GJ, Urata C, Miranda DF, Hozumi A. Large-scale and environmentally friendly synthesis of pH-responsive oil-repellent polymer brush surfaces under ambient conditions. ACS Appl Mater Interfaces 2014;6(15):11864–8.

[76] Dunderdale GJ, Urata C, Hozumi A. An underwater superoleophobic surface which can be activated/deactivated via external triggers. Langmuir 2014;30(44):13438–46.

[77] Miranda PB, Shen YR. Liquid interfaces: a study by sum-frequency vibrational spectroscopy. J Phys Chem B 1999;103(17):3292–307.

[78] Sander LC, Lippa KA, Wise SA. Order and disorder in alkyl stationary phases. Anal Bioanal Chem 2005;382(3):646–68.

[79] Miranda PB, Pflumio V, Saijo H, Shen YR. Conformation of surfactant monolayers at solid/liquid interfaces. Chem Phys Lett 1997;264(3–4):387–92.

[80] Miranda PB, Pflumio V, Saijo H, Shen YR. Chain-chain interaction between surfactant monolayers and alkanes or alcohols at solid/liquid interfaces. J Am Chem Soc 1998;120(46):12092–9.

Nanoarchitectonics of Inorganic Materials

Organic-Inorganic Hybrid Nanoarchitecture at Mesoscale

K. Katagiri
Hiroshima University, Higashihiroshima, Japan

CHAPTER OUTLINE

5.1 INTRODUCTION

Research on organic-inorganic hybrid materials is one of the most attractive fields of materials chemistry to this day [1]. Various types of material combinations and synthetic strategies have been developed for the hybrid materials up to the present time: (1) the encapsulation of functional organic molecules within the sol–gel-derived inorganic matrices; (2) the polymerization of functional organoalkoxysilanes; and (3) the organic functionalization of inorganic matters, such as nanofillers and nanoclays. However, the interface between the organic and inorganic components in these hybrids seems to be structurally ambiguous and difficult to control precisely. Therefore, the development of materials with organized structures in mesoscales is a fascinating research subject in the field of hybrid materials. The technique based on supramolecular chemistry, such as the self-assembly technique, is widely used as a powerful and versatile approach to achieve mesoscale materials with well-organized structure. Recently, this approach has also been

Supra-materials Nanoarchitectonics. http://dx.doi.org/10.1016/B978-0-323-37829-1.00005-5

employed for the development of organic-inorganic hybrid materials via nanoarchitectonics.

In this chapter, several approaches for the preparation of organic-inorganic hybrid materials by supramolecular organization are reviewed. Firstly, organic-inorganic hybrid hollow capsules were introduced. These capsules were prepared via layer-by-layer (LbL) assembly on colloid templates. This technique is suitable for nanoarchitectonics of organic-inorganic hybrids since the driving force of assembly is simple electrostatic interaction. Especially, UV and magnetically responsive capsules are described in detail. They have potentials for drug-delivery vehicles. Another topic is nanohybrid of magnetic nanoparticle clusters with polysaccharide nanogels. In this case, the key of self-assembly is hydrophobic interaction. It is also an important feature for construction of mesoscaled hybrids via nanoarchitectonics. This type of nanohybrid has potential for use as a contrast agent of magnetic resonance imaging and for use in magnetic hyperthermia therapy.

5.2 ORGANIC-INORGANIC HYBRID HOLLOW CAPSULES PREPARED BY COLLOID TEMPLATING

The preparation and study of core−shell colloids are of burgeoning interest, principally because such particles can display novel and enhanced properties (e.g., electrical, magnetic, catalytic, mechanical, chemical, and optical) when compared with those of the core particles [2,3]. Microcapsules have also attracted widespread interest because of their potential application in the areas of drug delivery, agriculture, and the food and cosmetics industries [4]. The core−shell particle and the hollow capsules formed with organic-inorganic hybrid materials are possible candidates for these applications since they can show both characteristics of organic and inorganic matter. A versatile method that affords microcapsules with control over their size, stability, loading, and release properties is the LbL assembly technique [3,5,6]. On the other hand, metal oxide precursors are widely used to form thin inorganic coatings on colloid surfaces. Herein, the preparation of organic-inorganic hybrid capsules by a combination of LbL assembly and sol−gel methods is reported. Polyelectrolyte capsules were fabricated by the alternate assembly of poly(sodium 4-styrenesulfonate) (PSS) and poly(-diallyldimethylammonium chloride) (PDDA) onto melamine formaldehyde (MF) particles, followed by removal of the MF particles by dissolution with HCl. An SiO_2 coating was deposited on the capsules by dispersing the polyelectrolyte-coated capsules in ethanol, adding tetraethylorthosilicate

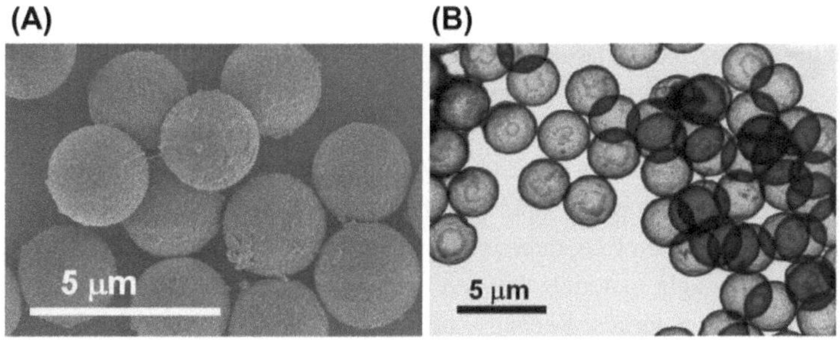

■ **FIGURE 5.1** Scanning electron microscope (SEM) (A) and transmission electron microscope (TEM) (B) images of $(PSS/PDDA)_5/SiO_2$ capsules.

(TEOS), ammonia, and water to cause hydrolysis and condensation of the alkoxide. Fig. 5.1 shows scanning electron microscope (SEM) and transmission electron microscope (TEM) images of hybrid hollow capsules formed with polyelectrolyte multilayers and SiO_2. From the SEM image, the particles have a spherical structure without any folds and creases. The numerous folds, creases, and flattening of the capsules are usually observed for SEM image of the hollow capsules formed with polyelectrolyte multilayer without SiO_2 [3]. Therefore, the siloxane network of surface SiO_2 strengthens the shell of the capsules. In addition, the size of the capsules well reflected the original size of the core templates, c.3.0 μm. The TEM image represents the hollow structure of the particles. The image also indicated that the shell has homogeneous thickness. The thickness is controllable by the reaction condition of sol−gel process. These techniques are applicable to the preparation of various kinds of hybrid capsules. Especially, the development of "smart capsules" with stimuli-responsive property is now the focus in several research groups. These systems will be aimed at applications in medical, cosmetic, and agricultural fields.

Optically addressable delivery systems [7−10] have been the subject of much research in recent years because irradiation with light promises to be a more direct means of releasing the active materials than an environmental trigger such as a change in pH, temperature, ionic strength, or enzyme concentration [11]. For example, Tao et al. [9] and Skirtach et al. [10] reported polyelectrolyte multilayer capsules doped with Congo red dye molecules and silver nanoparticles (NPs), respectively.

Despite these studies, there are considerable challenges in encapsulating and releasing low-molecular-weight molecules (including drugs) from LbL−engineered capsules. The stable encapsulation of small molecules in polyelectrolyte capsules is considerably more difficult than for macromolecules, because polyelectrolyte multilayers exhibit a high permeability to low-molecular-weight compounds [12]. This permeability can be reduced by the deposition of lipid coatings on the capsules [13], thus offering a pathway for encapsulating small molecules. However, such systems can be unstable under a variety of conditions, especially at different temperatures, as the permeability of the lipid (and hence the capsules) is highly temperature sensitive. Therefore, there is a need for stable capsules loaded with low-molecular-weight materials that can be triggered to controllably release the encapsulated materials. Recently, organic-inorganic hybrid capsules fabricated by a combination of LbL assembly and sol−gel methods was proposed. It shows that these capsules are suitable for the stable encapsulation of a low-molecular-weight dye (phenol red), which can be released upon UV irradiation. TiO_2 was selected as the UV−responsive component since it can decompose organic materials by photocatalytic reaction. Such a reaction was used to trigger rupturing of the polymer capsule walls, releasing the dye. The binary inorganic oxide system SiO_2−TiO_2 was selected since anatase TiO_2 nanocrystals can be obtained from SiO_2−TiO_2 gel films via treatment at less than 100°C [14]. It is also shown that release of the dye can be tuned through controlling the SiO_2:TiO_2 composition and UV-irradiation intensity. Fig. 5.2 shows the procedure used for preparing the organic-inorganic hybrid capsules by LbL assembly and sol−gel chemistry. Polyelectrolyte capsules were fabricated by the alternate assembly of PSS and PDDA onto MF particles, followed by removal of the MF particles by dissolution with HCl [3]. The capsules, terminated by a PSS layer, were subsequently coated with a cationic lipid bilayer of dioctadecyldimethylammonium bromide (DDAB). Phenol red was postloaded into the capsules by switching the lipid bilayer membrane permeability through the "gel" to "liquid crystalline" phase transition. An SiO_2 coating was deposited on the capsules by dispersing the DDAB−coated capsules in ethanol, adding TEOS, ammonia, and water to cause hydrolysis and condensation

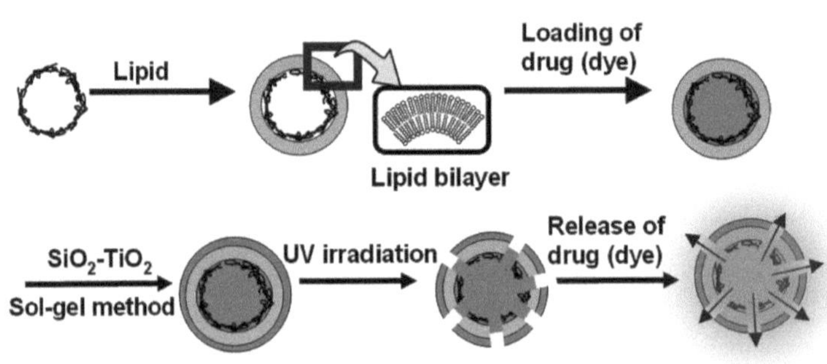

■ **FIGURE 5.2** Schematic drawing of preparation process of UV-responsive capsules.

of the alkoxide. In the case of the SiO_2-TiO_2 coating on the capsules, the precursor sol was prepared from TEOS and tetra-*n*-butylorthotitanate (TBOT). TEOS, TBOT, and acetylacetone, as a chelating agent, were dissolved in ethanol. A previous report by Wang and Caruso has shown that the sol–gel coating of polyelectrolyte multilayers assembled on particles results in uniform organic-inorganic coatings [15].

The effect of UV irradiation on the morphology of the capsules was examined by SEM. Fig. 5.3 shows SEM images of the organic-inorganic hybrid capsules before and after UV exposure. Prior to UV irradiation, the particles have a spherical structure for both SiO_2–polymer and SiO_2/TiO_2–polymer capsules. After UV irradiation, the SiO_2–polymer capsules remain spherical and intact, whereas the SiO_2/TiO_2–polymer capsules completely decomposed. In TiO_2, electron-hole pairs are generated through excitation with UV light, and the holes residing in the valence band of TiO_2 have high oxidative activity [16]. In the current study, the polyelectrolyte capsules are decomposed by the photocatalytic effect of the TiO_2 in the inorganic layer.

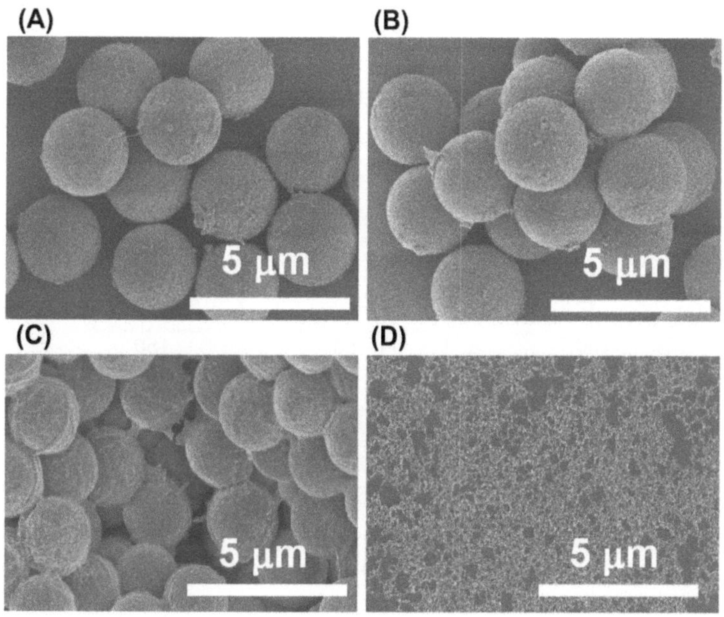

■ **FIGURE 5.3** Scanning electron microscope (SEM) images of $(PSS/PDDA)_5/PSS/DDAB/SiO_2$ capsules (A, B) and $(PSS/PDDA)_5/PSS/DDAB/SiO_2-TiO_2$ (75:25) capsules (C, D) before (A, C) and after (B, D) UV irradiation (20 mW cm^{-1}, 60 min).

The effects of UV-induced release of the encapsulated dye molecules were examined. Phenol red was selected for use in these experiments since it has been widely examined in release studies from colloidal materials, including liposomes. Phenol red can act as an indicator, changing color at different pH values [17]. The dye was encapsulated in the internal aqueous phase of the lipid-coated capsules at pH 7.0 above the phase transition temperature of the lipid membrane (c.40°C). At this condition, the molecular permeability of the membrane is high. After cooling to 20°C, the dye was sealed into the capsules because of the reduced permeability. The capsules were then washed and coated with SiO_2/TiO_2. The external aqueous phase was changed to pH 9.4. When the dye was released to the outer phase of the capsules, the color changed to purple. Therefore, the release behavior can be detected by an increase in absorbance at 550 nm (maximum absorbance of phenol red at pH 9.4). The control experiment was examined by using SiO_2–coated capsules. Fig. 5.4 shows the changes in absorbance (550 nm) with UV irradiation time. These plots are directly comparable since the same amount of encapsulated dye was employed in each experiment. The SiO_2/TiO_2–coated capsules retained the dye, with no dye release observed over a period of 24 h. When the SiO_2/TiO_2–coated capsules (SiO_2: TiO_2 = 75:25 and 95:5) were irradiated with UV light, the dye was released

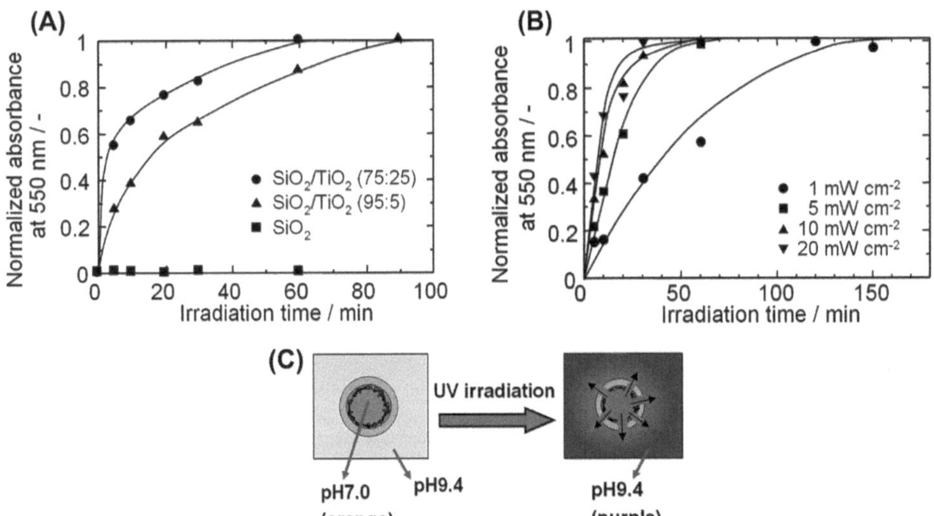

■ **FIGURE 5.4** Release of phenol red from the organic-inorganic hybrid capsules, monitored by changes in dye absorbance at 550 nm as a function of UV irradiation time. (A) Capsules coated with SiO_2 and SiO_2–TiO_2 (75:25 and 95:5) were examined. The UV intensity was fixed at 20 mW cm^{-2}. (B) Several UV intensities (1, 5, 10, and 20 mW cm^{-2}) were used to irradiate capsules coated with SiO_2–TiO_2 (75:25). (C) Schematic illustration of color change of the dispersion containing phenol red–loaded capsules by UV irradiation.

(Fig. 5.4A). The dye release rate was controllable through controlling the SiO_2:TiO_2 composition in the capsules, with quicker release observed for the capsules containing a higher TiO_2 content. In contrast, there was negligible increase in the absorbance after irradiation for 60 min for the SiO_2–coated capsules, reflecting virtually no dye release. These results indicate that the presence of TiO_2 in the capsule shells decomposes the polyelectrolytes, causing release of the dye. The release rate of the dye was also tunable by changing UV intensity. When UV intensities greater than 5 mW cm^{-2} were used to irradiate the SiO_2/TiO_2 (75:25)–coated capsules, the encapsulated dye was released with 60 min. In contrast, a UV intensity of 1 mW cm^{-2} resulted in considerably slower release (c.150 min) (Fig. 5.4B).

Light-triggered systems have disadvantages for actual medical applications. UV light cannot penetrate into the tissue and is toxic to human tissues. In the case of NIR light, most tissues show only weak absorption. However, a short-pulse NIR laser with a particular intensity is required for NIR–triggered systems [7,8,18]. On the other hand, a magnetic field represents a versatile candidate as an external force for an on-demand release system. Human tissues are sufficiently transparent to magnetic fields and are not damaged by them. Therefore, a magneto-responsive smart capsule was designed by a hybrid system using LbL–assembled polyelectrolytes, lipid bilayers, and magnetic nanoparticles [19]. These capsules are suitable for the stable encapsulation of low-molecular-weight compounds, which can be released upon irradiation with an alternating magnetic field. Membrane permeability of lipid bilayers changes with temperature through the "gel" to "liquid crystalline" phase transition [20]. Meanwhile, magnetic nanoparticles can generate heat in an alternating magnetic field due to hysteresis loss and/or Néel relaxations [21–23]. On the basis of these two phenomena, magneto-responsive microcapsules should be designable, functioning as follows. Irradiation with an alternating magnetic field will cause heating on the capsular shell of magnetic nanoparticles. A local temperature increase on the capsular shell will induce phase transition in the lipid membrane and then encapsulated molecules should be released from the interior of the capsules. Fe_3O_4 NPs were deposited on polyelectrolyte-hollow capsules using Pd catalysts. Fe_3O_4 NPs were deposited rather uniformly on the surfaces of polyelectrolyte-hollow capsules, and the deposited Fe_3O_4NPs had an average diameter of c.13 nm (Fig. 5.5).

The release behavior of the hybrid capsules formed with Fe_3O_4 NPs and the lipid bilayer was investigated by fluorescence microscopy. Fig. 5.6 shows fluorescence microscope images of the hybrid capsules formed with polyelectrolytes, Fe_3O_4 nanoparticles, and a lipid bilayer membrane loaded with fluorescence dye, calcein. Before alternating magnetic field irradiation,

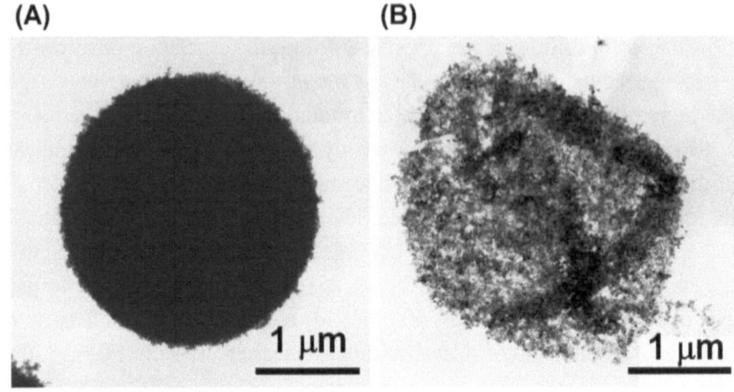

■ **FIGURE 5.5** Transmission electron microscope (TEM) images of (A) an Fe_3O_4 deposited MF particle coated with $(PSS/PAH)_5/Pd$ and (B) a Fe_3O_4/polyelectrolytes hollow capsule after removal of core.

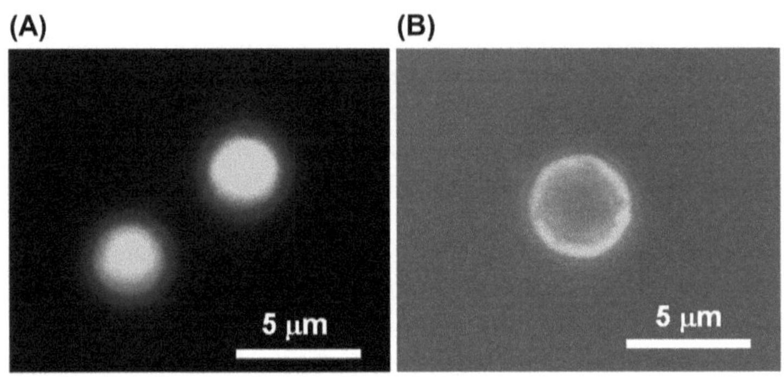

■ **FIGURE 5.6** Fluorescence microscope (A, B) images of hybrid capsules formed with polyelectrolytes, Fe_3O_4 nanoparticles, and a DDAC bilayer membrane loaded with calcein before (A) and after (B) alternating magnetic field irradiation at 360 kHz and 234 Oe for 30 min.

green fluorescence was observed only on the inside of capsules, indicating that calcein was stably enclosed within the lipid bilayer without a sign of leakage or release. In contrast, fluorescence was observed both inside and outside of the capsules after alternating magnetic field irradiation for 30 min. This result indicates that fluorescence dyes were effectively released from the capsule under magnetic stimuli.

The release behavior of encapsulated substances was quantitatively examined by using phenol red the same as UV−triggered system. Fig. 5.7 shows the changes in absorbance with time of exposure to the alternating magnetic field. A control experiment was carried out in the absence of the alternating magnetic field. The plots in Fig. 5.7 are directly comparable because the

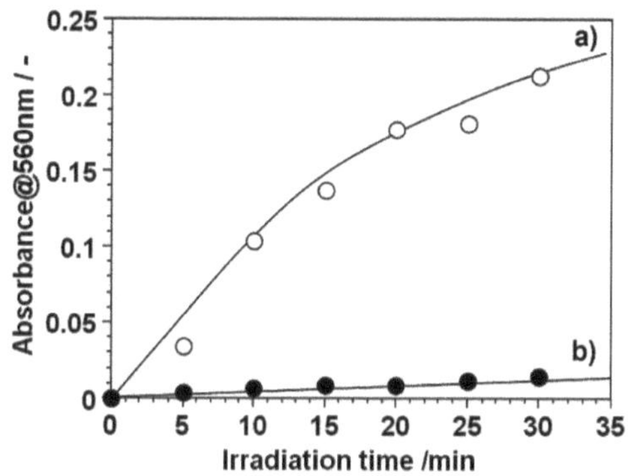

■ **FIGURE 5.7** Release of phenol red from the hybrid capsules, monitored by changes in dye absorbance at 560 nm as a function of irradiation time of alternating magnetic field (360 kHz and 234 Oe) (A). A control experiment was carried out by experiment without alternating magnetic field irradiation (B).

same amount of encapsulated dye was used in each experiment. When the hybrid capsules formed with polyelectrolytes, Fe_3O_4NPs and a lipid bilayer were irradiated with an alternating magnetic field, and dye was released (Fig. 5.7A). In contrast, there was a negligible increase in the absorbance after 30 min for capsules without the alternating magnetic field (Fig. 5.7B), indicating virtually no dye release. In addition, almost no dye release was confirmed when the hollow capsules formed without Fe_3O_4 NPs were irradiated with an alternating magnetic field. These results indicate that the presence of Fe_3O_4 NPs in the capsule shells induces phase transition of lipid membrane by heat in the presence of an alternating magnetic field, causing dye release.

5.3 SELF-ASSEMBLED NANOHYBRID OF MAGNETIC NANOPARTICLE CLUSTERS WITH POLYSACCHARIDE NANOGELS

The development of inorganic NPs has progressed over the past two decades [24–26]. They are widely used in various applications since they have specific physicochemical properties [27–29]. For example, semiconductor, noble metal, heterometallic, and metal oxide NPs have unique size-dependent properties that make them suitable for extensive applications. Fluorescence imaging benefits from the use of semiconductor NPs [e.g., quantum dots (QDs)], which have remarkable optical properties [30,31]. Compared with organic fluorescent dyes, QDs do not suffer from photobleaching. Magnetic NPs have also received a great deal of interest because of their potential applications in magnetic resonance imaging (MRI) and magnetic hyperthermia therapy (MHT) [32,33]. Current issues in the exploitation of inorganic NPs include the poor colloidal stability of the NPs in aqueous media and the difficulties associated with conjugating functional synthetic or biological molecules to the surface of inorganic NPs [34,35]. Development of organic-inorganic hybrids with mesoscale via nanoarchitectonics is one of the most versatile strategies for designing water-dispersible inorganic NPs. For example, coatings with lipids [36], polymers [37], and silica [38] have been widely examined. These materials should provide the NPs with optimal binding sites and improve water dispersibility, chemical functionality for further functionalization, and biocompatibility. Ideally, the coating process should be easy and environmentally friendly. Hydrogel nanoparticles (nanogels) are a promising class of biomaterials that has these properties [39,40]. Hydrophobized polysaccharides, such as cholesterol-bearing pullulan (CHP) (Fig. 5.8A), spontaneously form stable nanogels with a diameter of approximately 30 nm in water by self-association of hydrophobic groups, which form physical cross-links (Fig. 5.8B). CHP nanogels show high dispersibility and biocompatibility in aqueous media and have been widely used in drug-delivery systems [41,42]. Here a simple procedure to generate iron oxide cluster–polymer hybrids has been proposed for potential theranostic applications [43]. This method consists of mixing

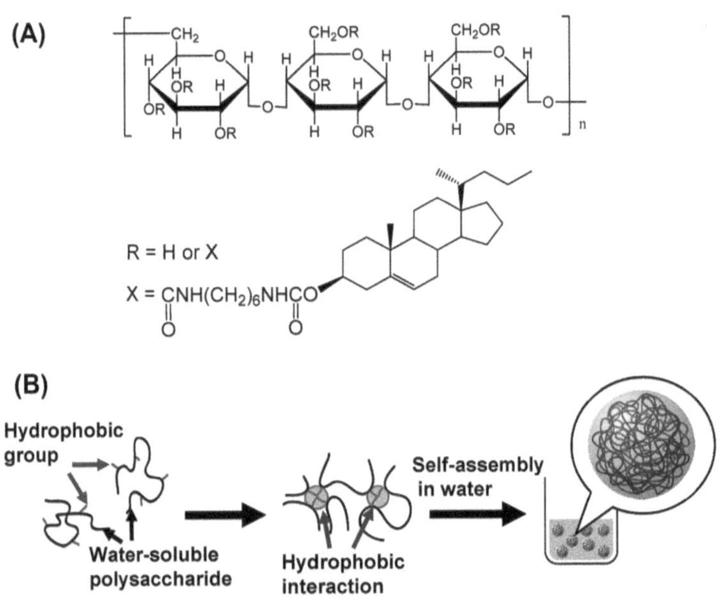

■ **FIGURE 5.8** Molecular structure of cholesterol-bearing pullulan (CHP) (A). Schematic representation of the formation of the CHP nanogel by self-association of cholesteryl groups in water (B).

self-assembled amphiphilic nanogels. Hydrophobized Fe_3O_4 NPs were initially synthesized by a conventional hydrothermal process. Then, the Fe_3O_4 NPs were complexed with CHP nanogels via hydrophobic interactions. The characteristics of the original CHP nanogels, especially their biocompatibility and high-colloidal stability, can be exploited to generate inorganic-organic hybrids containing Fe_3O_4 NPs. In addition, CHP nanogels have been accepted as a biomaterial with clinical applications because CHP nanogels can trap anticancer drugs, bioactive peptides, and proteins, for example, with the nanospace formed by the polymer networks, and have an artificial molecular chaperone function [44,45]. CHP nanogels can be conjugated with a variety of functional moieties and nanogels responsive to pH, temperature, and light have already been developed. For example, the accumulation of Fe_3O_4 NPs at sites of diseases is possible by conjugating the nanogel with an antibody or a specific peptide ligand. Therefore, the current approach has several advantages in terms of developing further functionality of nanogels comprising Fe_3O_4 NPs compared with other techniques to generate Fe_3O_4 NP clusters. In addition, the high-loading capacity of bioactive molecules, such as proteins and DNA/RNA, within nanogels allows us to expand the therapeutic effects beyond Fe_3O_4 NPs. Therefore, the self-assembled hybrid of CHP nanogels and Fe_3O_4 NPs is a promising multifunctional theranostic agent based on its applications in drug delivery, as a contrast agent in MRI, and for use in MHT.

Hydrophobized Fe_3O_4 NPs were synthesized by a conventional hydrothermal process. Oleic acid was used to control the size and hydrophobicity of Fe_3O_4 NPs. The particle size is approximately 12 nm and they have narrow particle size distribution. CHP nanogels (diameter = c.34.0 nm) formed via the self-association of hydrophobic cholesterol groups, which form physical cross-links. This indicates that certain hydrophobic cross-links are located in the interior of CHP nanogels. Therefore, hydrophobized Fe_3O_4 NPs can be incorporated into the hydrophobic nanospace formed by the polymer networks of CHP nanogels via hydrophobic interactions. Tetrahydrofuran (THF) was chosen as a dispersant for hydrophobized Fe_3O_4 NPs. Hydrophobized NPs easily disperse in THF, whereas THF is miscible in water. Here, a simple THF injection method was employed to prepare the CHP nanogel–Fe_3O_4 NP hybrid. Briefly, a THF dispersion containing the hydrophobized Fe_3O_4 NPs was rapidly injected into an aqueous dispersion of CHP nanogels through a fine needle. The force of the injection is usually sufficient to achieve complete mixing, so that the THF is almost instantly diluted in water, and the Fe_3O_4 NPs are homogeneously dispersed throughout the medium. The hydrophobized NPs were then spontaneously complexed with amphiphilic CHP nanogels because it is thermodynamically unfavorable

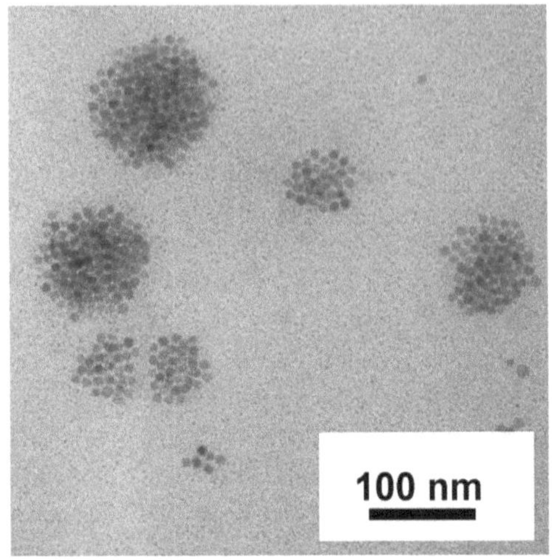

FIGURE 5.9 TEM images of CHP nanogel—Fe_3O_4 NP hybrids prepared using THF dispersions of Fe_3O_4 NPs at concentrations of 2.0 mg mL^{-1}. The concentration of the CHP nanogel was fixed to 0.1 mg mL^{-1}.

for the NPs to exist in a bulk aqueous phase. This process yielded a stable aqueous dispersion of the CHP nanogel—Fe_3O_4 NP hybrids. Fig. 5.9 shows a typical TEM image of the aqueous dispersion of the CHP nanogel—Fe_3O_4 NP hybrid. Aggregates of Fe_3O_4 NPs with diameters of 40—100 nm were observed. Because the samples were unstained, TEM only detected inorganic materials. Therefore, these aggregates should be surrounded by CHP nanogels. The size of the CHP nanogel—Fe_3O_4 NP hybrid increased along with the increase of Fe_3O_4 NPs concentration in THF.

The performance of the CHP nanogel—Fe_3O_4 NP hybrid as an MRI contrast agent was evaluated by measuring their relaxivities under a 1.5 T clinical MRI scanner. A phantom test was performed to investigate the MR signal-enhancing property of the hybrid using an agar phantom. Fig. 5.10A shows images obtained by T_2 MRI using various concentrations of the hybrid. This figure shows that the MR signal intensity (related to the T_2 relaxation time in T_2 images) is not identical for samples prepared using different concentrations of the hybrid. Of note, the MR signal (i.e., the negative signal on T_2 images) increased significantly with increasing concentration of the hybrid, similar to the results obtained using Resovist. The T_2 relaxation process occurs because of the exchange of energy between protons in water molecules. In the presence of an external magnetic field, superparamagnetic Fe_3O_4 NPs create inhomogeneity in the magnetic field that affects the microenvironment and dephases the magnetic moments of the

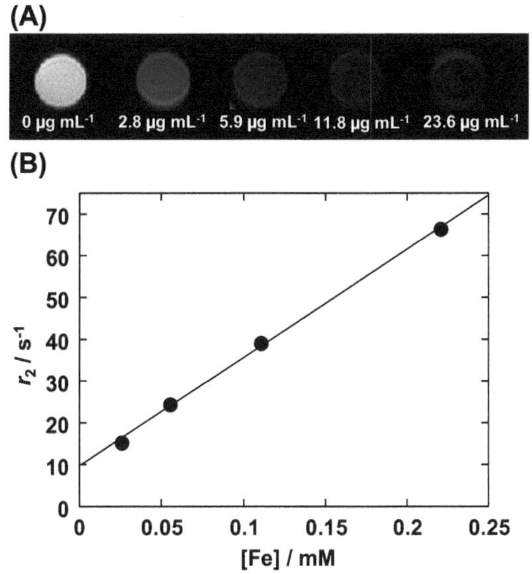

■ **FIGURE 5.10** T_2 MRIs of phantoms containing the CHP nanogel—Fe_3O_4 NP hybrid (A). The relaxivity of the CHP nanogel—Fe_3O_4 NP hybrid is linearly proportional to the Fe concentration (B).

protons, resulting in T_2 shortening. The results indicate that the hybrid generates a high magnetic field gradient near the surface of Fe_3O_4 NPs in CHP nanogels. MRI contrast agents shorten the T_2 (spin—spin) relaxation times, and their net effectiveness is expressed as T_2 relaxivity (r_2), which represents the reciprocal of the relaxation time per unit concentration of the iron ions. Fig. 5.10B shows the relaxation rate r_2 as a function of the iron concentration in the hybrid. At a magnetic field of 1.5 T, the CHP nanogel—Fe_3O_4 NP hybrid had a relatively high r_2 of 260 $mM^{-1}\ s^{-1}$, which is much higher than that of existing contrast agents, such as Resovist, MEIO, and Endorem (typically between 150 and 200 $mM^{-1}\ s^{-1}$) [28]. The r_2 increased dramatically with magnetic NP clustering. This enhancement of spin—spin relaxation is associated with the distortion of the local magnetic field by the magnetic NPs [46]. Similar results were reported for other copolymer/magnetic NP aggregates [32,47—49]. For example, Ai et al. reported that manganese ferrite NP clustering micelles have a r_2 of 270 $mM^{-1}\ s^{-1}$, which is much higher than single NP [47]. Therefore, the r_2 of CHP nanogel—Fe_3O_4 NP hybrid is comparable with that reported in these preceding papers. It was also reported that the aggregation of magnetic NPs transfers the size regimes of r_2 from the motional average regime (MAR) to the static dephasing regime (SDR), and that the maximum r_2 is reached in the SDR [49,50]. Here, the clusters of Fe_3O_4 NPs with a diameter of 10—13 nm have an

SDR of 60–110 nm. TEM showed that the hybrid formed aggregates with a diameter of 40–100 nm. Thus, the size distribution of the hybrid was within the SDR. Therefore, the r_2 of the CHP nanogel–Fe_3O_4 NP hybrid was much higher than that of existing contrast agents, such as Resovist.

To evaluate the potential applications of the CHP nanogel–Fe_3O_4 NP hybrid in MHT, the magnetic heating characteristics of the hybrid were determined using an induction heating system. Fig. 5.11A shows the temperature increases (ΔT) of aqueous dispersions of the CHP nanogel–Fe_3O_4 NP hybrid at various concentrations as a function of irradiation time. The temperature of the dispersion increased significantly following irradiation. The heating effect is caused by the absorption of energy from the alternating magnetic field and transformation into heat due to Néel and Brownian relaxations [21–23]. ΔT was proportional to the hybrid concentration. ΔT of an aqueous dispersion of the CHP nanogel–Fe_3O_4 NP hybrid was also determined as a function of alternating magnetic field irradiation time and magnetic field amplitude, and the results are shown in Fig. 5.11B. This experiment showed that the magnetic field amplitude had a profound effect on the maximum temperature and on the rate of ΔT. The maximum temperature achieved by the hybrid following irradiation at 168 Oe was only 32°C, which is insufficient for MHT. However, increasing the magnetic field amplitude to 307 Oe increased the maximum temperature 45°C, which is optimal for killing cancer cells.

In this subsection, nanoarchitectonics of mesoscaled hybrid materials consisting of a polysaccharide nanogel and Fe_3O_4 NPs were introduced. The hydrophobized Fe_3O_4 NPs were efficiently incorporated within CHP

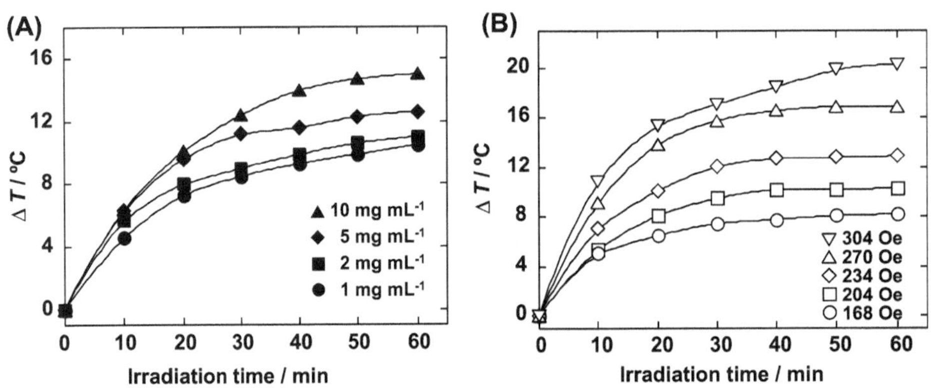

■ **FIGURE 5.11** Temperature profiles of aqueous dispersions of CHP nanogel–Fe_3O_4 NP hybrids as a function of irradiation time at an alternating magnetic field of 360 kHz. (A) Effects of varying the concentration of the CHP nanogel–Fe_3O_4 NP hybrid on ΔT. The magnetic field amplitude was fixed to 234 Oe. (B) Effects of varying the amplitude of the magnetic field on ΔT. The concentration of the CHP nanogel–Fe_3O_4 NP hybrid was 10 mg mL^{-1}.

nanogels through hydrophobic interactions between the cholesterol groups in the CHP nanogels and the oleate groups located on the surface of the Fe_3O_4 NPs. The hybrid showed good enhancement on T_2 MRI. The hybrid also provided higher relaxivity compared with existing contrast agents. In addition, the hybrid generated heat following irradiation with an alternating magnetic field. The heating temperature could be controllable by varying the concentration of the hybrid and the amplitude of the magnetic field, indicating that the hybrid is also applicable for MHT. In addition, other functional NPs modified with hydrophobic groups could also be produced using this method. Therefore, these nanoarchitectonics can be used to develop functional nanogel–NP hybrids not only for MRI and MHT but also for various applications.

5.4 SUMMARY

This chapter has highlighted some of the recent work in the area of organic-inorganic nanohybrid materials prepared via supramolecular nanoarchitectonics. Especially, organic-inorganic nanohybrids with spherical and capsular structures have been described in detail. Stimuli-responsive hybrid hollow spheres have been also developed by the colloid templating technique. These spheres should be applicable for the controlled delivery systems. The target external stimuli for the trigger of release can be controlled by selection of inorganic components of spheres. TiO_2 can be used for UV-responsive system by its photocatalytic activity. The release rate of this system is tunable by adopting an SiO_2–TiO_2 binary oxide system. Magneto-responsive microcapsules, comprising polyelectrolyte multilayers, lipid bilayers, and Fe_3O_4 nanoparticles, were also prepared via the colloid-templating technique. The magnetically induced release is attributed to the phase transition of the lipid membrane, caused by heat of Fe_3O_4 nanoparticles under magnetic stimuli, not to rupture of the capsules. In addition, hollow capsules prepared by LbL process have the ability to encapsulate various kinds of substances. Nanohybrids consisting of a polysaccharide nanogel and Fe_3O_4 NPs have also been prepared using a simple THF injection method. The hybrid showed excellent colloidal stability. The hybrid showed good enhancement on T_2 MRI. The hybrid also provided higher relaxivity compared with existing contrast agents. The hybrid generated heat following irradiation with an alternating magnetic field. The heating temperature could be controllable by varying the concentration of the hybrid and the amplitude of the magnetic field, indicating that the hybrid is also applicable for MHT. Other functional NPs modified with hydrophobic groups could also be produced using this method. Therefore, this method can be used to develop functional nanogel–NP hybrids not only for MRI and MHT but also for other biomedical applications, including bioimaging, photodynamic therapy, and neutron capture therapy. The author anticipates that, with further development, these nanoarchitectonics systems will make an important contribution not only to the development of organic-inorganic hybrid materials but also to the future application of bioinspired processing in nanoscience and nanotechnology.

REFERENCES

[1] Sanchez C, Ribot F. New J Chem 1994;18:1007.

[2] Caruso F. Adv Mater 2001;13:11−22.

[3] Caruso F, editor. Colloids and colloid assemblies: synthesis, modification, organization and utilization of colloidal particles. Weinheim: Wiley-VCH; 2003.

[4] Lim F. Biomedical applications of microencapsulation. Boca Raton, FL: CRC Press; 1984.

[5] Decher G, Hong J-D. Ber Bunsen Ges Phys Chem 1991;95:1430.

[6] Ariga K, Hill JP, Ji Q. Phys Chem Chem Phys 2007;9:2319.

[7] Radt B, Smith TA, Caruso F. Adv Mater 2004;16:2184.

[8] Angelatos AS, Radt B, Caruso F. J Phys Chem B 2005;109:3071.

[9] Tao X, Li J, Möhwald H. Chem Eur J 2004;10:3397.

[10] Skirtach AG, Antipov AA, Shchukin DG, Sukhorukov GB. Langmuir 2004;20:6988.

[11] Sershen S, Westcott SL, Halas NJ, West JL. J Biomed Mater Res 2000;51:293.

[12] Sukhorukov G, Brumen M, Donath E, Möhwald H. J Phys Chem B 1999;103:6434.

[13] Moya S, Donath E, Sukhorukov GB, Auch M, Baumler H, Lichtenfeld H, et al. Macromolecules 2000;33:4538.

[14] Katagiri K, Harada G, Matsuda A, Kogure T, Muto H, Sakai M. J Nanosci Nanotechnol 2006;6:1802.

[15] Wang D, Caruso F. Chem Mater 2002;14:1909.

[16] Ohko Y, Hashimoto K, Fujishima A. J Phys Chem A 1997;101:8057.

[17] Braganza LF, Blott BH, Coe TJ, Melville D. Biochim Biophys Acta 1983;731:137.

[18] Angelatos AS, Katagiri K, Caruso F. Soft Matter 2006;2:18.

[19] Katagiri K, Nakamura M, Koumoto K. ACS Appl Mater Interfaces 2010;2:768.

[20] Blok MC, van Deenen LL, De Gier J. Biochim Biophys Acta 1976;433:1.

[21] Rosensweig RE. J Magn Magn Mater 2002;252:370.

[22] Pankhurst QA, Connolly J, Jones SK, Dobson J. J Phys D Appl Phys 2003;36:R167.

[23] Mornet S, Vasseur S, Grasset F, Duguet E. J Mater Chem 2004;14:2161.

[24] Cushing BL, Kolesnichenko VL, O'Connor CJ. Chem Rev 2004;104:3893.

[25] Rao CNR, Vivekchand SRC, Biswas K, Govindaraj A. Dalton Trans 2007;34:3728.

[26] Gao Y, Tang Z. Small 2011;7:2133.

[27] Thanh NTK, Green LAW. Nano Today 2010;5:213.

[28] Na HB, Song IC, Hyeon T. Adv Mater 2009;21:2133.

[29] Cheon J, Lee J-H. Acc Chem Res 2008;41:1630.

[30] Yu WW, Chang E, Drezek R, Colvin VL. Biochem Biophys Res Commun 2006;348:781.

[31] Jamieson T, Bakhshi R, Petrova D, Pocock R, Imani M, Seifalian AM. Biomaterials 2007;28:4717.

[32] Hayashi K, Nakamura M, Sakamoto W, Yogo T, Miki H, Ozaki S, et al. Theranostics 2013;3:366.
[33] Ito A, Fujioka M, Yoshida T, Wakamatsu K, Ito S, Yamashita T, et al. Cancer Sci 2007;98:424.
[34] Erathodiyil N, Ying JY. Acc Chem Res 2011;44:925.
[35] Neoh KG, Kang ET. Polym Chem 2011;2:747.
[36] Gopalakrishnan G, Danelon C, Izewska P, Prummer M, Bolinger P-Y, Geissbühler I, et al. Angew Chem Int Ed 2006;45:5478.
[37] Nazli C, Ergenc TI, Yar Y, Acar FY, Kizilel S. Int J Nanomed 2012;7:1903.
[38] Yi DK, Selvan ST, Lee SS, Papaefthymiou GC, Kundaliya D, Ying JY. J Am Chem Soc 2005;127:4990.
[39] Akiyoshi K, Deguchi S, Moriguchi N, Yamaguchi S, Sunamoto J. Macromolecules 1993;26:3062.
[40] Kuroda K, Fujimoto K, Sunamoto J, Akiyoshi K. Langmuir 2002;18:3780.
[41] Sasaki Y, Akiyoshi K. Chem Rec 2010;10:366.
[42] Shimoda A, Sawada S, Akiyoshi K. Macromol Biosci 2011;11:882.
[43] Katagiri K, Ohta K, Sako K, Inumaru K, Hayashi K, Sasaki Y, et al. ChemPlusChem 2014;79:1631.
[44] Akiyoshi K, Sasaki Y, Sunamoto J. Bioconjugate Chem 1999;10:321.
[45] Nomura Y, Ikeda M, Yamaguchi N, Aoyama Y, Akiyoshi K. FEBS Lett 2003;553:271.
[46] Josephson L, Lewis J, Jacobs P, Hahn PF, Stark DD. Magn Reson Imaging 1988; 6:647.
[47] Lu J, Ma SL, Sun JY, Xia CC, Liu C, Wang ZY, et al. Biomaterials 2009;30:2919.
[48] Zhang B, Li Q, Yin P, Rui Y, Qiu Y, Wang Y, et al. ACS Appl Mater Interfaces 2012; 4:6479.
[49] Pöselt E, Kloust H, Tromsdorf U, Janschel M, Hahn C, Maßlo C, et al. ACS Nano 2012;28:1619.
[50] Min C, Shao H, Liong M, Yoon TJ. ACS Nano 2012;6:6821.

Nanoarchitectonic Metals

K. Eid[1,2,3], H. Wang[1] and L. Wang[1]
[1]*Zhejiang University of Technology, Zhejiang, P.R. China;* [2]*Changchun Institute of Applied Chemistry, Changchun, P.R. China;* [3]*University of Chinese Academy of Sciences, Beijing, P.R. China*

CHAPTER OUTLINE

6.1 INTRODUCTION

Metal nanocrystals (NCs) have unique physical and chemical properties enabling their utilization in a wide range of imperative applications such as catalysis, gas sensing, photonics, electronics, and biomedicine [1–5]. In principle, both properties and potential applications of metal NCs are mainly determined by different factors such as size, shape, crystallinity, composition, and architecture [4–8]. Controlling the morphology of metal NCs is one of the most pivotal parameters for production of various metals NCs with unusual properties [1,4,5,8,9]. For example, Pt-based NCs with high-index facets, porous morphologies, and multicompositions are superior to their solid counterparts and commercial Pt/C in both catalytic activity and durability toward oxidation and reduction reactions [10,11]. Likewise,

Supra-materials Nanoarchitectonics. http://dx.doi.org/10.1016/B978-0-323-37829-1.00006-7

anisotropic Ag NCs with multicorners and/or branches on their surface have a higher surface-enhanced Raman scattering activity than spherical Ag NCs, whereas hollow Au NCs possess surface plasmon resonance (SPR) near infrared relative to visible SPR of spherical Au NCs [12–14]. Thus, metal NCs with their exceptional shape-dependent properties (optical, magnetic, electrical, and catalytic, etc.) are intensively studied, not only for their importance in the basic science but also for their various potential applications ranging from medicine to catalysis. Consequently, in the past few decades, great efforts were dedicated to the precise control of the shape of metal NCs. These efforts culminated in the development of various physical, chemical, and biological synthetic approaches.

Among some of the most robust methods for the preparation of well-defined metal NCs are solution-based chemical routes such as polyol, seed-mediated growth, galvanic replacement (GR), and surfactant-based methods [5,15,16]. These methods entail several key parameters such as metal precursor concentration, capping agents, reductants, solvents, pH, and temperature, etc. Thus, the precise modulation of these parameters allowed for the preparation of wide varieties of profuse anisotropic morphologies such as rods, wires, cubes, dendrites, and cages with different sizes and composition. Despite the considerable advances achieved in this research area, many challenges still exist, such as multisteps, heating, and using organic solvents [5,15,16]. Additionally, it is a great challenge to obtain metal NCs with high surface area and active facets on their surface. This is because metal NCs with small size (<5 nm) are highly energetic and prone to grow into octahedral shapes with only {100} and {111} facets to reduce its total free energy [17]. There are massive efforts devoted not only to overcome these fences but also to develop new methods for the preparation of metal NCs. Although there are massive review papers, research articles, and books devoted to this topic, the link between synthetic methods of metal NCs and their mechanism is not highlighted enough [1,5–9]. Due to the vast progress in this research area, it is important here to discuss the latest significant developments in the synthesis of metal NCs. Therefore, this chapter is dedicated to synopsize the solution-based chemical methods in terms of seed-mediated growth, polyol, GR, and surfactant-based methods for controlling synthesis of nanoarchitectonic metals with special emphasis on their mechanism. Additionally, several immaculate examples of different metal NCs are also introduced. The composition- and morphology-dependent catalytic performance of metal NCs are also investigated roughly. The chapter concludes with a brief summary of the current challenges and future perspective on the development of metal NCs.

6.2 SEEDED-MEDIATED GROWTH METHOD

The seed-mediated growth method is a robust route for the preparation of metal NCs with different shapes and compositions ranging from mono- to multimetallic [17–19]. It involves the initial preparation of metal NCs with small sizes as starting seeds that subsequently grow in the reaction solution or drive the growth of other metal atoms into anisotropic shapes. Due to the high reduction rate resulting from the reducing agents and autocatalytic-mediated processes, the metal precursors nucleate promptly. Then, the nucleated metal atoms serve as catalytic sites for further reduction of the other metal precursors and create favorable sites for the atomic addition during particle growth [17]. Low lattice mismatching between seeds and other metal atoms leads to the epitaxial growth and vice versa. The essence of this method is the isolation of nucleation and growth stages, enabling the precise control over the size, shape, and crystallinity of the metal NCs that cannot be achieved otherwise. This intrinsically drives forward the formation of anisotropic shapes with high-index facets that are highly desired in various applications.

In the early 1990s Natan and coworkers developed the seed-mediated growth for the preparation of mono-dispersed Au NCs [20,21]. This was based on the colloidal gold surface-catalyzed reduction of gold precursors by hydroxylamine. Afterward, Murphy and coworkers used seed-mediated growth for controlling the morphology of Au and Ag NCs [22]. They obtained Au and Ag nanorods with a high aspect ratio via a three-step seeding route [23,24]. Interestingly, Au nanorods exposed penta-twinned structures that induce the elongation of the NCs. In 2002 Nikoobakht and El-Sayed [25] made a breakthrough in the seed-mediated growth of Au nanorods. They found that the addition of $AgNO_3$ to the growth solution [Au seeds and cetyltrimethylammonium bromide (CTAB)] drives the formation of monocrystalline Au nanorods with high yields nearly 100%. The concentrations of $AgNO_3$ determine the aspect ratio of Au nanorods (from 1.5 to 10) as well as their plasmon band (600 and 1300 nm) [26]. Additionally, higher concentrations of CTAB along with benzyldimethylhexadecylammonium chloride were favored for producing longer Au nanorods. Later Murphy and coworkers found that the concentrations of seed, surfactant, and reactant [Au^{3+} and ascorbic acid (AA)] molecules have a considerable effect on tuning the shape of Au NCs. Based on these parameters, they synthesized a wide range of Au NCs with different morphologies, including rods, rectangles, hexagons, cubes, triangles, branched, and star-like (Fig. 6.1) [27]. Following that, various efforts were devoted to understanding the mechanism of seed-mediated growth and its related factors. For instance, Xu and coworkers demonstrated that the size and crystallinity of seeds have a

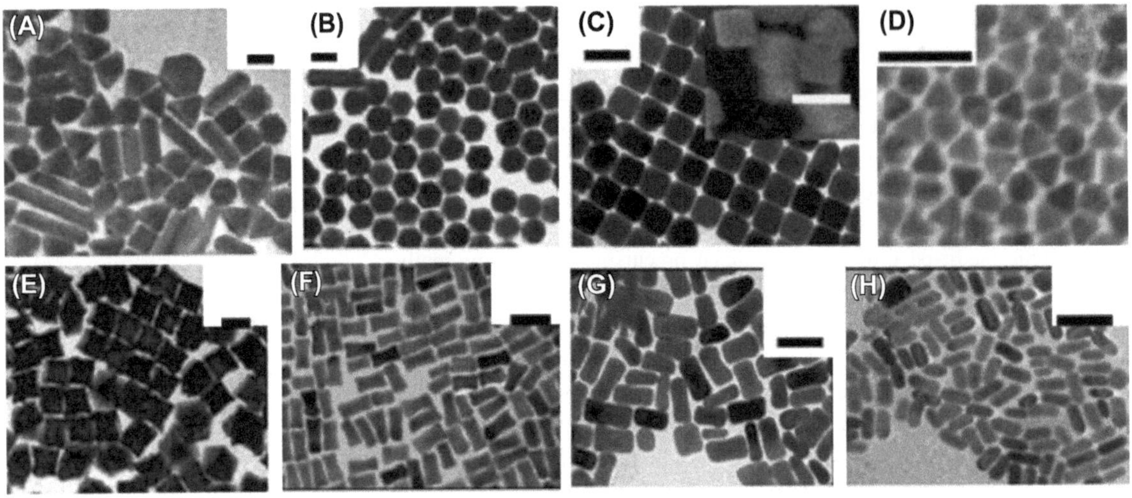

■ **FIGURE 6.1** (A—D) Transmission electron microscopy (TEM) of Au NCs produced under different conditions. AA concentration increased from (A) to (C) while seed concentration increased from (C) to (D). (E and F—H) TEM images of Au nanocube and nanorods formed in the presence of Ag NO_3 and low AA concentrations, respectively. CTAB concentration increased from (E) to (H). Au^{3+} concentration decreased from (F) to (H), while seeds increased from (G) to (H). The indicated *scale bars* in (A—H) are 100 nm. *Adapted with permission from Sau TK, Murphy CJ. J Am Chem Soc 2004;126:8648. Copyright 2004 American Chemical Society.*

significant effect on the growth of Au NCs [28]. The results clearly revealed that monocrystalline seeds with larger size were preferred for the production of a high yield of monocrystalline Au NCs. In particular, cubic Au NCs were obtained with yields of 95.2%, 26.1%, and 4% via using of 41.3, 1.5, and 3 nm of monocrystalline Au seed capped with cetylpyridinium chloride (CPC), CTAB, and citrate, respectively (Fig. 6.2) [28]. Au seeds with small sizes cannot overcome twinning during the growth process, which leads to subsequent production of multicrystalline Au NCs [29]. In addition to the seed size, the selective synthesis of monocrystalline metal NCs is determined by surfactant type/concentration, growth kinetics, temperature, reduction rate, and adsorbates type/concentration [28,29]. Controlling these parameters led to the preparation of single-crystalline Au NCs with different shapes such as cubic, octahedral, and rhombic dodecahedral NCs. Similarly, using large size of cubic Pd seeds, CTAB as a surfactant, KI as an additive, and AA as a reductant led to fabrication of various shapes of metal NCs such as rhombic dodecahedral, cubic, and octahedral Pd NCs at different temperatures (Fig. 6.3) [30—32]. In particular, a monocrystalline Pd {100} nanocube is obtained in the absence and/or at very low concentrations of KI. In the absence of KI, Br^- ions from CTAB adsorb onto the surface of Pd seeds and guide their growth into nanocubes with {100} fates. Meanwhile, Pd {110} rhombic dodecahedral and Pd {111} octahedral NCs are obtained

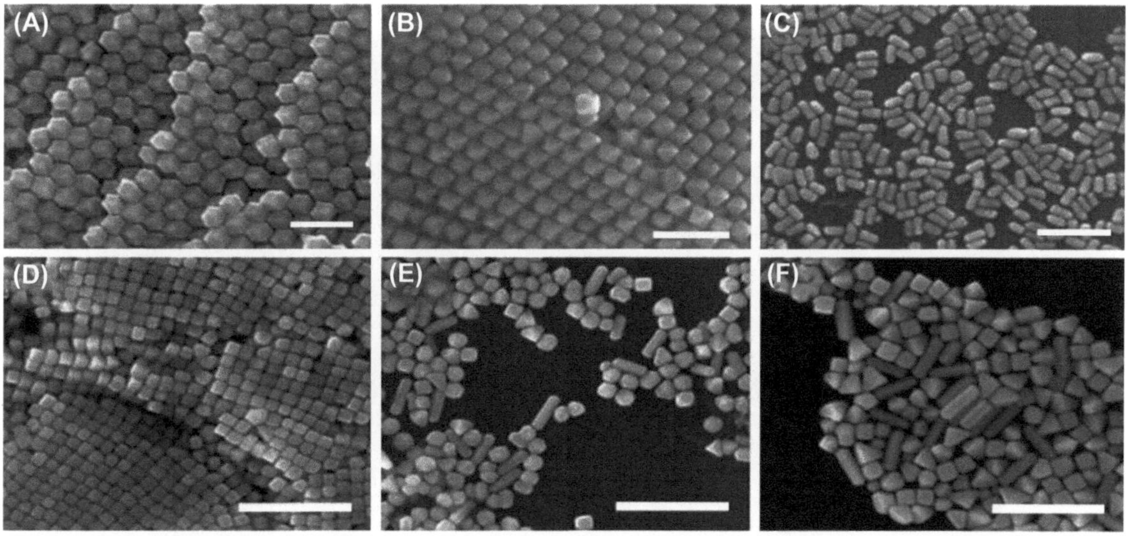

■ **FIGURE 6.2** (A) Rhombic dodecahedral Au NCs formed by using 10 mM CPC-capped Au seeds, (B) octahedral Au NCs by 100 mM CPC-capped Au seeds, and (C) Au nanorods by CPC-capped Au seeds with 10 mM $AgNO_3$. The indicated *scale bars* in (A), (B), and (C) are 200, 100, and 200 nm, respectively. (D) SEM images of Au NCs prepared by using CPC-capped Au seeds (41.3 nm), (E) citrate-capped Au seeds (3 nm), and (F) CTAB-capped Au seeds (1.5 nm). The indicated *scale bars* are 500 nm. *Adapted with permission from Niu W, Zheng S, Wang D, Liu X, Li H, Han S, et al. J Am Chem Soc 2008;131:697. Copyright 2008 American Chemical Society.*

at high and low temperatures, respectively, in the presence of medium concentrations of KI. This is because of, I^- ions from KI adsorbing on the surface of the Pd seeds and promptly producing rhombic dodecahedral with {110} facets that are more stable. Additionally, at a high temperature, the atomic deposition of Pd atoms takes place over {111} facets and guides the formation of octahedral Pd NCs.

Solvents are another important factor in the seed-mediated growth. In this regard, Carbó-Argibay and coworkers studied the growth of monocrystalline and penta-twinned Au nanorods in the presence of dimethylformamide (DMF) solvent and poly(vinyl pyrrolidone) (PVP) surfactant [33,34]. Using monocrystalline Au nanorods and penta-twinned Au nanorods drove the production of Au {111} octahedrons and Au {111}/{110} truncated decahedrons, respectively. This was due to the effect of PVP on modulating the surface energies of Au NCs in the order of $\{111\} < \{110\} < \{100\}$ facets in DMF solvent along with the highly anisotropic nature of Au nanorods [33–36]. This result infers that the final product of Au NCs preserved the initial crystalline structure of their preformed seeds. Interestingly, Au nanorods prepared by seed-mediated growth are successfully used for the fabrication of different bimetallic NCs such as Pd on Au and Ag on Au as

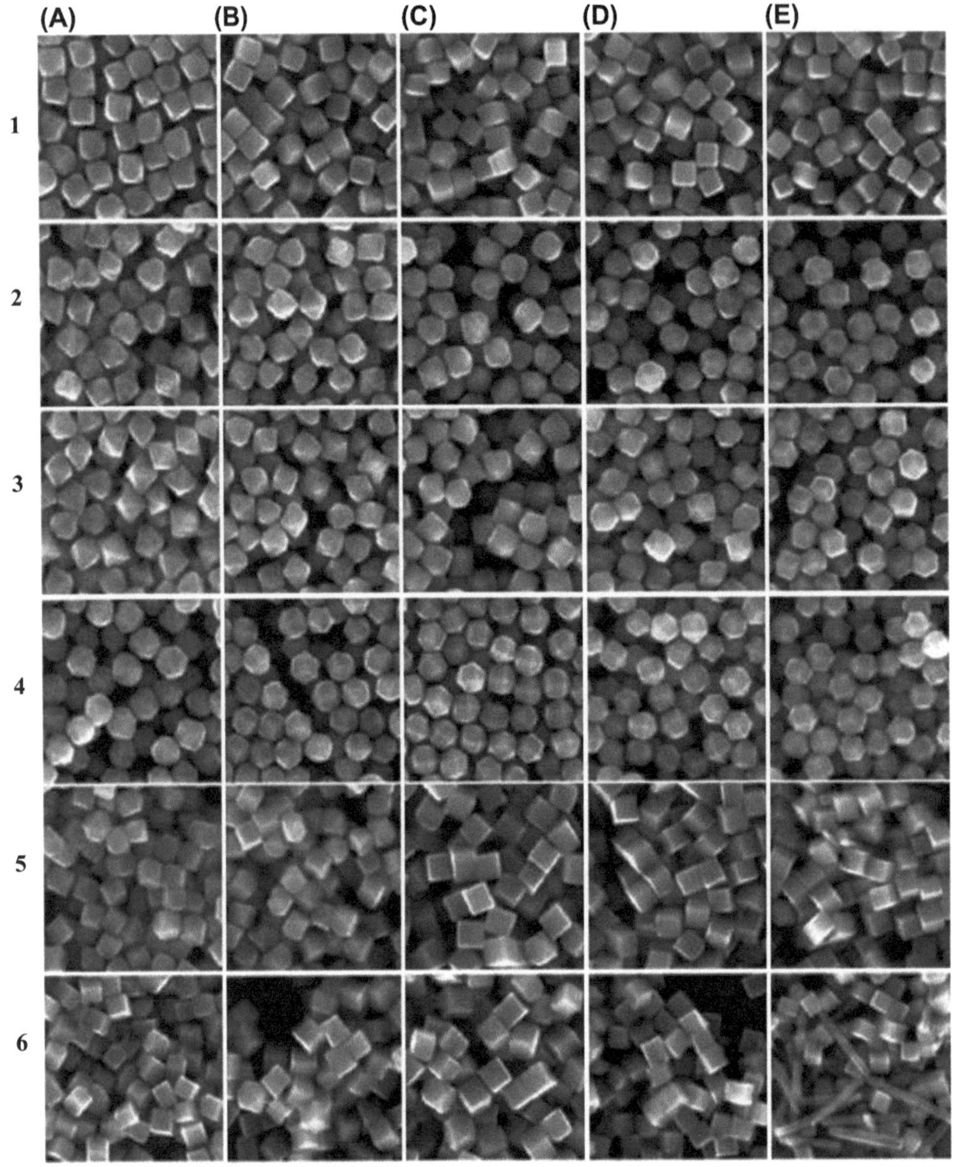

■ **FIGURE 6.3** Scanning electron microscope (SEM) images of polyhedral Pd NCs prepared under different conditions. In columns (A—E) the samples were synthesized at 30°C, 40°C, 50°C, 60°C, and 80°C, respectively. In rows 1—5 the samples formed by changing the concentration of KI. In row 6, samples were prepared without KI. The *indicated scale* bar is 200 nm. *Adapted with permission from Niu W, Zhang L, Xu G. ACS Nano 2010;4:1987. Copyright 2010 American Chemical Society.*

core–shell structures [35,37]. The morphology and crystallinity of the final metal NC products are determined by the structure of the starting seeds. Wide varieties of Au, Ag, and Pd can be obtained with different morphologies such as tetrahexahedral concave cubes and concave trisoctahedral with high-index facets by seed-mediated growth [38–42]. For example, Huang and coworkers synthesized Au-Pd core–shell heterostructures with various morphologies by using Au nanocubes as starting seeds [43]. Tetrahexahedral Au-Pd core–shell NCs were obtained by hiring a proper concentration of Au seeds at a high temperature, while concave octahedral and octahedral Au-Pd core–shell NCs were obtained at low temperatures and a long reaction time (Fig. 6.4A–C). The obtained concave metal NCs resulted from the oxidation of Pd atoms by O_2 with the assistance of Cl^- ions from cetyltrimethylammonium chloride (CTAC). The same group also synthesized various morphologies of Au-Pd core–shell NCs, including octahedral, truncated octahedral, cuboctahedral, truncated cubes, and concave cubes by controlling the amount of the Au octahedral NCs seeds in the growth solution (Fig. 6.4D–H) [44]. Xia and coworkers used Pd

■ **FIGURE 6.4** (A) Scanning electron microscope (SEM) images of the tetrahexahedral Au-Pd core–shell NCs formed by using Au nanocube seed, CTAC surfactant, H_2PdCl_4, and AA, at 31°C for 30 min; (B) concave cubes Au-Pd formed at 10°C for 4 h; and (C) octahedral Au-Pd core–shell NCs formed at 10°C for 2 h; (D) SEM images of octahedral, (E) truncated octahedral, (F) cuboctahedral, (G) truncated cubes, and (H) concave cubes Au-Pd core–shell NCs obtained by using 450, 400, 350, 300, and 200 μL of Au seeds, respectively. The *indicated scale* bars are 100 nm. *(A–C) Adapted with permission from Lu C-L, Prasad KS, Wu H-L, Ho J-A, Huang MH. J Am Chem Soc 2010;132:14546 Copyright 2010 American Chemical Society; (D–H) Adapted with permission from Yang C-W, Chanda K, Lin P-H, Wang Y-N, Liao C-W, Huang MH. J Am Chem Soc 2011;133:19993. Copyright 2011 American Chemical Society.*

NCs with different shapes, including truncated octahedral, cube, octahedron, and thin plate for the fabrication of Pd-Rh nanodendrites with the assistance of PVP (Fig. 6.5) [45]. The shape of Pd seeds and the concentration of Rh precursor drove the formation of Pd-Rh nanodendrites. Intriguingly, the degree to which Rh branches grew over the Pd core was determined by the concentration of Rh salt. The shape of Pd core did not change after growth of Rh branches. Binding the capping agent on the {100} and {111} facets drove the formation of concave cubes and concave octahedrons, respectively. For example, Pt-Rh concave cube was produced by using Pt nanocubes as seeds with controlling the injection rate of Rh salt in the presence of KBr, because Br^- ions block the {100} facets and enabling overgrowth at corners and edges [46].

Despite the great advances present in the seed-mediated growth method, reported morphologies were mostly nonporous. The preparation of porous metal NCs is highly desired for various applications due to their unique physical and chemical properties. In pursuit of this aim, Xia and coworkers prepared Pd-Pt nanodendrites with high-index facets upon usage of Pd seeds and PVP as a structure-directing agent [17]. The resulted porous dendrite morphology was ascribed to the strong reduction rate of AA and autocatalytic-mediated process. These led to prompt nucleation of Pt onto the surface of Pd seeds, which facilitated the autocatalytic process for further

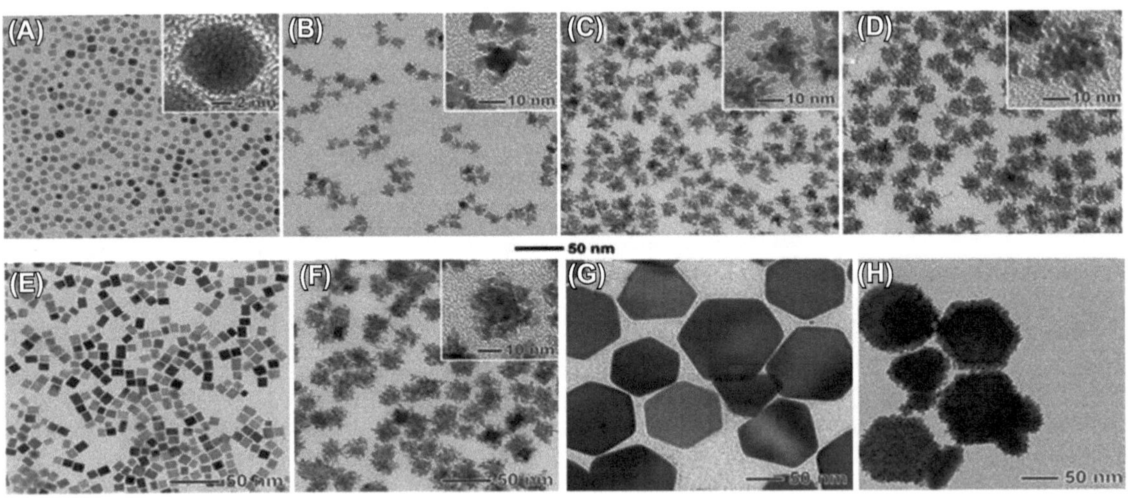

■ **FIGURE 6.5** (A) Transmission electron microscope (TEM) image of truncated octahedral Pd NCs seeds. (B—D) TEM images of Pd-Rh nanodendrites prepared by reducing of Na_3RhCl_6 by AA in the presence of Pd seeds. The concentration of Na_3RhCl_6 increased from (B to D). (E) TEM images of cubic Pd NCs seed. (F) cubic, octahedral, and (G) plate-like of Pd-Rh NCs prepared by using cubic, octahedral, and plate-like Pd seeds. The insets in (A—D), (E) are the HRTEM images. *Adapted with permission from Kobayashi H, Lim B, Wang J, Camargo PH, Yu T, Kim MJ, et al. Chem Phys Lett 2010;494:249. Copyright 2010 Elsevier.*

reduction of Pt precursor and resulted in the formation of favorable sites for the atomic addition of Pt atom. Then, the epitaxial growth occurred over the Pt nuclei instead of Pd seed, due to the low lattice mismatching factor (0.77% for Pt-Pd). Later, Yang and coworkers fabricated a wide range of porous hollow-structured metallic NCs with assistance of oxidative etching of bis(*p*-sulfonatophenyl) phenylphosphane legend [47], which bound strongly with Ag(I)/Ag(0) to allow the complete removal of Ag in 24–28 h, leaving behind an organosol of hollow or cage-bell structured metallic NCs. Interestingly, this strategy allowed the preparation of single, binary, and ternary hollow metallic NCs such as Ru, Os, Ir, PtRu, and PtRuOs with various porous shapes (Fig. 6.6). The aforementioned results should trigger researchers to exert more research endeavors in extending the seed-mediated growth process toward preparing novel porous metallic NCs with high-index facets, given their importance in different applications.

6.3 GALVANIC REPLACEMENT METHOD

The galvanic replacement (GR) is a facile and versatile method for fabrication of hollow metallic NCs with tunable pore structure and compositions through replacing the sacrificial templates with other metal species [48–61]. It is a corrosive process that results from the dissimilar redox potentials of the sacrificial templates and the metal required for the final product. This leads to the oxidative dissolution of the sacrificial templates to form porous structure. The shape, pore size, composition, and crystallinity of hollow metallic NCs are determined by the sacrificial templates. Xia and coworkers reported for the first time on the replacement reaction between Au^{3+}, Pt^{2+}, and Pd^{2+} precursors and Ag NCs or nanowires [48]. Right after that, this route was extended to produce other noble metals, semiconductors, and hybrid nanostructure [49–52]. For instance, upon usage of Ag NCs as sacrificial templates, wide varieties of porous metal nanostructures, such as spherical nanoshells, nanodots, nanocages, nanotubes, nanoframes, and nanodendrites, were obtained with different compositions [53–58].

Xia and coworkers studied the mechanism of the GR reaction between Ag NCs and $HAuCl_4$ in an aqueous solution (Fig. 6.7) [48]. Particularly, Ag NCs oxidized promptly to Ag^+ upon mixing with aqueous solution of $HAuCl_4$ due to the difference in the redox potentials among Au^+/Au 0.99 V and Ag^+/Ag 0.8 V versus standard hydrogen electrode, i.e.,

$$3Ag_{(s)} + AuCl_4{}^-{}_{(aq)} = Au_{(s)} + 3Ag^+{}_{(aq)} + 4Cl^-{}_{(aq)}$$

This reaction occurred on the facets with the highest surface energy. Then, a porous nanostructure with smooth Au-Ag alloyed shell was formed during

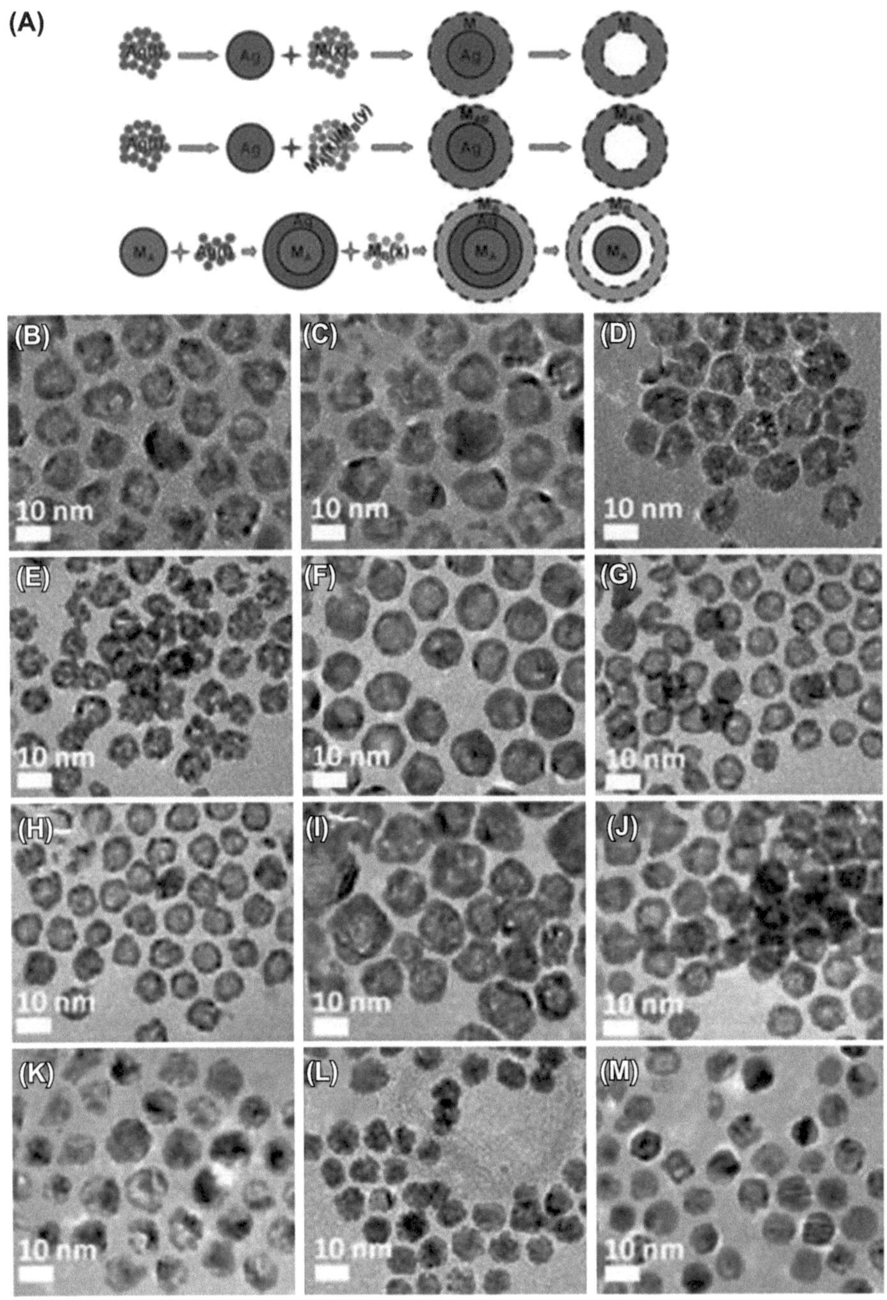

■ **FIGURE 6.6** (A) Scheme illustrates the preparation of various shapes of porous metallic NCs by using Ag NCs as starting seeds for preparation of various porous. TEM images of hollow (B) Ru, (C) Rh, (D) Os, (E) Ir, (F) Pt, (G) PtRu, (H) PtRh, (I) PtOs, and (J) PtRuOs. (K) TEM images of cage-bell structured of Pt-Ru, (L) Pt-Os, and (M) Pt—Pt NCs. *Adapted with permission from Kobayashi H, Lim B, Wang J, Camargo PH, Yu T, Kim MJ, et al. Chem Phys Lett 2010;494: 249. Copyright 2012 American Chemical Society.*

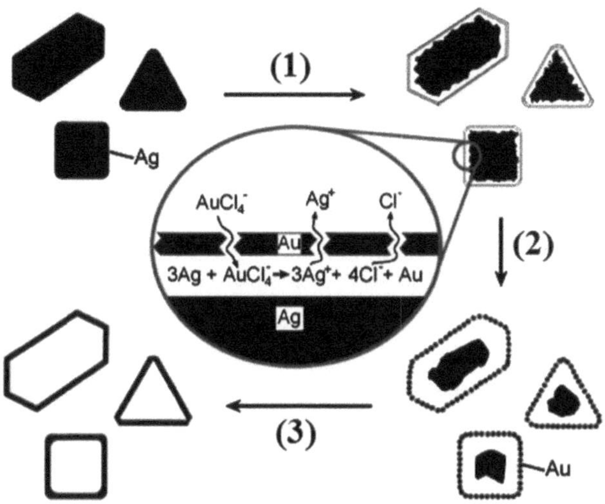

■ **FIGURE 6.7** Scheme shows the preparation of Au NCs with different shapes from Ag templates based on the galvanic replacement (GR). The reaction steps, including (1) initiation of the GR by addition of $HAuCl_4$ to of Ag NCs solution, (2) continuous the GR between $HAuCl_4$ and Ag NCs, and (3) dissolution of Ag and annealing of the resultant shells. *Adapted with permission from Sun Y, Mayers BT, Xia Y. Nano Lett 2002;2:481. Copyright 2002 American Chemical Society.*

combination of replacement reaction with alloying. Once the concentration of Au atoms reached the critical value, nuclei started to form followed by growth into small clusters, and then into complete shell-like covered Ag template via Ostwald ripening. Simultaneously, Ag atoms diffused into the Au shell to form seamless porous structures composed of Au-Ag shell. Finally, selective dissolution of Ag atoms from the alloyed wall occurred to allow shape reconstruction into porous morphology.

Several key factors have a substantial effect on the morphology, composition, and crystallinity of the final product, including metal precursor type/ concentration, reaction temperature, and the added reagents [59–61]. In this regard, Xia and coworkers observed tuning of the surface roughness of the obtained porous Au nanostructures via reaction temperature. Porous Au nanostructures with smooth and rough surfaces were obtained at high and low temperatures, respectively. This is because the AgCl (formed as a byproduct during the replacement reaction) solubility at different temperatures has a substantial effect on the precipitates as well as interdiffusion rates between Au precursors and Ag template [60,62]. They also reported that monocrystalline Pd-Ag nanoboxes with smooth exterior were obtained via reaction of Ag nanocubes with Na_2PdCl_4, whereas, multicrystalline Pt-Ag nanoboxes with a rough surface were produced upon usage of Na_2PtCl_4

[63]. This was ascribed to the difference in the solid–solid interdiffusion between Pt and Pd precursors on the Ag surface, resulting from their dissimilarity in chemical valences. This subsequently drove the uniformity of alloyed structures formed in the initial stage of the replacement reaction. In other words, the surface roughness decreased when the lattice matching factor increased and vice versa. This was clearly evident during the reaction of Au precursors with different chemical valences (e.g., $AuCl_2^-$ and $AuCl_4^-$) with cubic Ag nanostructures [64]. The results obviously warrant that, at the initial stage of reaction, a thick Au-Ag nanoshell is obtained when $AuCl_2^-$ is used and with partial dealloying of Ag from Au at the final stage relative to Au^{3+} (Fig. 6.8) [64]. Therefore, the discrepancies in the valence state among Au precursors lead to different reaction stoichiometry. Similarly, Yu and coworkers found that Pt nanotubes are formed upon reaction of H_2PtCl_6 with Te nanowires as a sacrificial template, while Pt and Pd nanowires are produced when using Pt^{2+} and Pd^{2+} under the

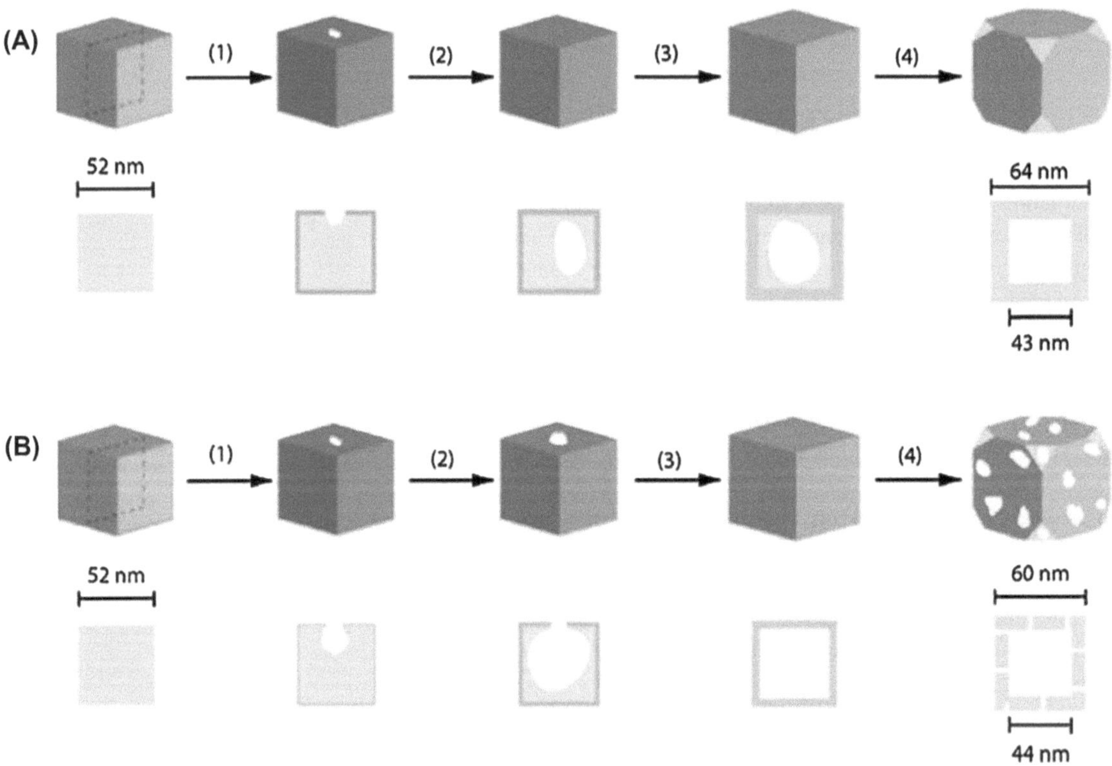

■ **FIGURE 6.8** (A) Scheme illustrates the effect of $AuCl_2^-$ and (B) $AuCl_4^-$ on the morphological evolution throughout the GR using Ag nanocubes as sacrificial template. *Adapted with permission from Au L, Lu X, Xia Y. Adv Mater 2008;20:2517. Copyright 2008 Wiley-VCH.*

same reaction conditions. They attributed this to the dissimilarity in the reaction stoichiometry [65].

Another important key factor in the replacement reaction is the additional reagents that alter the redox potential of the metal salts and, consequently, tune the replacement reaction directions. Specifically, Millstone and co-workers obtained mixed-metal nanoframes via reaction of Au nanoprism with OH^--substituted $[PtCl_6]^{2-}$ [66]. Using I^- ion, on the other hand, drives the formation of Ag@Au core–shell nanoplates and hollow Rh nanotubes [67]. The surface facets of the sacrificial templates drive the generation of NCs with desired facets. Particularly, during the GR reaction, the selective dissolution of template atoms occurs on the facets with the highest energy, while the deposition of other metallic atoms happens on the facets with the lowest free energy. The free energy of noble metal facets decreases on the order of (110) > (100) > (111). The surface free energy of the specific facets can be controlled upon usage of structure-directing agents and/or atomic species for selective fabrication of complex porous nanostructures [67]. For example, Ag cub octahedrons, enclosed by six (100) and eight (111) facets, was used as a sacrificial template for selective preparation of novel nanocages [68,69]. Owing to the involvement of PVP, the dissolution of Ag atoms started from (111) facets, while the deposition of Au atoms occurred on the (100) facets stabilized by PVP [68,69]. Likewise, during the replacement reaction between $PtCl_6^{2-}$ and a Pd nanocube enclosed by (100) facets that were adsorbed with Br^- ion led to the growth of Pd-Pt concave cubes and eventually octapods. Due to the presence of Br^- ions, Pd atoms on the (100) facets of the nanocubes were oxidized and removed. In this sense, surfactants and/or atomic species are also capable of altering the redox potentials of metallic precursors along with tuning the surface free energy. This facilitates the production of active metallic NCs at the early stage of the reaction, which can be used as a template to fabricate porous metallic nanostructures in one-step method [70]. For example, cubic $PtCu_3$ nanocages are obtained by the solvothermal method with the assistance of CTAB. This is achieved by initial mixing of $H_2PtCl_6 \cdot 6H_2O$, Cu(a-cac)$_2$, and CTAB in oleylamine (OAm) followed by heating [71,72]. Although the reduction potential of Cu precursor (0.34 V) is lower than Pt precursor (1.18 V), Cu NCs are initially formed and act as a template for Pt atoms enabling GR reaction to produce $PtCu_3$ nanocages. This is attributed to the presence of CTAB surfactant, which reverses the reduction kinetics of Pt and Cu species. Similarly, mixing Pt and Cu salts with glycine in the presence of PVP led to the formation of various shapes of PtCu NCs, including dendrites, yolk-cages, boxes, rhombic dodecahedral, and nanoframes. This is because glycine alters the reduction order of Cu and Pt precursors [73,74]. Additionally, using NaI, ethanolamine, and glycine in the presence of PVP drives the synthesis of octahedral Pt-Cu nanoframes [75]. This approach can be used for producing other multimetallic NCs with different porous structures upon hiring proper concentrations of surfactant and other reagents.

GR can be inhibited or regulated through the addition of reducing or oxidizing agents, which subsequently tune the reduction kinetics [76]. For example, Qin and coworkers prepared Ag@Au core−shell nanocubes by using Ag nanocubes as a template in the presence of AA (Fig. 6.9A) [77]. Interestingly, the morphology of the final product was determined by the GR rate (R_{gal}) relative to the AA reduction rate (R_{red}). This was achieved via mixing $HAuCl_4$ with Ag nanocubes, AA, and PVP in the presence of HCl (pH = 2.63). Both Ag nanocubes and AA facilitated the reduction of $AuCl_4^-$ and drove the formation of porous structures via GR. In this case, the R_{gal} was higher than the R_{red}, attributed to the poor reduction power of AA at this pH, whereas the reduction by AA led to the growth of a conformal Au shell on the Ag nanocubes. Likewise, PdPt nanocages, PdPt dendrites, and Pd@Pt nanocubes with dendritic shells were obtained by a similar approach [78]. Interestingly, using AA and hydrazine drove the formation of well-defined mesoporous Pt nanostructures and Pd-Ni-Pt core-sandwich shell NCs, respectively, via fine inhibition of the GR reaction [79,80]. On the other hand, using strong reductants such as hydroxylamine with NaOH inhibited the replacement reaction between Ag templates and Au precursor, and led to formation of Ag@Au core−shell nanoprisms [80]. The strong reduction power herein drives the formation of porous Ag@Au nanoprism via reduction instead of GR. Intriguingly, Xia and coworkers used Ag nanocubes as a template for fabrication of Au-Ag nanocages upon usage of $Fe(NO_3)_3$ as an etchant to isolate the dealloying of Ag from the deposition of Au (Fig. 6.9B) [56]. Initially Au-Ag alloy nanoboxes formed followed by the selective dissolution of Ag by $Fe(NO_3)_3$ because R_{gal} was lower than R_{red}. Similarly, other etchants, such as H_2O_2 and NH_4OH, were also used for the fabrication of novel porous metallic nanostructures through different dissolution mechanism [81−83].

GR method is very flexible and can be easily combined with other synthetic approaches for predation of multicomplex nanostructures [84−87]. For example, Sun and coworkers combined between seed-mediated growth and GR for the synthesis of Au NCs bimetallic nanoshell dimmers (Fig. 6.10) [88]. The synthetic process starts with seeding Au NCs on the surface of Fe_3O_4@SiO_2 followed by the deposition of another layer of SiO_2 on the surface of Fe_3O_4@SiO_2@Au. Then seed-mediated growth of Ag over the surface of Au seeds was followed by subsequent GR among

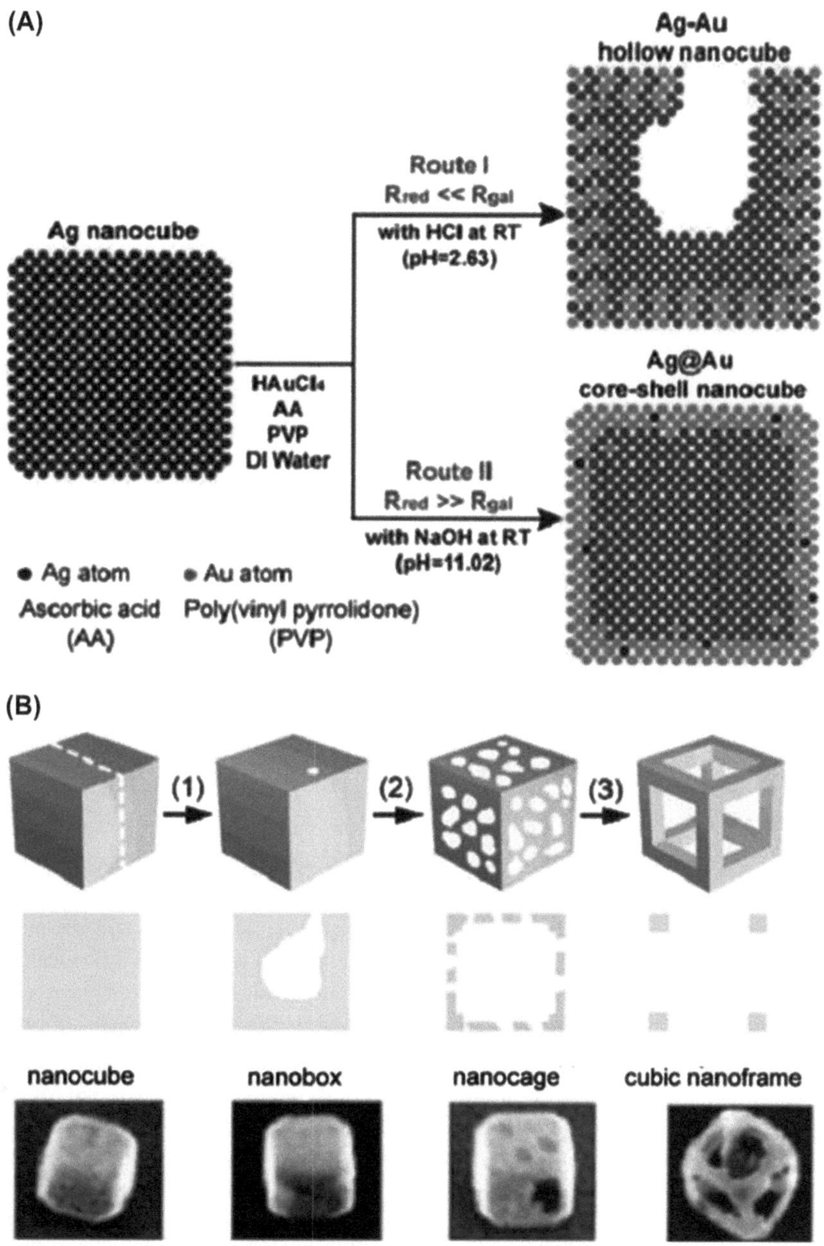

■ **FIGURE 6.9** (A) Scheme represents the effect of reduction kinetic by AA at different pH on the GR between Ag nanocubes and $HAuCl_4$. (B) Scheme shows the preparation of Au/Ag nanocage and Au cubic nanoframe by selectively etching of the Ag atoms from Au/Ag alloy nanobox by $Fe(NO_3)_3$ or NH_4OH solutions. *(A) Adapted with permission from Yang Y, Liu J, Fu Z-W, Qin D. J Am Chem Soc 2014;136:8153. Copyright 2014 American Chemical Society; (B) Adapted with permission from Lu X, Au L, McLellan J, Li Z-Y, Marquez M, Xia Y. Nano Lett 2007;7:1764. Copyright 2007 American Chemical Society.*

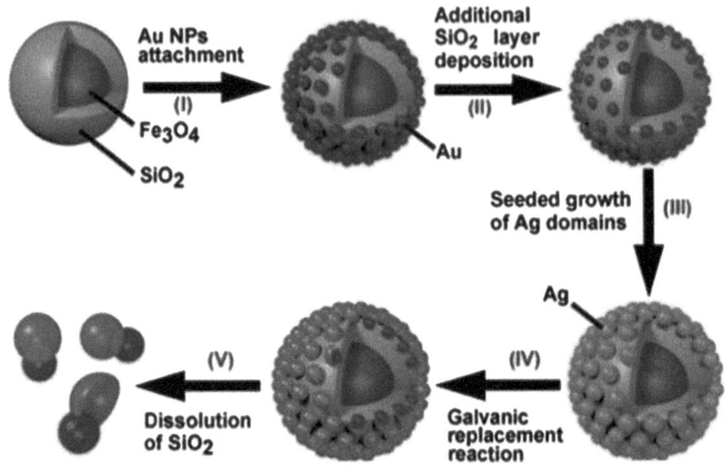

■ **FIGURE 6.10** Scheme shows the production of Au NCs-bimetallic nanoshell dimmers by seed-mediated growth and GR. *Adapted with permission from Hu Y, Liu Y, Li Z, Sun Y. Adv Funct Mater 2014;24: 2828. Copyright 2014 Wiley-VCH.*

Au and Ag. Finally, selective chemical etching of SiO_2 layers proceeds to form Au NCs bimetallic nanoshell dimmers. There are many nonprecious metals, including Ni, Co, Cu, and Mg, and Zn, that were used as sacrificial templates in the GR method for the preparation of multicomplex hollow nanostructures [89–97]. One-dimensional Te nanowires were used for the preparation of various single and binary noble metals with different structure such as nanotube and core-shells [65,98–100]. Metal oxides such as Cu_2O and (Mn_3O_4) were successfully used for the preparation of mesoporous Au NCs and porous Pt nano structures were obtained, respectively [101,102]. Although there have been great advances in the GR method, many challenges still exist, such as a limitation to the chemical transformation of metallic nanostructures. In pursuit of this aim, Hyeon and coworkers prepared hollow Fe_2O_3 nanocages upon GR between Mn_3O_4 nanoboxes and iron (11) perchlorate [103]. This method is generalized to synthesize another porous metal oxide NCs, including Co_3O_4/SnO_2 and Mn_3O_4/SnO_2. A well-defined 3D porous Pt_3Ni nanoframe with Pt-skin is obtained by interior erosion of $PtNi_3$ NCs [104]. Hiring a thermal treatment leads to decreasing the conversion time from 2 weeks to 12 h.

On the other hand, Murray and coworkers reported the antigalvanic replacement reaction (AGR) between Ag salt and Au nanoclusters $[Au_{25}(SC_2H_4Ph)_{18}]^{-1}$ [105]. The results demonstrate that, the AGR preferred small nanoclusters capped with thiolates. Right after that, Zhikun Wu studied the AGR among Au NCs and different metals such as Cu, Ag,

Fe, Ni, and Pb. The results showed that Au NCs can act as a template for Ag and Cu only [106]. This phenomenon may be a facile and versatile route for controlling shape and composition of metal alloy NCs.

6.4 POLYOL METHOD

Polyol method is a robust strategy for preparation of well-defined metal NCs in terms of size, shape, composition, and crystallinity. Typical synthesis entails the reduction of metal precursor by polyol in the presence of appropriate capping agent at an elevated temperature. In 1988 Figlarz et al. [107] developed the polyol method to synthesize metal powders, including Co, Ni, and Cu particles. They mixed the metal salts in ethylene glycol (EG) at different ratios followed by addition of sulfuric acid (H_2SO_4) and heating until boiling. The obtained particles were highly uniform and well defined. Furthermore, increasing the ratio of the starting metal precursor to EG while decreasing the reaction temperature led to increasing the particle size. The metal particles were formed through a homogeneous nucleation separated from the growth step. Ever after, the polyol route attracted great attention for controlled synthesis of metal NCs with different sizes, shapes, compositions, and crystallinity [108–112]. This is based on several key parameters, including polyol solvents, capping agents, reducing agents, metal precursor concentrations, and atomic species. The typical polyols are EG, 1, 2-propylene glycol, and 1, 5-pentanediol with viscosity (η) values within the range of $\eta = 6.1$, 40.4, and 140 mPa s, respectively, based on the length of the polyols' hydrocarbon chain [110]. The precise choice of polyols drives both nucleation and growth of metal NCs [113,114]. This is because at elevated temperatures, polyols are oxidized to various aldehyde and ketone species enabling the reduction of metal precursors.

The capping agent is an indispensable factor for stabilization of the formed nuclei at the initial stage of reaction and for directing morphology of the metal NCs. PVP is the most typical capping agent in the polyol method, enabling production of a wide range of metal NCs such as Ag, Pt, Au, Pd, Cu, and Rh in addition to their alloys via adsorption on the metal surface through the carbonyl group of the pyrolidone ring [107–111,114–119]. The PVP-metal legend facilitates the reduction on specific crystal faces, while preventing reduction onto others, enabling anisotropic growth through the formation of metal NCs with high-index facets. This process is not only selective but also prefers binding on the crystal facets at low energy in the {111} and {100} planes rather than {110} facet. The mechanism of binding PVP on specific crystal facets is vague and still under consideration, thus it is rather difficult to resolve the role of PVP in the polyol process. However, empirically the morphology, size, and uniformity of metal NCs are mainly determined by the PVP chain length. PVP with an average molecular weight ($M_w \geq 55{,}000$), was preferred for fabrication of various metal NCs [114–119]. Furthermore, the reaction temperature, reactant concentration, injection rate, and adsorbates have an unambiguous effect on tuning particle size and morphology.

Yang and coworkers prepared Pt NCs with different shapes via using EG polyol and PVP as a structure-directing agent with assistance of $AgNO_3$ [116]. Intriguingly, the addition of $AgNO_3$ into reacting solution alters the shape of Pt NCs from irregular nanocubes into uniform cubes, cuboctahedra, and octahedral Pt NCs (Fig. 6.11A–D) [116]. This is because the silver ions enhance crystal growth rate along {100} facets of Pt nuclei rather than {111}, which determine the surface structure of the Pt NCs. They also synthesized Ag NCs with various morphologies, including cubes, truncated cubes, cuboctahedra, truncated octahedral, and octahedral through controlling the injection rate of $AgNO_3$into a solution of hot pentanediol in the presence of PVP (Fig. 6.11H–M) [114]. Likewise, reduction of $HAuCl_4$ by 1, 5-pentanediol in the presence of PVP and $AgNO_3$ promoted production of Au NCs with the shape of octahedra, truncated octahedral, cuboctahedra, cubes, and higher polygons (Fig. 6.11E–G) [120]. Noticeably, using high and low concentration of $HAuCl_4$ led to production of tetrahedra and icosahedra Au NCs, respectively, due to the sensitivity of the polyol process to the $HAuCl_4$ concentration. Xia and coworkers empirically generalized the polyol method for morphology control in various metal NCs. Particularly, they prepared monodispersed Ag nanocubes enclosed by {100}, {110}, and {111} facets upon reduction of $AgNO_3$ in EG in the presence of PVP [115]. The ratios of PVP/$AgNO_3$ determine the size and geometric morphology of the final product. Moreover, the as-prepared Ag nanocubes were successfully used as sacrificial templates to produce monocrystalline nanobox and hollow polyhedral of Au NCs with six {100} and eight {111} facets. This approach is also used for fabrication of metal, metal oxide, and semiconductor NCs with various shapes, such as nanowires, nanorods, nanobars, and cubes. Xia group achieved another breakthrough in the polyol method via employment of the ionic species during synthesis of metal NCs, enabling production of unique morphologies [112,113,121,122]. Specifically, they added NaCl during reduction of $AgNO_3$ by EG in the presence of PVP with and without oxygen. The results depicted that the number of twinned NCs decreased gradually, while the number of single-crystalline NCs increased in the presence of air and NaCl (Fig. 6.12A–D) [112]. The as-formed twinned seeds were attacked by the etchant and dissolved promptly into the solution due to their instability in the oxidative medium

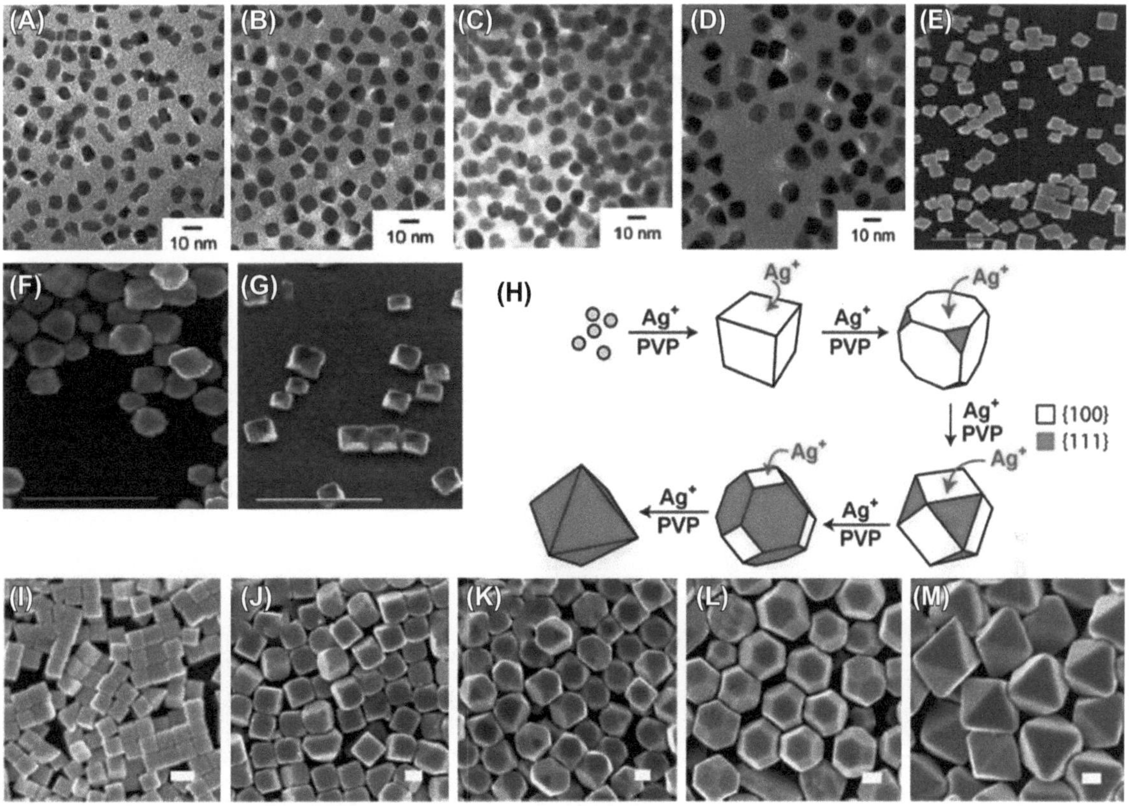

■ **FIGURE 6.11** (A) Transmission electron microscope (TEM) image of the Pt NCs prepared by reduction of H_2PtCl_6 by EG in the presence of PVP without addition of $AgNO_3$. (B, C, and D) TEM image of Pt cubes, cuboctahedron, and octahedral synthesized in the presence of 1.1, 11, and 32 mol % of $AgNO_3$ respectively, under typical conditions. (E, F, G) Scanning electron microscope (SEM) images of Au cubes, truncated octahedral, and octahedral NCs prepared by reduction of $HAuCl_4$ by 1, 5-pentanediol in the presence of PVP and using different concentrations of Ag NO_3. (H) Scheme shows the growth of Ag octahedron with {111} facet NCs in pentanediol by the sequential additions of $AgNO_3$ and PVP. (I—M) SEM images of Ag cubes, truncated cubes, cuboctahedra, truncated octahedral, and octahedra, respectively. The indicated *scale bars* are 100 nm. *(A—D) Adapted with permission from Song H, Kim F, Connor S, Somorjai GA, Yang P. J Phys Chem B 2005;109:188. Copyright 2005 American Chemical Society; (E—G) Adapted with permission from Seo D, Park JC, Song H. J Am Chem Soc 2006;128:14863. Copyright 2006 American Chemical Society; (H—M) Adapted with permission from Tao A, Sinsermsuksakul P, Yang P. Angew Chem Int Ed 2006; 45:4597. Copyright 2006 Wiley-VCH.*

(Fig. 6.12A). Conversely, the single-crystal seeds are not vulnerable to the oxidative etching, which resulted in their continuous growth to form monodispersed twinned decahedron and icosahedron NCs (Fig. 6.12B), while replacement of NaCl with KCl resulted in formation of polydispersed NCs with the same shape (Fig. 6.12C). Interestingly, in the absence of oxygen, the twinned seeds grew rapidly to produce highly uniform Ag nanowires (Fig. 6.12D). Later, Xia prepared Pd nanocubes, and nanorods by reduction of Pd salt by EG or diethylene glycol (DEG) in the presence of PVP, KBr,

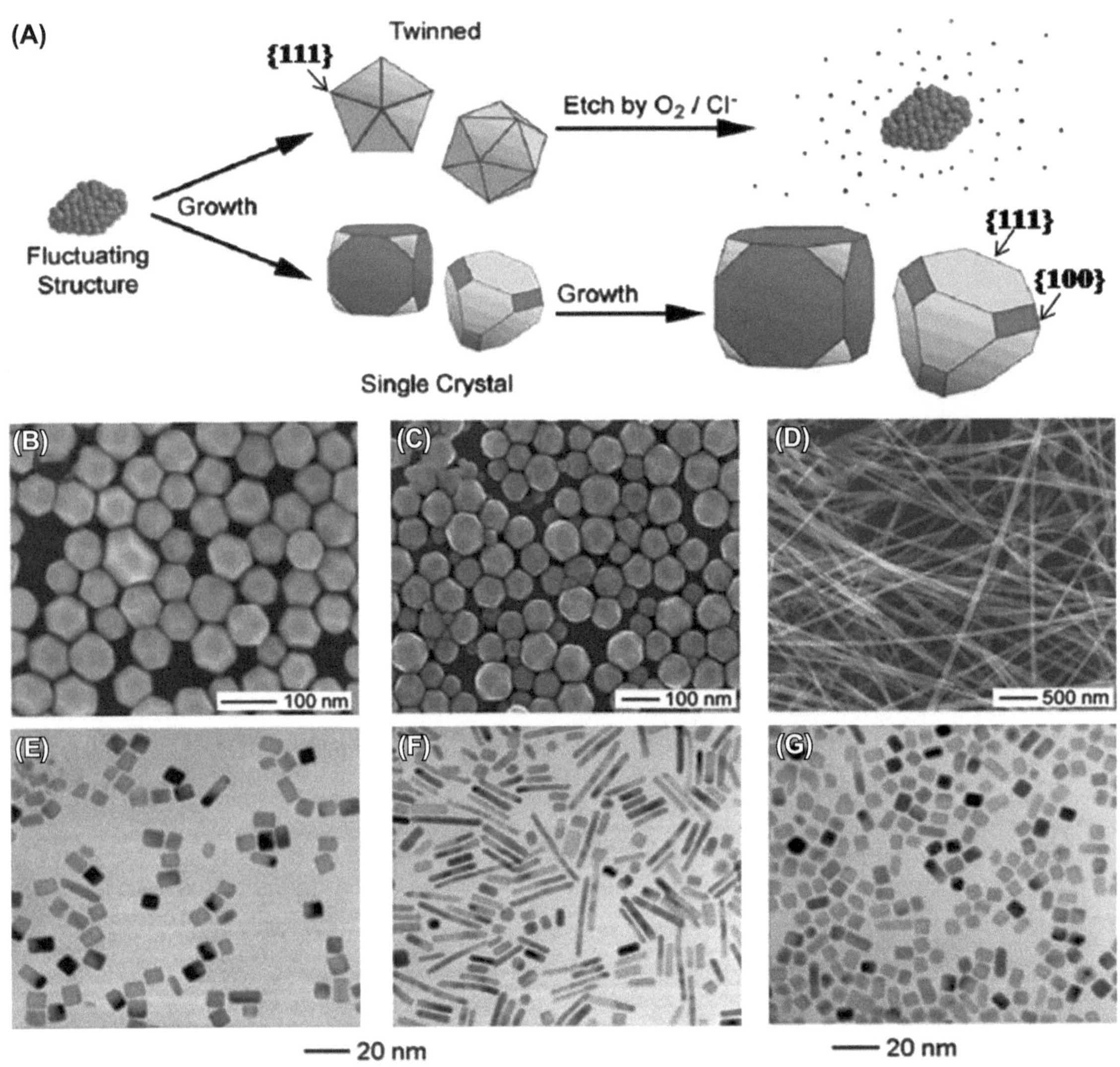

■ **FIGURE 6.12** (A) Scheme represents the dissolution and growth of twinned Ag NCs in the presence and absence of oxygen, respectively. (B) Scanning electron microscope (SEM) image of truncated Ag NCs obtained after 46 h in the presence of NaCl under oxygen, (C) Transmission electron microscope (TEM) images of Ag NCs formed after 54 h, in the presence of KCl under oxygen, (D) SEM image of Ag nanowires formed after 90 min in the presence of NaCl and absence of oxygen. (E) TEM images of Pd NCs formed by using DEG at 100°C, (F) by using EG at 120°C in the presence of a fixed molar ratio of KBr/Na_2PdCl_4, and PVP. (G) TEM images of Pd NCs prepared under the same condition in (E) except using 0.13 M citric acid to retard the oxidative etching effect. *(A—D) Adapted with permission from Wiley B, Herricks T, Sun Y, Xia Y. Nano Lett 2004;4:1733. Copyright 2004 American Chemical Society; (E and F) Adapted with permission from Xiong Y, Cai H, Wiley BJ, Wang J, Kim MJ, Xia Y. J Am Chem Soc 2007;129:3665. Copyright 2007 American Chemical Society.*

and oxygen (Fig. 6.12E and F) [113]. The oxidative etching effect of Br^- ion and oxygen led to the anisotropic growth of Pd NCs. In contrast, using NaCl with citric acid or citrate ions retarded the oxidative etching effect and produced multiple-twinned Pd nanocubes and short nanobars (Fig. 6.12G). Similarly, Yang and coworkers prepared various morphologies of pod-like Ag NCs with eight distinct arms by selective etching of monodispersed octahedral Ag NCs with a mixture of NH_4OH and H_2O_2 [12]. The etching agents promote the etching on the {100} facets that drive the formation of concave features and an octapod shape.

Although there has been massive progress achieved in the polyol method, it is still a great challenge for rational designing and controlling the shape of metal NCs along with their properties. This is ascribed to the absence of quantitative factors that can be used to control the nucleation and growth of metal NCs. This barrier can be solved upon optimization of the reduction rate of the metal precursors as indicated recently by Xia group. They reduced Pd precursor by different polyols, including EG, DEG, and triethylene glycol in the presence of PVP at various reaction temperatures. That led to formation of Pd NCs with unique twin structures, including single crystal, multiplies twinned, and stacking-fault-lined (Fig. 6.13A–F) [123]. This is clearly determined by the type of polyol and the reaction temperature. These findings pave a new way toward the controlled synthesis of metal NCs with well-defined twin structures.

Other organic solvents such as DMF can be also used in the polyol synthesis for the fabrication of various metal NCs. For instance, several shapes of Au and Ag, including nanoprisms, decahedra, tetrahedra, and nanoplates, nanocubes, rhombic dodecahedral, squashed dodecahedra, and octahedral are obtained through DMF in the presence of PVP [124–128]. Notably, in case of DMF the modulation of the surface energy of metal seeds by PVP is different from EG, for example, Ag with {111} facet is more stable in DMF relative to {100} facet [115,129].

6.5 SURFACTANT-MEDIATED METHOD

Surfactant-mediated synthesis is a facile, versatile, and effective route to synthesize metallic NCs with various morphologies and composition. It involves the reduction of metal precursors by using appropriate reductants in the presence of surfactants, which cap the tiny metal crystallites and serve as templates to facilitate their coalescence into different shapes [50,130]. The surfactant type, concentration, pH, its interaction with metal salts, and reduction kinetics are the main parameters that determine the shape of metal NCs. In particular, surfactant type is an indispensable factor for tuning the

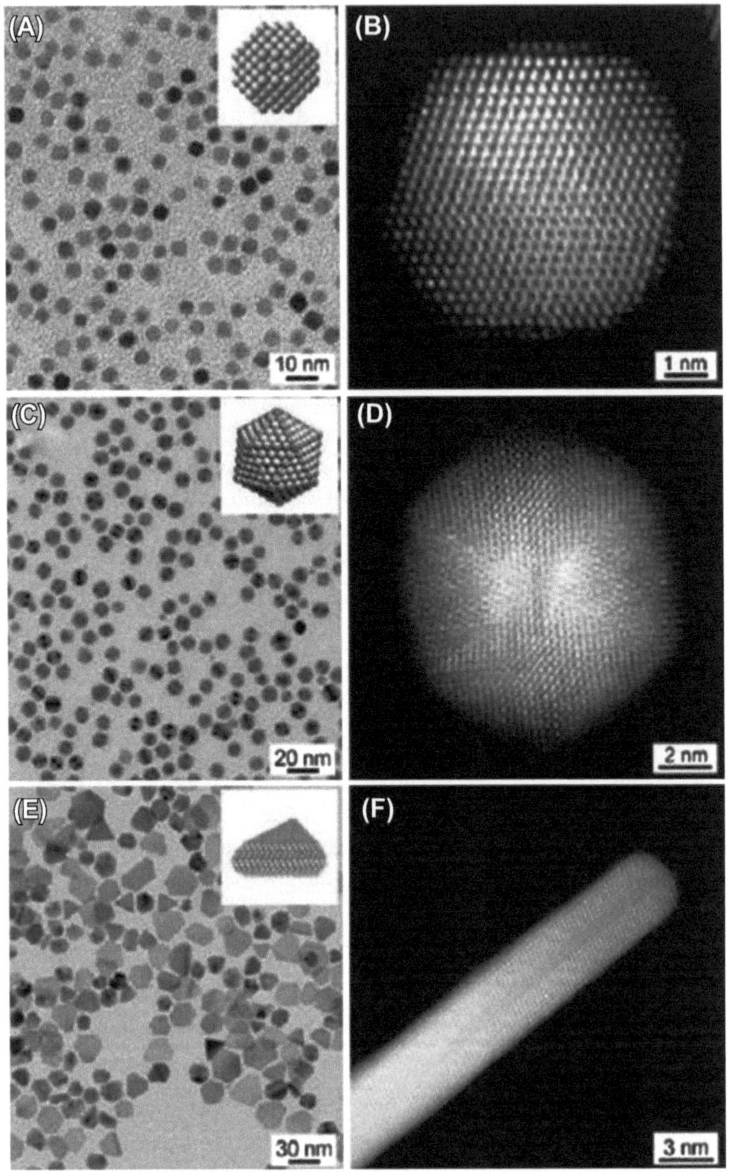

■ **FIGURE 6.13** (A, C, E) Transmission electron microscope (TEM) and (B, D, F) HAADF-STEM images of Pd NCs prepared under different reaction conditions. (A—B) single-crystal truncated octahedral in EG at 140°C, (C—D) multiple-twinned icosahedra in DEG at 140°C, and (E—F) stacking fault-lined nanoplates in DEG at 75°C. The insets in (A—E) represent the corresponding models of the Pd NCs. *Adapted with permission from Wang Y, Peng H-C, Liu J, Huang CZ, Xia Y. Nano Lett 2015;15:1445. Copyright 2015 American Chemical Society.*

morphology of metal NCs due to their inherent self-assembly properties. For example, block copolymers spontaneously self-assemble into a wide range of porous structures because of their hydrophobic effect. Specifically, block copolymers are composed of two or more incompatible linear polymers joined end to end causing a long range of repulsive and a short range of attractive forces. This, subsequently, facilitates their spontaneous self-assembly into a wide range of porous structures such as spheres, cylinders, lamellae, cubic, hexagonal, and layered architectures [131–133]. The concentration of block copolymers should be greater than the critical micelle concentration and hydrophobic force interaction's efficient self-assembly of surfactants with metal precursors. For instance, Shelnutt and coworkers synthesized various dendritic metal NCs using seeding and autocatalytic reduction of metal precursors by AA in aqueous surfactant solution (Fig. 6.14A–D) [134]. The metal NCs' shape and ultimate size are determined by the surfactant type and concentration of seed particles, respectively. Particularly, liposome drove the fabrication of dendritic platinum sheets in the form of thin circular disks or solid foam-like NCs with assistance of tin-porphyrin photocatalyst, while using Brij-35 micelles promoted the formation of spherical dendritic Pt NCs. The photocatalyst facilitates the

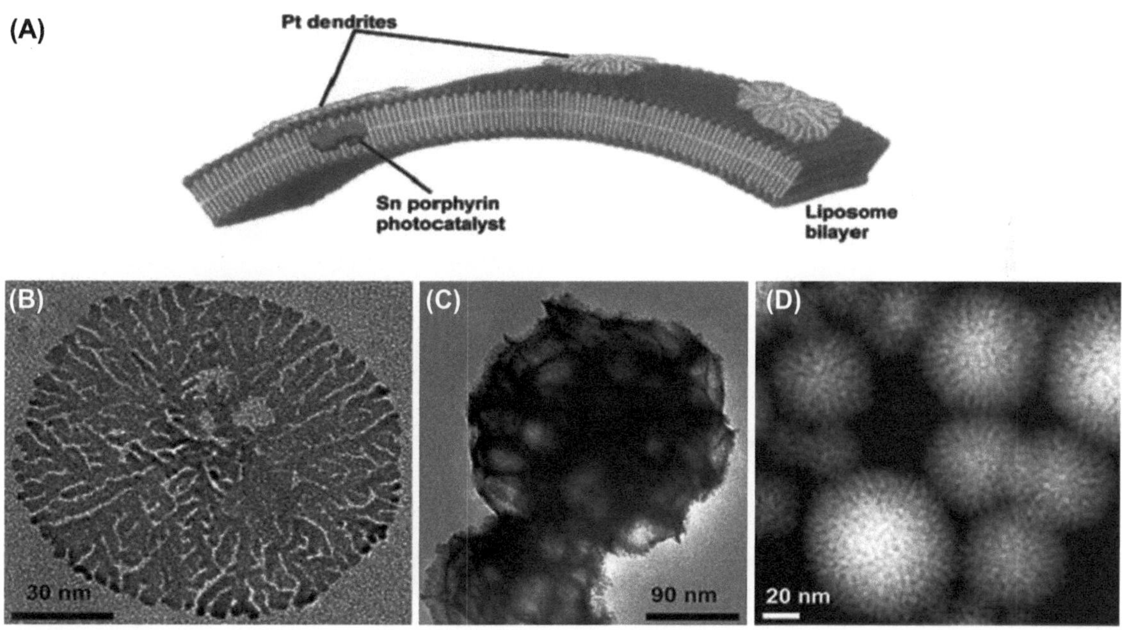

■ **FIGURE 6.14** (A) Scheme represents the growth of Pt nanosheets on the liposomal surface. (B) Transmission electron microscope (TEM) image of dendritic Pt nanosheet and (C) foam-like balls prepared by liposome in the presence of a photocatalyst. (D) TEM image of Pt nanodendrites formed by Brij-35. *Adapted with permission from Song Y, Yang Y, Medforth CJ, Pereira E, Singh AK, Xu H, et al. J Am Chem Soc 2004;126:635. Copyright 2004 American Chemical Society.*

initial production of a huge population of catalytic growth centers to enable assembly surfactant with metal seeds into uniform porous dendritic structures. This method paves the way for the preparation of metal NCs with various porous morphologies. Right after this pioneer work, different surfactants were successfully used for preparation of a wide range of metal NCs with various morphologies and compositions. For instance, Zhang and coworkers fabricated diverse 3D porous network-like metal NCs, including Pt, Pd, Au, Ag, Pt-Pd, Pd-Ru, Pd-Au, Pd-Cd, and Pt-Cu, through surfactant-mediated synthesis based on modulation of the synergism of interparticle electrostatic repulsion and heat-induced fusion during nucleation and growth (Fig. 6.15A–G) [135]. This was achieved through reduction of metal salts by EG and formaldehyde in the presence of dodecyl trimethyl ammonium bromide (DTAB) at 150°C for 8 h. DTAB adsorbed onto the surface of the as-prepared metal nuclei for modulation of the electrostatic repulsion that drives the growth process by Ostwald repining. The charged surfactant, including cationic (e.g., CTAB), anionic (e.g., dioctylsulfosuccinate sodium), or charging molecules (e.g., aminoacetic acid) was indispensable for the metal production and/or their alloy. Furthermore, polyol solvent

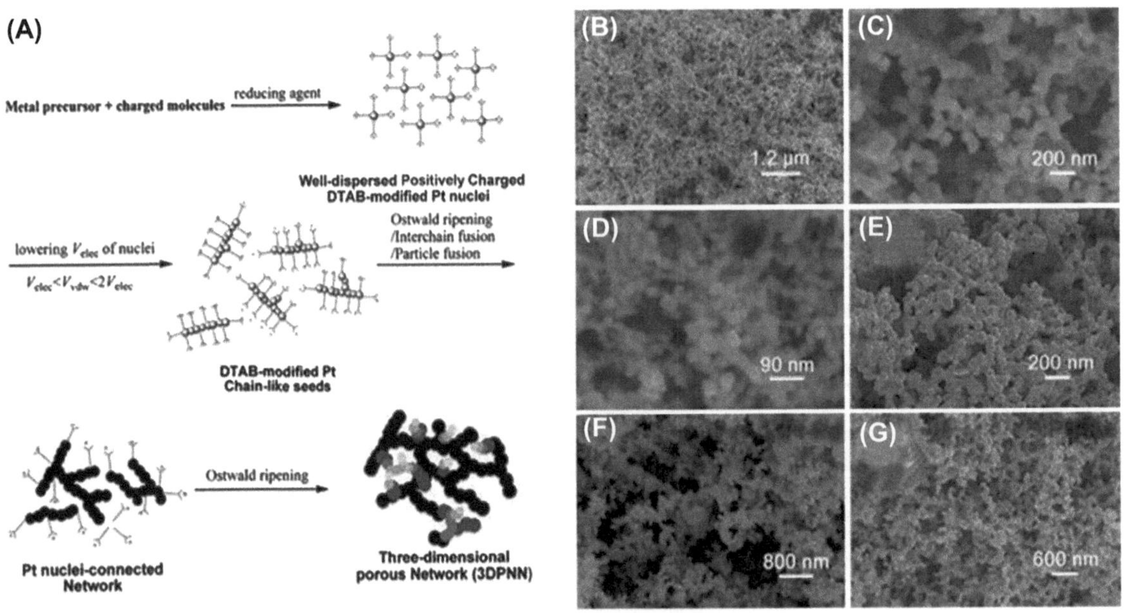

■ **FIGURE 6.15** (A) Scheme shows the synthesis of a 3D porous metal network by correlating heat-induced particle-fusion with the modulation of the electrostatic repulsion of charged nuclei. (B and C) Scanning electron microscope (SEM) images of 3D porous Pt, (D) Pd, (E) Au, (F) $Pt_{0.14}$-$Pd_{0.86}$, and (G) $Pd_{0.90}$-$Ru_{0.10}$ NCs. *Adapted with permission from Cui J, Zhang H, Yu Y, Liu Y, Tian Y, Zhang B. J Mater Chem 2012;22:349. Copyright 2012 The Royal Society of Chemistry.*

is also necessary for anisotropic nucleation and growth of metal NCs via promotion of the dipolar interaction and van der Waals attraction. Moreover, the size and composition of 3D porous network metal NCs can be easily adjusted by varying the metal precursor concentration. This facile method can be extended to fabricate other metals, alloys, and even metal oxide network-like NCs.

Reduction kinetics is an important factor in the surfactant-mediated synthesis method. Particularly fast reduction of metal precursors is required to generate myriad tiny crystallites in the nucleation stage. These crystallites act as building blocks to construct the final product in the growth step. In this regard, adjusting reduction kinetics in the presence of nonionic surfactants with different molecular structures, including Pluronic F127, Brij 700, Tetronic 1107, PVP, and poly(1-vinylpyrrolidone-co-vinyl acetate) (PVP-co-VA), leads to formation of various porous dendritic metal NCs. Noticeably, AA and formic acid that have a mild reduction power were used as reductants in fabricating those morphologies [136,137]. Wang and Yamauchi prepared a wide range of porous dendritic Pt-based NCs with different compositions such as Pt, Pd@Pt, Au@Pd@Pt, and Au@Pt by reduction of metal precursors by AA in presence of block copolymer, including Pluronic F127 or Pluronic F123 (Fig. 6.16A–C) [136,138,139]. The essence in this method is the spontaneous separation between nucleation and growth step based on the difference in the redox potentials of metal species. AA facilitates prompt nucleation of metal with the highest redox potential that serve as a seed for the subsequent depositions of other metals Pd or Pt during the atomic deposition of other metals. Pluronic F127 was adsorbed on the surface of metal crystallites with its hydrophobic poly(propylene oxide) group and formed cavities, which facilitated the assembly of the dendritic Pt outer shell.

It was believed that the exact molecular structure of surfactant molecules is not necessary to obtain porous dendritic metal NCs due to the similarity between the obtained dendritic structures via using different surfactant. Recently, Huang and coworkers synthesized novel porous dendrites Pd NCs that consist of nearly parallel branches to form perpendicular pore channels through using hexadecylpyridinium chloride as structure-directing agents (Fig. 6.16D) [140]. Most recently we successfully prepared porous PtPdRu nanodendrites via simultaneous coreduction of three metal precursors by AA in aqueous Pluronic F127 or PVP solution [141,142]. The mechanism is based on the automatic phase separation resulting from the difference in the redox potential of three metal precursors. We could not obtain dendritic shape in the absence of PVP and or Pluronic F127 indicating their role in the synthesis process. Han and coworkers synthesized

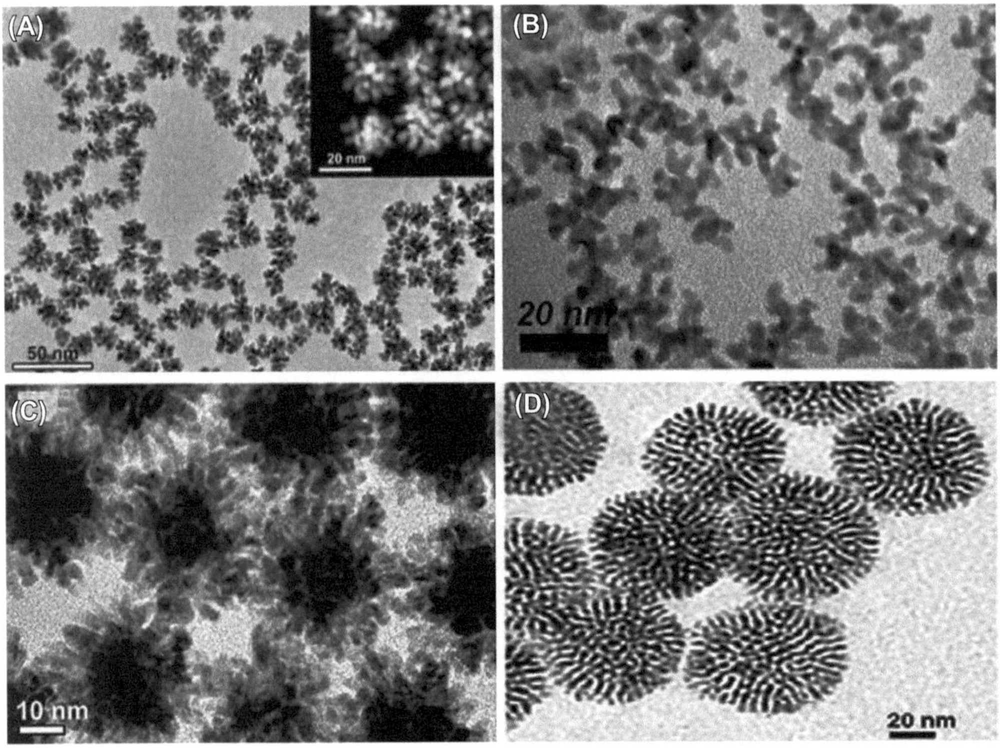

■ **FIGURE 6.16** (A) Pt nanodendrites by Pluronic F127. (B) Pd-Pt nanodendrites by Pluronic F123. (C) Au@Pd@Pt nanodendrites by Pluronic F127. (D) Porous Pd NCs prepared by hexadecylpyridinium chloride. *(A—C) Adapted with permission from Wang L, Yamauchi Y. J Am Chem Soc 2009;131:9152. Copyright 2009 American Chemical Society; Wang L, Nemoto Y, Yamauchi Y. J Am Chem Soc 2011;133:9674. Copyright 2011 American Chemical Society; Wang L, Yamauchi Y. J Am Chem Soc 2010;132:13636. Copyright 2010 American Chemical Society; (D) Adapted with permission from Huang X, Li Y, Chen Y, Zhou E, Xu Y, Zhou H, et al. Angew Chem Int Ed 2013;125:2520. Copyright 2013 Wiley-VCH.*

Au_{oct}@PdPt NCs by reduction of three metal precursors by AA and hydrazine followed by heating to 95°C in presence of CTAC as a structure-directing agent [143]. Heating was fully required to avoid aggregation of CTAC at room temperature. The as-obtained particles have an octahedral shape with dendritic shell. The interaction of CTAC with Pd via amino group rather than Pt is the main key to obtain well-defined NCs facilitated by AA. Interestingly, when CTAC is replaced with sodium dodecyl sulfate, F127, P123, and Brij58, highly aggregated particles are obtained because of the absence of the amino group in these surfactants. The as-prepared Au_{sph}@PdPt has a spherical core when CTAC is used as reductant and structure-directing agent without addition of any reductants.

Other morphologies such as nanowires can be obtained by the surfactant-mediated synthesis through controlling 1D coalescence. This can be conducted via using particular facet capping agent. For example, Yang and coworkers synthesized Pt-Ag alloy nanowires by using oleic acid (OA) and OAm as a structure-directing agent (Fig. 6.17A–E) [144]. OA/OAm adsorbed on the {111}facets with the lowest energy and drove the anisotropic growth into nanowires. The nanowires growth started with the prompt formation of small primary particles and some short rod-like nanostructures within 2 min (Fig. 6.17B–E). Then through continuous reaction, worm-like shape formed and eventually nanowires appeared after 1 h. Han and coworkers synthesized octapod Au-Pd NCs via reduction of $NaAuBr_4$ and K_2PtCl_4 by AA in presence of CTAC (Fig. 6.17F–H) [145]. This unique structure is produced through selective etching of {100} facets by in situ Br^- ions generated from the reduction of gold precursor. The results revealed that octahedral Au-Pd NCs formed, within 10 min. By this time, octahedral Au-Pd NCs formed through etching in the {100} facets, which alter to truncated NCs with concave features, and finally octapodal within 4 h. This is due to the progressive generation of Br ions during reduction process facilitating the formation of octapod structure. Noticeably, addition of NaBr into the reaction solution yielded Au-Pd NCs with different shapes due to the strong binding affinity on the {100} facets. Hoefelmeyer and coworkers prepared Ru NCs with various morphologies through decomposition of Ru salt in the presence of hexadecylamine and tri(n-octyl) amine [146]. The results show that decompositions at a low temperature produce highly branched Ru NCs, while spherical Ru NCs are formed at an elevated temperature. The formation of branched Ru is attributed to the aggregation-directed growth mechanism. Liu and coworkers synthesized PtRh alloy 3D porous nanostructures by controlled aggregation of nanoparticles in OAm [147]. Specifically, they mixed metal precursors in OAm followed by sequential heating cycles. Using OAm as reductants and as a structure-directing agent led to the initial formation of PtRh NCs coated with OAm. That is naturally aggregated into close-packed clusters at high temperatures to decrease their total free energy. This is due to the rapid decomposition of the metal precursors at the elevated temperatures in a strong redox reaction. The particle composition can be easily controlled by adjusting the initial ratio of Pt/Rh salts.

6.6 ELECTROCATALYTIC APPLICATIONS

Nanoarchitectonic metals with their unique physical and chemical properties such as highly accessible surface area, active surface facets, and high electrical conductivity enable their successful utilization in various electrocatalytic applications [1,2,8,9]. It is well recognized that the size, morphology, and composition are the key factors that determine the catalytic

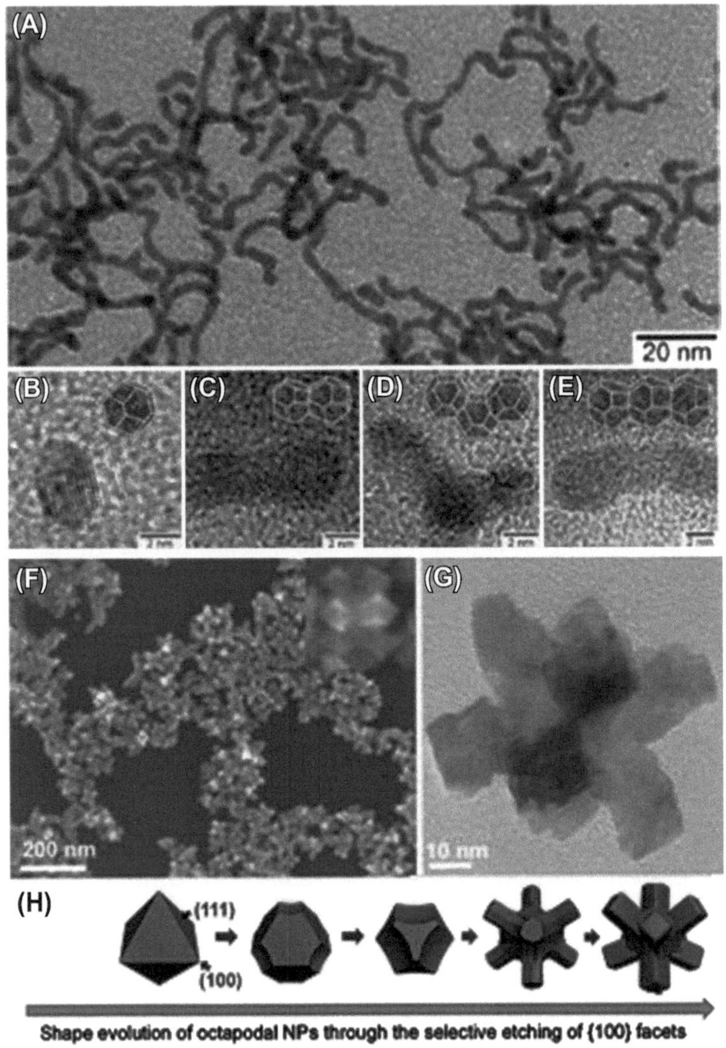

■ **FIGURE 6.17** (A) Transmission electron microscope (TEM) image of worm-like nanowires $Pt_{53}Ag_{47}$ NCs prepared by reduction of metal precursors in OA, OAm, 1,2-hexadecanediol, and diphenyl ether at high temperature. (B—E) TEM images and models show growth of nanowires. (F) Scanning electron microscope (SEM), (G) TEM image Au-Pd octapodal NCs prepared by coreduction of metal precursors by AA in the presence of CTAC, and (H) Scheme shows the shape evolution of metal NCs by the selective etching process. *(A—E) Adapted with permission from Peng Z, You H, Yang H. ACS Nano 2010;4:1501. Copyright 2010 American Chemical Society; (F—H) Adapted with permission from Hong JW, Lee YW, Kim M, Kang SW, Han SW. Chem Commun 2011;47:2553. Copyright 2011 The Royal Society of Chemistry.*

properties and potential application of metal NCs [5,10,11]. In particular, the catalysts at the nanoscale have a high Fermi level resulting from changing their electronic structure. This subsequently decreases the reduction potential of the metal on the surface of the nanocatalyst and increases the number of chemically unsaturated and thermodynamically high-energy active surface atoms [148,149], which has a significant effect on improvement of the catalytic performance of metal NCs. For example, spherical metal NCs with small particle size show a substantial catalytic performance toward ORR and alcoholic methanol oxidation reactions (e.g., methanol, ethanol, formic). This is due to the high accessible surface area of metal NCs facilitating catalytic reaction kinetic and preventing the Ostwald repining pathway. Therefore, the commercial electrocatalysts such as Pt and Pd are always provided with size in the range of 3–5 nm. Shoemaker and coworkers reported that the mass activity and specific activity of Pt increased by twofold and fourfold from 1.3 to 2.2 nm (Fig. 6.18A) [150]. This is

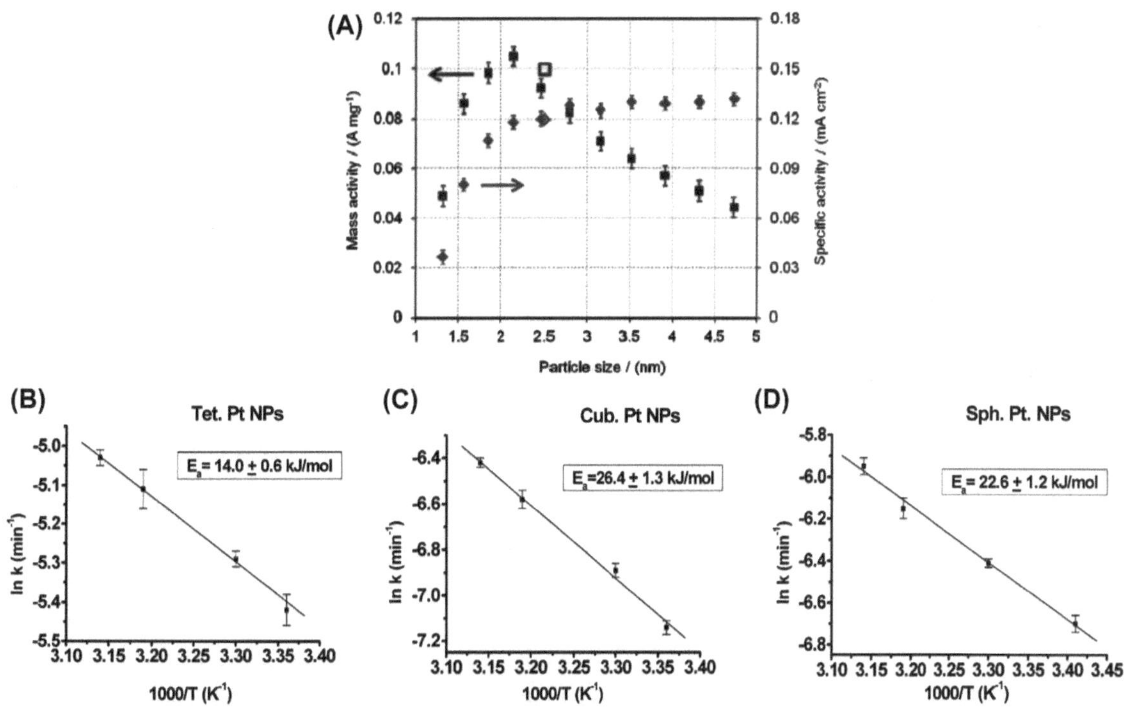

■ **FIGURE 6.18** (A) Size dependence of specific activity and mass activity of Pt/C for ORR at 0.93 V versus RHE. (B) The activation energies and catalytic activities of tetrahedral Pt NCs, (C) cubic Pt NCs, and (D) spherical Pt NCs. *(A) Adapted with permission from Shao M, Peles A, Shoemaker K. Nano Lett 2011;11: 3714. Copyright 2011 American Chemical Society; (B–D) Adapted with permission Narayanan R, El-Sayed MA. Nano Lett 2004;4:1343. Copyright 2004 American Chemical Society.*

correlated to the binding energy of oxygen molecules on the different Pt sites that decreased with increasing particle size.

Controlling the shape of metal NCs shape played a pioneer role in improving the catalytic activity of metal NCs. Particularly, metal NCs with specific surface features including high-index facets, corners, sharp edges, and porosity have a substantial effect on enhancement of the catalytic activity. For instance, El-sayed and coworkers investigated the relation between the shape of Pt NCs and the catalytic activity for the electron transfer reaction between hexacyanoferrate (III) and thiosulfate ions to form hexacyanoferrate (II) and tetrathionate ions [151]. The results revealed that the catalytic activity is highly correlated to the atoms present at the corners and edges of Pt NCs. Particularly, the catalytic activity of tetrahedral Pt NCs is superior to both spherical and cubic Pt NCs (Fig. 6.18B–D) [151]. More recently we reported that PtPdRu nanocages exhibited superior catalytic activity and durability toward ORR than PtPdRu nanodendrites [142]. Specifically, the mass and specific activity of PtPdRu nanocages was 2.08 and 1.81 times of PtPdRu nanodendrites. This is attributed to the existence of nanocages with porous dendritic surface providing huge active catalytic sites for reactant species and allowing reactions to occur over the highly stable cavities in the core. Colliere and coworkers showed that sponge-like Ru NCs revealed higher catalytic activity and stability than spherical Ru NCs in the hydrogenation of benzene and quinolone [152].

Metal NC facets also tune the catalytic activity and stability because of their ability to tune the adsorption rate of reactant molecules. For example, star-like Pt NCs with high-index facets decrease the activation energy for the reduction of ferricyanide by 1.6 times relative to Pt tetrahedrons bound with low-index facets (Fig. 6.19) [153]. Tilley and coworkers reported that branched Pd with high-index facets exhibited higher catalytic activity in the hydrogenation of nitrobenzene to aniline relative to commercial Pd catalyst [154]. Furthermore, the catalytic selectivity can be easily tuned by the surface facets of metal NCs [155]. Regarding this, Pt NCs with {111} facets exhibit higher catalytic activity than Pt NCs with {100} toward dehydrocyclization of *n*-heptane, meanwhile, the opposite is revealed for the isomerization of isobutene [156].

Metal NC composition also has a considerable effect on improvement of the catalytic activity. Multimetallic NCs exhibit superior catalytic activity and stability compared with their counterpart mono- and bimetallic NCs. This is ascribed to the alloy, electronic, and strain effect. For example, Au-Pt dendritic NCs revealed higher catalytic oxidation activity for CO oxidation in H_2 gas than Pt catalysts [157]. Most recently, we prepared porous PtCu

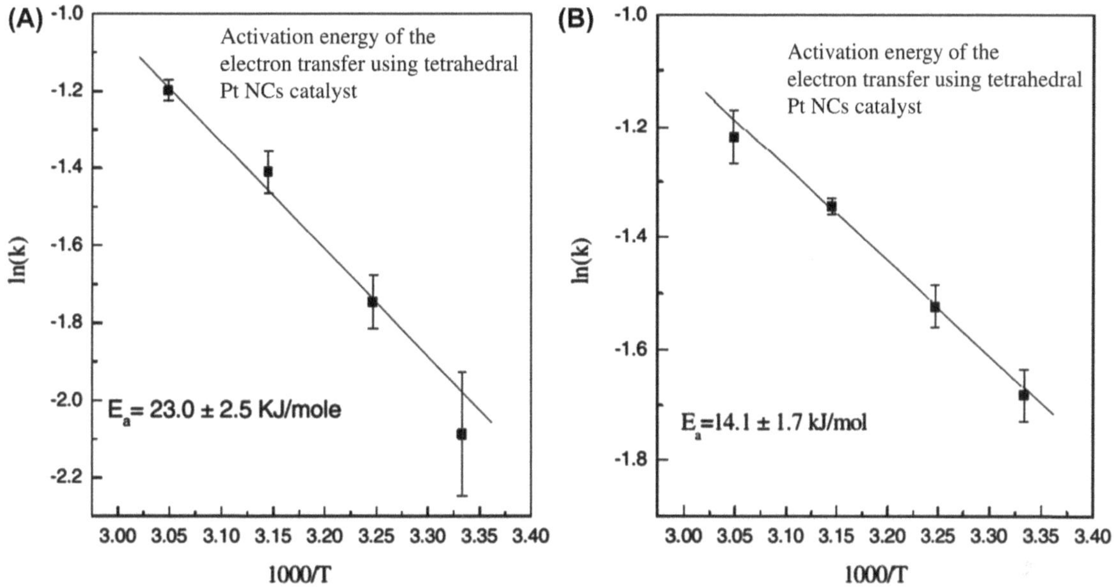

■ **FIGURE 6.19** (A) The activation energy for the reduction of ferricyanide by thiosulfate in the presence of tetrahedral Pt NCs and (B) nanostar Pt NCs. *Adapted with permission from Mahmoud MA, Tabor CE, El-Sayed MA, Ding Y, Wang ZL. J Am Chem Soc 2008;130:4590. Copyright 2008 American Chemical Society.*

NCs with porous dendritic shell, which exhibit superior electrocatalytic activity toward MOR relative to Pt/C due to the electronic and alloying effect [158]. Sun and coworkers reported that Au/CuPt core−shell NCs show significant enhancement in the catalytic activity toward both ORR and MOR relative to commercial Pt [159]. This is due to the alloy effect and core/shell interactions. More recently we revealed that, PtPdRu nanodendrites depicted superior mass and specific activity toward ORR and MOR comparative to binary PtPd flowers and Pt/C (Fig. 6.20) [141]. This is attributed to the synergism between Pt and Pd, which increases the adsorption rate of reactant species; meanwhile strain effect between Pt and Ru retard adsorption of poisoning intermediate species. This subsequently counterbalances among strong adsorption for reactants and decreases the binding energies of the intermediates, which leads to a substantial increment in the catalytic activity.

6.7 CONCLUSIONS

In this chapter we discussed the controllable synthesis of nanoarchitectonic metals through wet-chemical methods, including four general methods: seed-mediated growth, galvanic replacement approach, polyol synthesis, and surfactant-mediated route. These methods allow the control of the

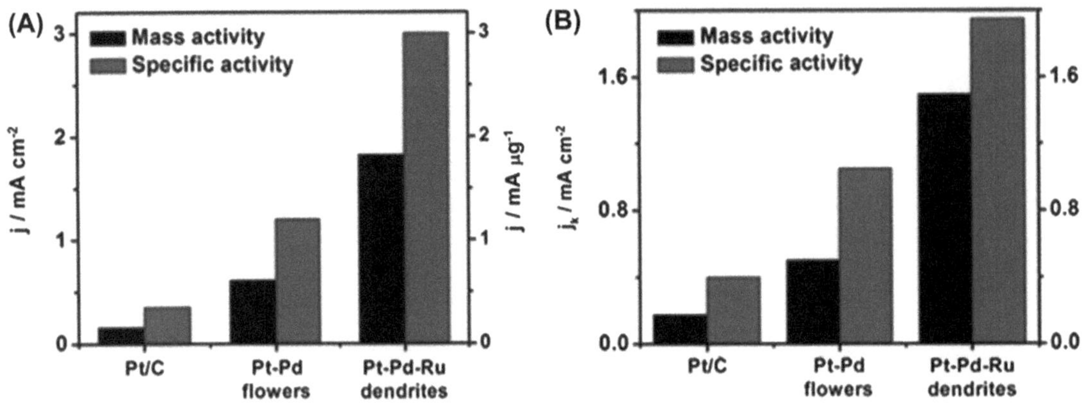

■ **FIGURE 6.20** (A and B) Comparisons of the mass activities and specific activities of the three materials at 0.6 V versus Ag/AgCl toward methanol oxidation reaction and oxygen reduction reaction, respectively. *Adapted with permission from Eid K, Malgras V, He P, Wang K, Aldalbahi A, Alshehri SM, et al. RSC Adv 2015;5:31147. Copyright 2015 The Royal Society of Chemistry.*

nucleation and growth process that are very important for synthesis of nanoarchitectonic metals. The rational use of the preformed seed to direct the particle growth is the key factor for the seed-mediated growth. The difference in the metallic redox potential is a driving force in the galvanic replacement approach. The reasonably used surfactant is very important for polyol synthesis and surfactant-mediated route. Controlled synthesis of metallic nanoarchitectonics is important for tailoring their properties and potential functions. This is ascribed to the strong relationship between properties of metal NCs and their sizes, shapes, and compositions. It still remains a great challenge to develop facile wet-chemical methods for precise control of the shapes, sizes, and compositions of nanoarchitectonic metals with desired functions.

ACKNOWLEDGMENT

This work was supported by the National Natural Science Foundation of China (No. 21273218). K. Eid greatly appreciates the UCAS-TWAS President's Fellowship.

REFERENCES

[1] Liu H-l, Nosheen F, Wang X. Chem Soc Rev 2015;44:3056.
[2] Zhou Z-Y, Tian N, Li J-T, Broadwell I, Sun S-G. Chem Soc Rev 2011;40:4167.
[3] Ariga K, Ji Q, Nakanishi W, Hill J-P, Aono M. Mater Horiz 2015;2:406.
[4] Xia Y, Xia X, Wang Y, Xie S. MRS Bull 2013;38:335.
[5] Zhu C, Du D, Eychmüller A, Lin Y. Chem Rev 2015;115:8896.
[6] Guo S, Wang E. Acc Chem Res 2011;44:491.

[7] Quan Z, Wang Y, Fang J. Acc Chem Res 2012;46:191.
[8] Ariga K, Vinu A, Yamauchi Y, Ji Q, Hill J-P. Bull Chem Soc Jpn 2012;85:1.
[9] Watt J, Cheong S, Tilley RD. Nano Today 2013;8:198.
[10] Xu Y, Zhang B. Chem Soc Rev 2014;43:2439.
[11] Wu J, Yang H. Acc Chem Res 2013;46:1848.
[12] Mulvihill MJ, Ling XY, Henzie J, Yang P. J Am Chem Soc 2009;132:268.
[13] Skrabalak SE, Chen J, Sun Y, Lu X, Au L, Cobley CM, et al. Acc Chem Res 2008; 41:1587.
[14] Cobley CM, Chen J, Cho EC, Wang LV, Xia Y. Chem Soc Rev 2011;40:44.
[15] Niu W, Xu G. Nano Today 2011;6:265.
[16] Gu J, Zhang Y-W, Tao FF. Chem Soc Rev 2012;41:8050.
[17] Lim B, Jiang M, Camargo PH, Cho EC, Tao J, Lu X, et al. Science 2009;324: 1302.
[18] Murphy CJ, Sau TK, Gole AM, Orendorff CJ, Gao J, Gou L, et al. J Phys Chem B 2005;109:13857.
[19] Peng H-C, Park J, Zhang L, Xia Y. J Am Chem Soc 2015;137:6643.
[20] Brown KR, Natan MJ. Langmuir 1998;14:726.
[21] Brown KR, Lyon LA, Fox AP, Reiss BD, Natan MJ. Chem Mater 2000;12:314.
[22] Jana NR, Gearheart L, Murphy CJ. Chem Commun 2001:617.
[23] Jana NR, Gearheart L, Murphy CJ. J Phys Chem B 2001;105:4065.
[24] Johnson CJ, Dujardin E, Davis SA, Murphy CJ, Mann S. J Mater Chem 2002;12: 1765.
[25] Nikoobakht B, El-Sayed MA. Chem Mater 2003;15:1957.
[26] Kou X, Zhang S, Tsung CK, Yang Z, Yeung MH, Stucky GD, et al. Chem Eur J 2007;13:2929.
[27] Sau TK, Murphy CJ. J Am Chem Soc 2004;126:8648.
[28] Niu W, Zheng S, Wang D, Liu X, Li H, Han S, et al. J Am Chem Soc 2008;131: 697.
[29] Elechiguerra JL, Reyes-Gasga J, Yacaman MJ. J Mater Chem 2006;16:3906.
[30] Niu W, Zhang L, Xu G. ACS Nano 2010;4:1987.
[31] Niu W, Li Z-Y, Shi L, Liu X, Li H, Han S, et al. Cryst Growth Des 2008;8:4440.
[32] Zhang L, Niu W, Xu G. Nanoscale 2011;3:678.
[33] Carbó-Argibay E, Rodríguez-González B, Pacifico J, Pastoriza-Santos I, Pérez-Juste J, Liz-Marzán LM. Angew Chem Int Ed 2007;46:8983.
[34] Carbó-Argibay E, Rodríguez-González B, Pastoriza-Santos I, Pérez-Juste J, Liz-Marzán LM. Nanoscale 2010;2:2377.
[35] Xiang Y, Wu X, Liu D, Jiang X, Chu W, Li Z, et al. Nano Lett 2006;6:2290.
[36] Xiang Y, Wu X, Liu D, Feng L, Zhang K, Chu W, et al. J Phys Chem C 2008;112: 3203.
[37] Xiang Y, Wu X, Liu D, Li Z, Chu W, Feng L, et al. Langmuir 2008;24:3465.
[38] Ming T, Feng W, Tang Q, Wang F, Sun L, Wang J, et al. J Am Chem Soc 2009; 131:16350.
[39] Li J, Wang L, Liu L, Guo L, Han X, Zhang Z. Chem Commun 2010;46:5109.
[40] Zhang J, Langille MR, Personick ML, Zhang K, Li S, Mirkin CA. J Am Chem Soc 2010;132:14012.
[41] Yu Y, Zhang Q, Lu X, Lee JY. J Phys Chem C 2010;114:11119.
[42] Wu H-L, Kuo C-H, Huang MH. Langmuir 2010;26:12307.

[43] Lu C-L, Prasad KS, Wu H-L, Ho J-A, Huang MH. J Am Chem Soc 2010;132: 14546.
[44] Yang C-W, Chanda K, Lin P-H, Wang Y-N, Liao C-W, Huang MH. J Am Chem Soc 2011;133:19993.
[45] Kobayashi H, Lim B, Wang J, Camargo PH, Yu T, Kim MJ, et al. Chem Phys Lett 2010;494:249.
[46] Zhang H, Li W, Jin M, Zeng J, Yu T, Yang D, et al. Nano Lett 2010;11:898.
[47] Liu H, Qu J, Chen Y, Li J, Ye F, Lee JY, et al. J Am Chem Soc 2012;134:11602.
[48] Sun Y, Mayers BT, Xia Y. Nano Lett 2002;2:481.
[49] Xia X, Wang Y, Ruditskiy A, Xia Y. Adv Mater 2013;25:6313.
[50] Liu Y, Goebl J, Yin Y. Chem Soc Rev 2013;42:2610.
[51] Zhang H, Jin M, Liu H, Wang J, Kim MJ, Yang D, et al. ACS Nano 2011;5:8212.
[52] Alia SM, Yan Y, Pivovar B. Catal Sci Technol 2014;4:3589.
[53] Gilroy KD, Farzinpour P, Sundar A, Hughes RA, Neretina S. Chem Mater 2014;26: 3340.
[54] Song HM, Anjum DH, Sougrat R, Hedhili MN, Khashab NM. J Mater Chem 2012; 22:25003.
[55] Wang C, Wang Y, Xu L, Shi X, Li X, Xu X, et al. Small 2013;9:413.
[56] Lu X, Au L, McLellan J, Li Z-Y, Marquez M, Xia Y. Nano Lett 2007;7:1764.
[57] Chen Z, Waje M, Li W, Yan Y. Angew Chem Int Ed 2007;46:4060.
[58] Wan D, Xia X, Wang Y, Xia Y. Small 2013;9:3111.
[59] Zhang G, Sun S, Cai M, Zhang Y, Li R, Sun X. Sci Rep 2013;3:1526.
[60] Sun Y, Xia Y. J Am Chem Soc 2004;126:3892.
[61] Sun Y, Xia Y. Nano Lett 2003;3:1569.
[62] Lu Y, Chen W. J Phys Chem C 2010;114:21190.
[63] Chen J, Wiley B, McLellan J, Xiong Y, Li Z-Y, Xia Y. Nano Lett 2005;5:2058.
[64] Au L, Lu X, Xia Y. Adv Mater 2008;20:2517.
[65] Liang HW, Liu S, Gong JY, Wang SB, Wang L, Yu SH. Adv Mater 2009;21:1850.
[66] Straney PJ, Marbella LE, Andolina CM, Nuhfer NT, Millstone JE. J Am Chem Soc 2014;136:7873.
[67] Bi Y, Lu G. Chem Commun 2008:6402.
[68] Chen J, McLellan JM, Siekkinen A, Xiong Y, Li Z-Y, Xia Y. J Am Chem Soc 2006;128:14776.
[69] Cobley CM, Xia Y. Mater Sci Eng R 2010;70:44.
[70] Fang Z, Wang Y, Liu C, Chen S, Sang W, Wang C, et al. Small 2015;11:2593.
[71] Xia BY, Wu HB, Wang X, Lou XW. J Am Chem Soc 2012;134:13934.
[72] Dhavale VM, Kurungot S. ACS Catal 2015;5:1445.
[73] Nosheen F, Zhang Z-c, Zhuang J, Wang X. Nanoscale 2013;5:3660.
[74] Zhang Z, Yang Y, Nosheen F, Wang P, Zhang J, Zhuang J, et al. Small 2013;9: 3063.
[75] Ding J, Zhu X, Bu L, Yao J, Guo J, Guo S, et al. Chem Commun 2015;51:9722.
[76] Jing H, Wang H. Chem Mater 2015;27:2172.
[77] Yang Y, Liu J, Fu Z-W, Qin D. J Am Chem Soc 2014;136:8153.
[78] Hong JW, Kang SW, Choi B-S, Kim D, Lee SB, Han SW. ACS Nano 2012;6:2410.
[79] Li C, Jiang B, Imura M, Malgras V, Yamauchi Y. Chem Commun 2014;50:15337.
[80] Sneed BT, Young AP, Jalalpoor D, Golden MC, Mao S, Jiang Y, et al. ACS Nano 2014;8:7239.

[81] Zhang Q, Cobley CM, Zeng J, Wen L-P, Chen J, Xia Y. J Phys Chem C 2010;114: 6396.
[82] Jang H, Min D-H. ACS Nano 2015;9:2696.
[83] McEachran M, Keogh D, Pietrobon B, Cathcart N, Gourevich I, Coombs N, et al. J Am Chem Soc 2011;133:8066.
[84] Mahmoud M, El-Sayed M. Langmuir 2012;28:4051.
[85] Sun Y, Wiley B, Li Z-Y, Xia Y. J Am Chem Soc 2004;126:9399.
[86] Ridelman Y, Singh G, Popovitz-Biro R, Wolf SG, Das S, Klajn R. Small 2012;8:654.
[87] Yang J, Shen D, Zhou L, Li W, Li X, Yao C, et al. Chem Mater 2013;25:3030.
[88] Hu Y, Liu Y, Li Z, Sun Y. Adv Funct Mater 2014;24:2828.
[89] Wang M, Zhang W, Wang J, Wexler D, Poynton SD, Slade RC, et al. ACS Appl Mater Interfaces 2013;5:12708.
[90] Shan A, Chen Z, Li B, Chen C, Wang R. J Mater Chem A 2015;3:1031.
[91] Ni W, Wu HB, Wang B, Xu R, Lou XWD. Small 2012;8:3432.
[92] Adams S, Thai D, Mascona X, Schwartzberg AM, Zhang JZ. Chem Mater 2014;26: 6805.
[93] Vasquez Y, Sra AK, Schaak RE. J Am Chem Soc 2005;127:12504.
[94] Chen Z, Ye S, Wilson AR, Ha Y-C, Wiley BJ. Energy Environ Sci 2014;7:1461.
[95] Mohl M, Dobo D, Kukovecz A, Konya Z, Kordas K, Wei J, et al. J Phys Chem C 2011;115:9403.
[96] Li X, Liu H, Yang J, Qiao S-Z, Du X-W. RSC Adv 2014;4:1185.
[97] Niu KY, Kulinich SA, Yang J, Zhu AL, Du XW. Chem Eur J 2012;18:4234.
[98] Guo S, Dong S, Wang E. Chem Commun 2010;46:1869.
[99] Hong W, Wang J, Wang E. Small 2014;10:3262.
[100] Zhu C, Guo S, Dong S. Adv Mater 2012;24:2326.
[101] Kim KW, Kim SM, Choi S, Kim J, Lee IS. ACS Nano 2012;6:5122.
[102] Ma A, Xu J, Zhang X, Zhang B, Wang D, Xu H. Sci Rep 2014;4:4849.
[103] Oh MH, Yu T, Yu S-H, Lim B, Ko K-T, Willinger M-G, et al. Science 2013;340:964.
[104] Chen C, Kang Y, Huo Z, Zhu Z, Huang W, Xin HL, et al. Science 2014;343: 1339.
[105] Choi J-P, Fields-Zinna CA, Stiles RL, Balasubramanian R, Douglas AD, Crowe MC, et al. J Phys Chem C 2010;114:15890.
[106] Wu Z. Angew Chem Int Ed 2012;51:2934.
[107] Fievet F, Lagier J, Blin B, Beaudoin B, Figlarz M. Solid State Ion 1989;32:198.
[108] Wiley B, Sun Y, Mayers B, Xia Y. Chem Eur J 2005;11:454.
[109] Wiley B, Sun Y, Xia Y. Acc Chem Res 2007;40:1067.
[110] Tao AR, Habas S, Yang P. Small 2008;4:310.
[111] Xia Y, Xiong Y, Lim B, Skrabalak SE. Angew Chem Int Ed 2009;48:60.
[112] Wiley B, Herricks T, Sun Y, Xia Y. Nano Lett 2004;4:1733.
[113] Xiong Y, Cai H, Wiley BJ, Wang J, Kim MJ, Xia Y. J Am Chem Soc 2007;129: 3665.
[114] Tao A, Sinsermsuksakul P, Yang P. Angew Chem Int Ed 2006;45:4597.
[115] Sun Y, Xia Y. Science 2002;298:2176.
[116] Song H, Kim F, Connor S, Somorjai GA, Yang P. J Phys Chem B 2005;109:188.
[117] Jun Y, Choi w, Cheon Js. J Angew Chem Int Ed 2006;45:3414.
[118] Kim F, Connor S, Song H, Kuykendall T, Yang P. Angew Chem Int Ed 2004;43: 3673.

[119] Montejano-Carrizales JM, Rodríguez-López JL, Pal U, Miki-Yoshida M, José-Yacamán M. Small 2006;2:351.
[120] Seo D, Park JC, Song H. J Am Chem Soc 2006;128:14863.
[121] Xiong Y, Chen J, Wiley B, Xia Y, Yin Y, Li Z-Y. Nano Lett 2005;5:1237.
[122] Xiong Y, McLellan JM, Yin Y, Xia Y. Angew Chem Int Ed 2007;46:790.
[123] Wang Y, Peng H-C, Liu J, Huang CZ, Xia Y. Nano Lett 2015;15:1445.
[124] Pastoriza-Santos I, Liz-Marzán LM. Adv Funct Mater 2009;19:679.
[125] Gao Y, Jiang P, Song L, Wang J, Liu L, Liu D, et al. J Cryst Growth 2006;289:376.
[126] Wang D, Liu Y, You T. CrystEngComm 2010;12:4028.
[127] Pastoriza-Santos I, Liz-Marzán LM. Nano Lett 2002;2:903.
[128] Jeong GH, Kim M, Lee YW, Choi W, Oh WT, Park Q-H, et al. J Am Chem Soc 2009;131:1672.
[129] Tsuji M, Maeda Y, Hikino S, Kumagae H, Matsunaga M, Tang X-L, et al. Cryst Growth Des 2009;9:4700.
[130] Deng Y, Wei J, Sun Z, Zhao D. Chem Soc Rev 2013;42:4054.
[131] Bucknall DG, Anderson HL. Science 2003;302:1904.
[132] Förster S, Konrad M. J Mater Chem 2003;13:2671.
[133] Hillmyer MA, Bates FS, Almdal K, Mortensen K. Science 1996;271:976.
[134] Song Y, Yang Y, Medforth CJ, Pereira E, Singh AK, Xu H, et al. J Am Chem Soc 2004;126:635.
[135] Cui J, Zhang H, Yu Y, Liu Y, Tian Y, Zhang B. J Mater Chem 2012;22:349.
[136] Wang L, Yamauchi Y. J Am Chem Soc 2009;131:9152.
[137] Wang L, Wang H, Nemoto Y, Yamauchi Y. Chem Mater 2010;22:2835.
[138] Wang L, Nemoto Y, Yamauchi Y. J Am Chem Soc 2011;133:9674.
[139] Wang L, Yamauchi Y. J Am Chem Soc 2010;132:13636.
[140] Huang X, Li Y, Chen Y, Zhou E, Xu Y, Zhou H, et al. Angew Chem Int Ed 2013; 125:2520.
[141] Eid K, Malgras V, He P, Wang K, Aldalbahi A, Alshehri SM, et al. RSC Adv 2015; 5:31147.
[142] Eid K, Wang H, Malgras V, Alothman ZA, Yamauchi Y, Wang L. J Phys Chem C 2015;119:19947.
[143] Kang SW, Lee YW, Park Y, Choi B-S, Hong JW, Park K-H, et al. ACS Nano 2013; 7:7945.
[144] Peng Z, You H, Yang H. ACS Nano 2010;4:1501.
[145] Hong JW, Lee YW, Kim M, Kang SW, Han SW. Chem Commun 2011;47:2553.
[146] Vanden Brink M, Peck MA, More KL, Hoefelmeyer JD. J Phys Chem C 2008;112: 12122.
[147] Zhang Y, Janyasupab M, Liu CW, Li X, Xu J, Liu CC. Adv Funct Mater 2012;22: 3570.
[148] Wagner FT, Lakshmanan B, Mathias MF. J Phys Chem Lett 2010;1:2204.
[149] Porter NS, Wu H, Quan Z, Fang J. Acc Chem Res 2013;46:1867.
[150] Shao M, Peles A, Shoemaker K. Nano Lett 2011;11:3714.
[151] Narayanan R, El-Sayed MA. Nano Lett 2004;4:1343.
[152] Pelzer K, Vidoni O, Philippot K, Chaudret B, Colliere V. Adv Funct Mater 2003; 13:118.
[153] Mahmoud MA, Tabor CE, El-Sayed MA, Ding Y, Wang ZL. J Am Chem Soc 2008;130:4590.

[154] Watt J, Cheong S, Toney MF, Ingham B, Cookson J, Bishop PT, et al. ACS Nano 2009;4:396.
[155] Bratlie KM, Lee H, Komvopoulos K, Yang P, Somorjai GA. Nano Lett 2007;7: 3097.
[156] Somorjai G, Zaera F. J Phys Chem C 1982;86:3070.
[157] Zhou S, McIlwrath K, Jackson G, Eichhorn B. J Am Chem Soc 2006;128:1780.
[158] Eid K, Wang H, He P, Wang K, Ahamad T, Alshehri SM, et al. Nanoscale 2015. http://dx.doi.org/10.1039/c5nr04557f.
[159] Sun X, Li D, Ding Y, Zhu W, Guo S, Wang ZL, et al. J Am Chem Soc 2014;136: 5745.

Functional Layered Compounds for Nanoarchitectonics

N. Miyamoto and S. Yamamoto
Fukuoka Institute of Technology, Fukuoka, Japan

CHAPTER OUTLINE

7.1 INTRODUCTION

Inorganic layered materials are composed of stacked nanometer-thick inorganic crystalline sheets, which are weakly bound by electrostatic, van der Waals, or hydrogen bonding interactions. These materials are obtained from natural soil or are synthesized by hydrothermal heating or calcination at high temperature from mixture of raw chemicals. Depending on the chemical composition, layered materials possess functionalities such as conductivity, semiconductivity, magnetic property, luminescence property, and photocatalytic activity. In this section, recent development of nanoarchitectonics with layered materials as achieved by intercalation and exfoliation technique are reviewed as summarized in Fig. 7.1.

Supra-materials Nanoarchitectonics. http://dx.doi.org/10.1016/B978-0-323-37829-1.00007-9

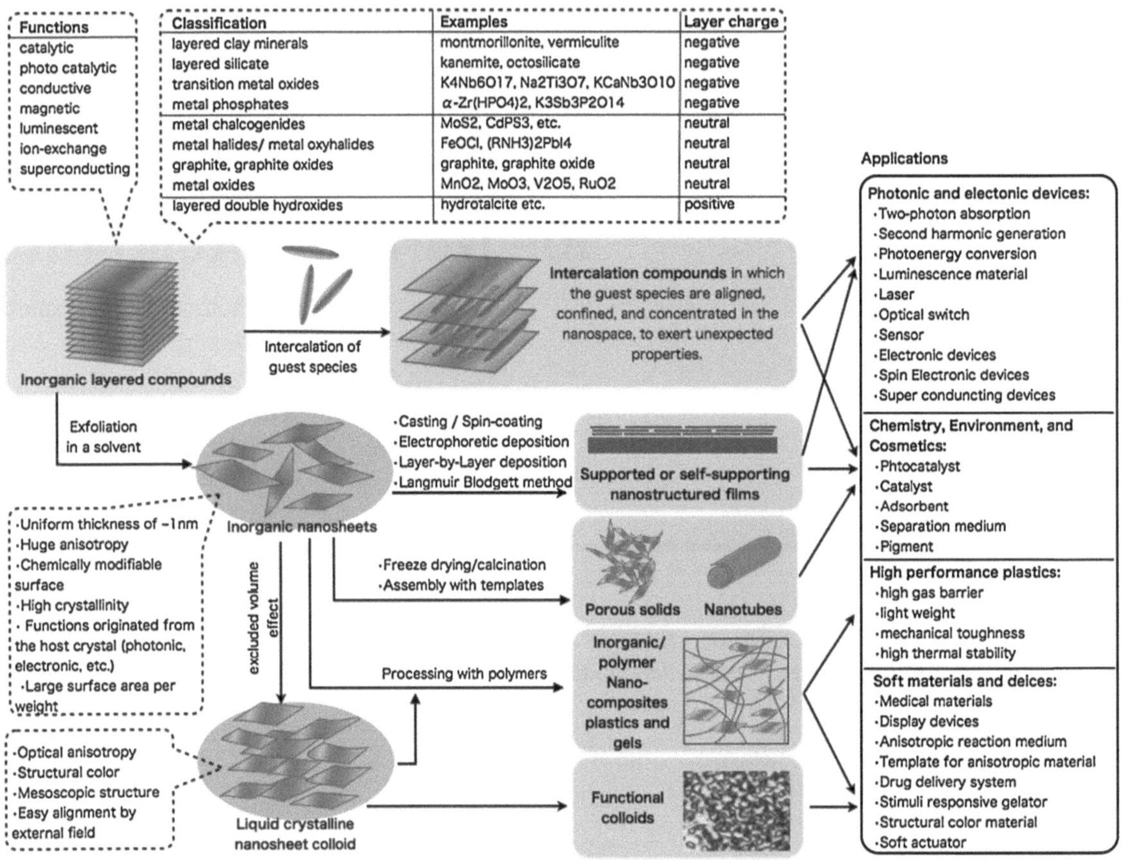

■ **FIGURE 7.1** Overview of nanoarchitectonics starting from inorganic layered materials.

7.2 PHOTOFUNCTIONS OF INTERCALATION COMPOUNDS

Intercalation reaction, insertion of guest species into the interlayer space of the host layered solids, is an effective way to further functionalize layered materials. Intercalation reaction proceeds by replacing or coadsorbing the originally presenting interlayer species with guest species. Various types of guest species such as ions, solvent molecules, and polymers can be intercalated into various host materials. By intercalation reactions, the properties of the guest species are expected be modified through guest–guest and guest–host interactions in the nanoscopic two-dimensional interlayer space. In particular, organic dyes have been selected as the guest species [1] to fabricate novel photofunctional materials as well as for fundamental studies to understand host–guest interactions and nanostructures of intercalation compounds. Whereas many applications for catalyst, adsorbents, drug-delivery system, pigments, optics, photoenergy conversion, etc. are possible with intercalation compounds, here I highlight some of recently reported dye-intercalated compounds for photofunctional materials.

The simplest one is the application of dye-intercalated compounds for color materials; The color of an organic dye is tunable through the confinement, guest−guest interaction, and host−guest interactions. Improved thermal and photostability of the dye is also expected by shielding from surrounding atmosphere. Since only a limited number of dyes are approved by government for use in cosmetics and foods, modification of the dye property through intercalation is desired in these fields. Furutani et al. reported effective color tuning of rhodopsin RSB-11 by intercalation into four different types of natural montmorillonite [2]. In the human eye, RSB-11 is present as complexes with different types of proteins, giving λ_{max} of 500, 530, and 560 nm for the blue, green, and red photoreceptor cells, respectively. The RSB-11 incorporated in the montmorillonites (Bengel Bright 11, Mikawa, and Bengel A) showed λ_{max} of 472, 503, and 532 nm, respectively, covering the similar wavelength range of the natural system. Although the detailed mechanism is not clear, different degrees of protonation of the dye in the different type of interlayer are thought to tune the color. Hata et al. [3] investigated the intercalation of an acidic dye Brilliant Blue FCF into fluortetrasilicic mica preintercalated with polyalylamine (PAA). Due to overcompensation of the negative layer charge with PAA, the acidic dye can be incorporated into the interlayer. In this system, the pH of the dye solution for intercalation reaction as well as the amount of intercalated PAA and dye affected the color, giving powder materials with the color tuned in the full-color range (Fig. 7.2). The color change is explained by partial protonation of the dye molecule through the interaction with the coadsorbed PAA.

Intercalation compounds of fluorescent dye lead to the materials for lightning, laser, and sensing. Incorporation of fluorescent dye into layered materials usually gives rise in fluorescence quantum yield because radiationless deactivation process from excited state is inhibited by confinement and immobilization of the dye in the narrow nanospace. However, in many cases, it is troublesome that dye molecules form nonemissive aggregates when incorporated and concentrated in the interlayer. Sasai et al. overcame this problem by cointercalation of a typical laser dye Rhodamine-6G (R6G) with alkyltrimethylammonium (CnTMA), which inhibits aggregation of R6G. The emission intensity increased with the increase in the carbon number of the CnTMA (Fig. 7.3) and is also dependent on the amount of CnTMA and R6G. In the optimum condition, the quantum yield of up to 80% is attained.

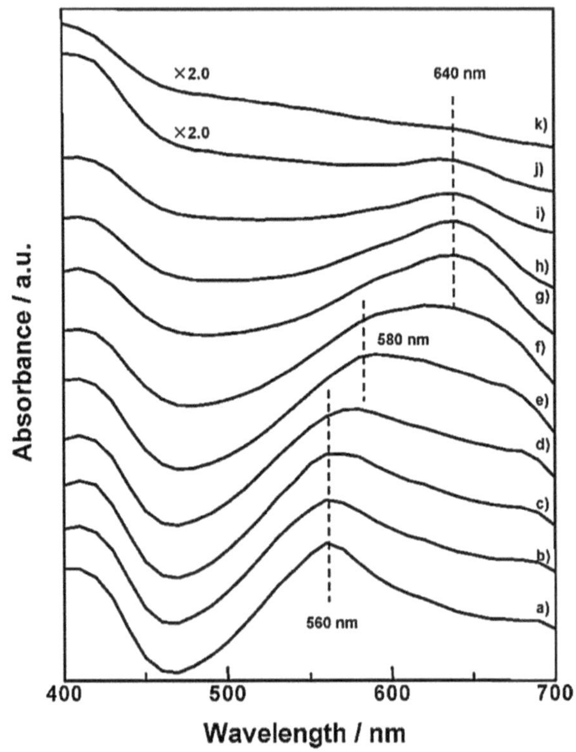

■ **FIGURE 7.2** UV—vis absorption spectra of (B) diffused reflectance spectra of B1/PAA/Na-TSM (1/5/1) at pH (a) 1.5, (b) 3.0, (c) 7.0, (d) 8.0, (e) 9.0, (f) 9.5, (g) 10.0, (h) 10.5, (i) 11.0, (j) 11.5, and (k) 12.0 [3].

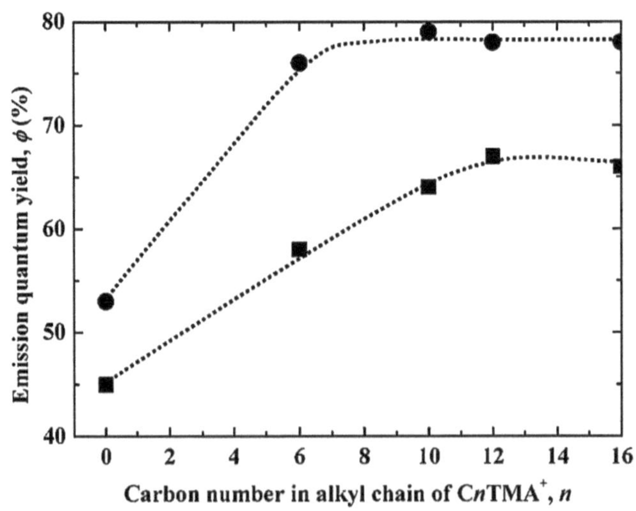

■ **FIGURE 7.3** Dependence of emission quantum yield, φ, on n (carbon number in alkyl chain) of $CnTMA^+Br^-$. (●) 0.1 and (■) 0.5 mol % R6G molecules versus CEC of Lap. n) 0 means R6G/Lap powder. [CnTMA+]) c. 70% CEC [48].

Intercalation compounds with nonlinear optical property is also promising. Two-photon absorption (TPA) is the nonlinear optical phenomenon that occurs when two photons are simultaneously absorbed by a material. By using a focused laser, TPA occurs only on a very small confocal spot in the material, allowing applications for 3D microfabrication and optical data storage system. Suzuki et al. reported enhancement of two-photon absorption efficiency when cationic organic dyes are intercalated into saponite. Among four kinds of dyes investigated, TMPyP showed the largest enhancement with the factor of 13 times (Fig. 7.4). Enhanced planarity of the dye in the interlayer space causes extension of pi-conjugate system and reduction in the detuning energy. Other nonlinear optics materials of

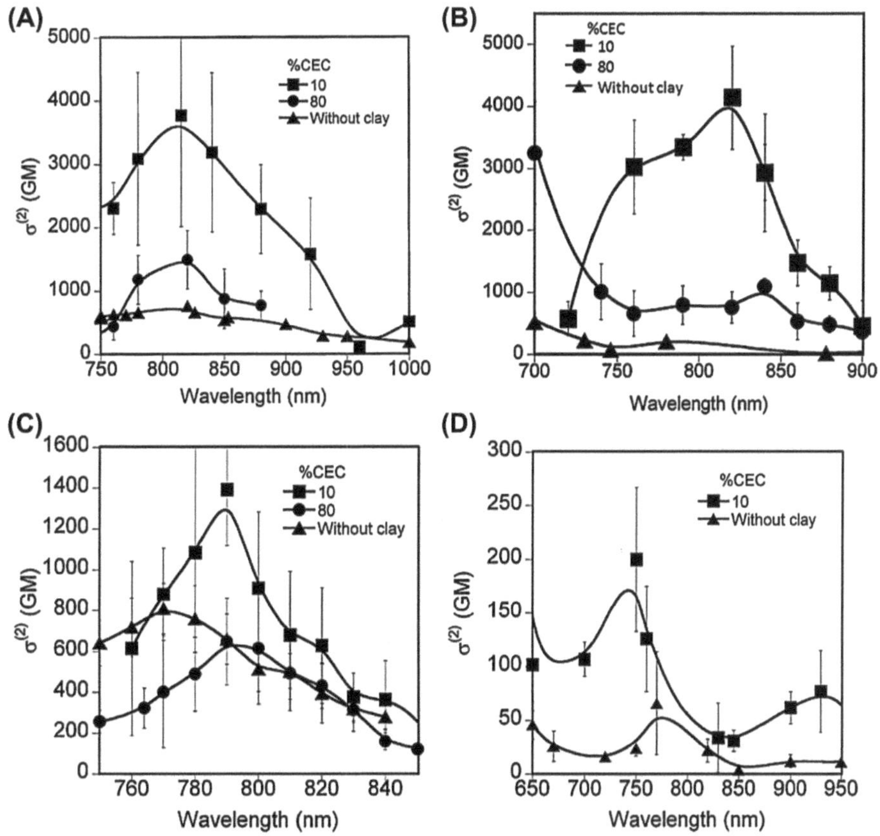

■ **FIGURE 7.4** TPA spectra of (A) MPPBT, (B) TMPyP, (C) FL, and (D) PIC in DMSO and composite films. The DMSO solution concentration was 3.0×10^{-3} mol L^{-1} for MPPBT, 6.0×10^{-3} mol L^{-1} for TMPyP, 2.5×10^{-3} mol L^{-1} for FL, and 2.0×10^{-2} mol L^{-1} for PIC. The curve overlapping the circles is a guide for the eye only. Error bars are attached to each experimental data [49].

second harmonic generation and upconversion emission have also been reported as summarized in a recent review paper [4].

Since an intercalation compound can further incorporate other guest species such as solvent molecules, application for a sensor or a smart adsorbent [5] is also possible. Shinozaki et al. [6] reported that Rhodamine 6G/$K_4Nb_6O_{17}$ intercalation compounds show reversible change in absorption and fluorescence spectra, triggered by adsorption of small amounts of water vapor. It is considered that coadsorption of water molecules causes reversible aggregation/deaggregation of the dye in the interlayer. Sasai et al. [7] reported that Rhodamine 3B/decyltrimethylammonium/$Ti_{1.825}O_4$ intercalation compounds show large change in the emission intensity depending on relative humidity. Basal spacing also varies with relative humidity, indicating that coadsorption of water molecules causes changes of the dye orientation and interaction with host and alkylammonium, resulting in varied emission intensity.

7.3 RECONSTRUCTION OF NANOSHEETS INTO FUNCTIONAL NANOSTRUCTURES

7.3.1 Exfoliation of Layered Solids Into Nanosheets

Another important strategy for nanoarchitectonics with layered material is to exfoliate layered materials into inorganic nanosheets. Inorganic layered materials are swollen by intercalation of solvent molecules into their expandable interlayer space. Infinite swelling of the interlayer space leads to *exfoliation* (or delamination) of layered materials to individual inorganic nanosheets. The obtained nanosheets are regarded as novel nanomodules characterized by (1) extremely large aspect ratio with large lateral size up to several μm and uniform thickness of several nm; (2) electric, optical, and magnetic properties and defined structure with high crystallinity, which originate from intrinsic property of host materials; and (3) capability of supramolecularly organizing guest species on the sheet. These features open applications of the inorganic nanosheets as nanomodules for functional nanohybrid solids and novel colloid materials.

7.3.2 Nanostructured Films

Since the exfoliated layered solids possess highly anisotropic shape, highly oriented multilayered nanostructured films are obtained by simply casting or spin coating a nanosheet sol on a substrate, followed by evaporation of the solvent. The thin films have been prepared by this method from the layered materials such as MPS_3 (M = Mn, Cd) [8], $K_4Nb_6O_{17}$ [9], $Cs_xTi_{2-x/4}\cdot_{x/4}O_4$ [9b,10], and smectite clays [11]. Due to the large anisotropy of the

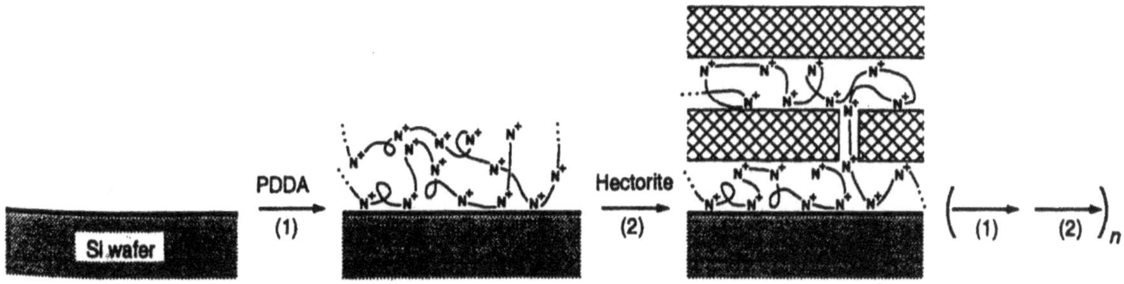

■ **FIGURE 7.5** Layer-by-layer deposition of PDDA-Laponite film [13a].

nanosheets, the nanosheets are oriented along the surface of substrates. The films thus obtained are highly transparent and anisotropic, in contrast with the powders with the same composition and nanostructure. These films are suitable for photo [4] and electronic devices [12].

Layer-by-layer (LBL) deposition is a valuable method for fabrication of heterostacked multilayered films. Kleinfeld and Ferguson [13] reported stepwise formation of multilayered films by alternately adsorbing a cationic polyelectrolyte, poly(diallyldimethyl-ammonium chloride) (PDDA), and anionic nanosheets of hectorite. Fig. 7.5 shows the deposition process: first, aqueous solution of PDDA was dropped on the substrate (hydroxylated silicon wafer), followed by rinse with water and drying. Subsequently, aqueous dispersion of Laponite was dropped, rinsed, and dried. Repeating this cycle 200 times resulted in the films with thickness of 0.2 μm. Constant increase of ellipsometric thickness with increasing number of the cycles and observation of the *00l* reflections by X-ray diffraction (XRD) evidenced the formation of a multilayered structure. LBL film deposition has been applied to a wide range of materials: α-$Zr(HOPO_3)_2$ [14], niobates ($KTiNbO_5$, $KCa_2Nb_3O_{10}$, and $KCa_2NaNb_4O_{13}$) [14c,15], titanate [16], and layered clay minerals [17].

The great merit of LBL film deposition is potential for fabricating heterogeneous stacking of various components in desired sequence, as proposed by Kellar et al. [14a]. For example, the heterostructure was obtained by sequential deposition of anionic nanosheets of α-ZrP and $K_2Nb_6O_{17}{}^{2-}$ with a polycation poly(allylamine)hydrochloride; increment of the ellipsometric thickness corresponded to the thickness of each component, indicating the formation of the heterogeneous superstructure. By heterogeneously assembling electron acceptor, sensitizing dyes, and nanosheets of titanoniobate or zirconium phosphate, the model system of photoinduced electron transfer was fabricated [14b,18]. As another example of applications, Cassagneau et al. fabricated high-density (1232 mAh g^{-1} of graphitic carbon) rechargeable lithium ion battery cathode composed of heterogeneously stuck polyethyleneoxide, PDDA, and graphite oxide.

Multilayered films are also deposited electrophoretically. The advantage of this technique compared to the casting and layer-by-layer depositions is high uniformity of resulting film, easiness in thickness control, applicability to large-scale deposition, and deposition to complex-shaped substrate. Sugimoto et al. reported electrophoretic deposition of exfoliated $H_2Ti_4O_9$ nanosheets (Fig. 7.6) [19]. The counter cations of the titanate, the exfoliating agent tetrabutylammonium cations, are collected around the cathode, whereas the negatively charged nanosheets are deposited onto the anode, resulting in the formation of $H_2T_4O_9$ film as confirmed by XRD. In contrast, the electrophoretic deposition of the films from the suspended fine powders of $Cs_xTi_{2\text{-}x/4}\cdot_{x/4}O_4$ [20] and $K_4Nb_6O_{17}$ [21], which are not completely exfoliated, results in the film of alkylammonium-intercalated forms. Thermal treatment of the deposited films led to an oriented $TiO_2(B)$ film, which is valuable due to electrochemical and catalytic properties compared to other TiO_2 polymorphs, with the *(0k0)* planes lying perpendicular to the substrate.

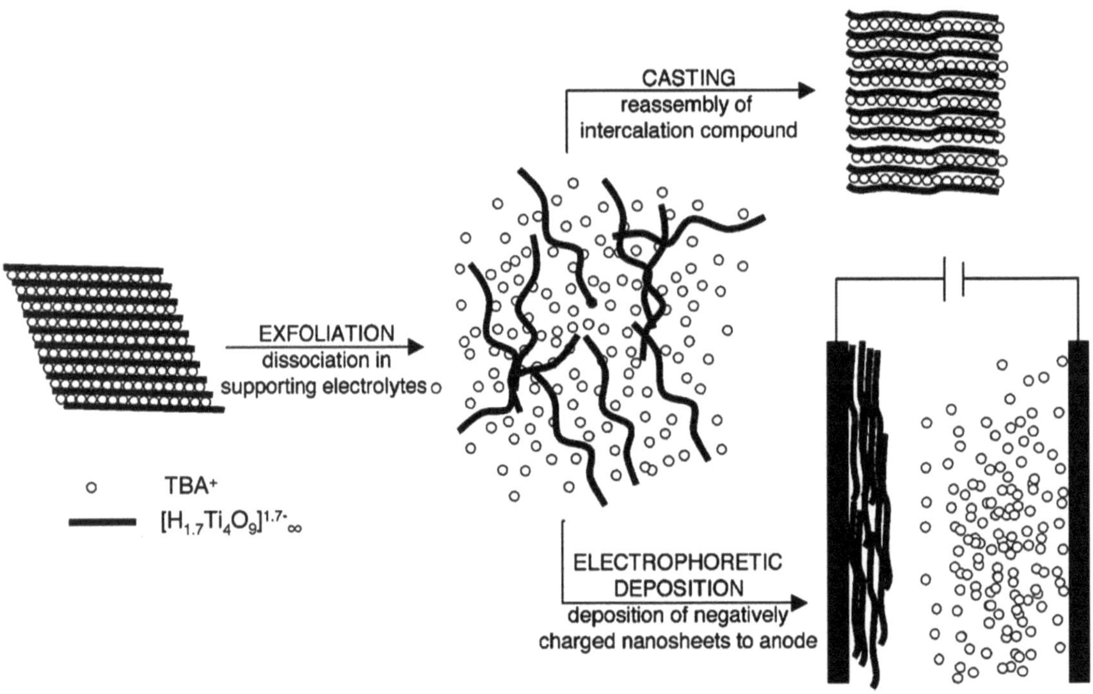

■ **FIGURE 7.6** Schematic of the exfoliation and deposition of $H_2Ti_4O_9$ H_2O.

7.3.3 **Porous Solids**

Inorganic nanosheets obtained by exfoliation can be building units for porous materials. Sasaki et al. reported fabrication of porous titanium dioxide powders by reassembling exfoliated layered titanate $Cs_xTi_{2-x/4}\cdot_{x/4}O_4$ (Fig. 7.7) [22]. The sol containing the titanate nanosheets was freeze-dried to yield "voluminous white solid with cotton-like appearance." By heating this solid at 400–700°C, the material was recrystallized into titanium dioxide with pores. The final product was a flaky powder with

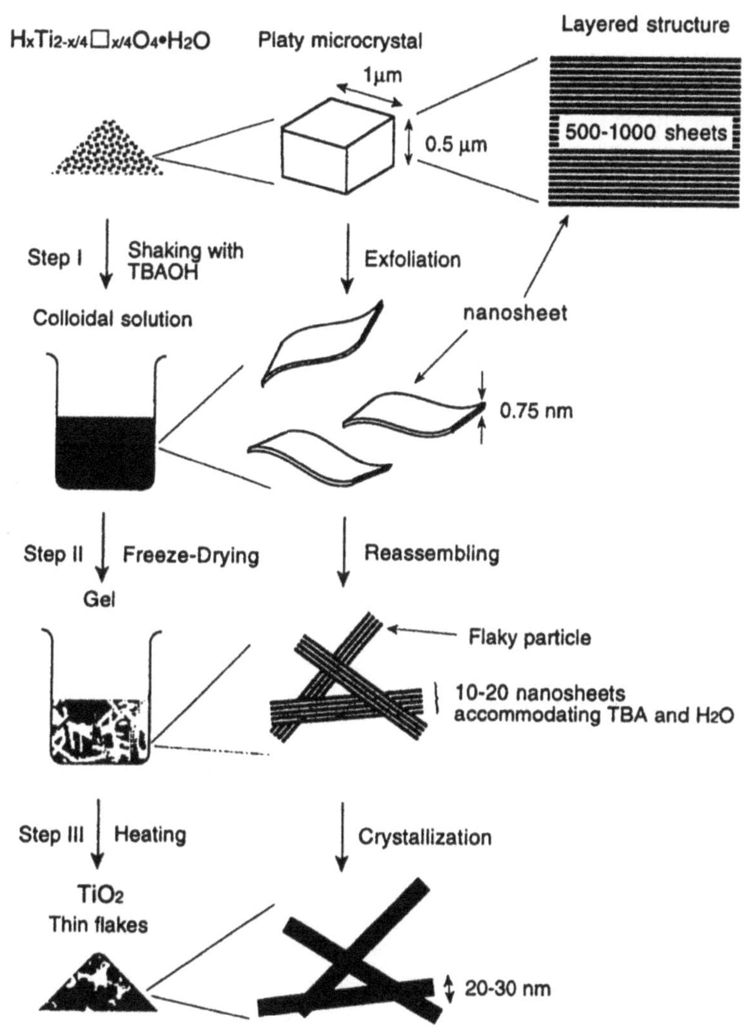

■ **FIGURE 7.7** Schematic illustration of synthetic process of the titanium dioxide thin flakes [22].

the Brunauer-Emmett-Teller (BET) surface area of 47−104 $m^2 g^{-1}$ and mode pore size of 23−51 nm depending on the temperature of heating.

Reassembling of exfoliated nanosheets with pillars also leads to porous solids. Exfoliated nanosheets of $Cs_xTi_{2-x/4}\cdot_{x/4}O_4$ [23] and $K_4Nb_6O_{17}$ [24] were reassembled with metal oxide fine particles in the interlayer spaces; the potential use of these materials as photocatalysts was examined in these reports. Even without freeze-drying and pillaring, acidification of $HTiNbO_5$ nanosheet colloids also led to aggregation and precipitation of the powders, which worked as stronger solid acid catalyst than the bulk solid of $HTiNbO_5$ [25].

Another strategy of fabricating porous solids from exfoliated nanosheets is to curl up nanosheets into tubules. Saupe et al. found that exfoliated niobate nanosheets are curled up as well-defined tubules [26]. The formation of tubules was observed by TEM; the nanosheets tended to be curled at higher colloid concentration and lower pH. It was also inferred that curling occurred only when the niobate is exfoliated to single layers as evidenced by AFM. From these observations, it was suggested that the curling is driven by the inherent asymmetricity of the exfoliated nanosheets (one side is less sterically crowded than the other). Drying the colloid containing the tubules resulted in porous powders with the BET surface area of 37 $m^2 g^{-1}$ and mean pore size of 10 nm, which was comparable to the diameter of the tubules observed by TEM.

Camerel et al. [27] extended the concept of Saupe et al. in a unique way to obtain mesoporous niobate-aluminosilicate composite with unidirectionally oriented pores. The colloid of the tubules was added with the solutions of aluminum sulfate and sodium silicates, followed by hydrothermal treatment. The tubules were self-organized with their main axis in same directions due to the formation of nematic liquid crystalline state driven by excluded volume effect between the tubules. The obtained composite possessed BET surface area of 102 $m^2 g^{-1}$ and mean pore size of 14.6 nm, indicating that the method presented is valuable for synthesizing composite mesoporous materials.

Nematic liquid crystalline phase of V_2O_5 nanosheet ribbons (the details of the liquid crystalline nature of V_2O_5 system are described in the following section) was used as a template, not as a module, to fabricate mesoporous composite with aligned pores on the cm scale [27]. Hydrolysis of tetramethoxysilane in the presence of liquid crystalline V_2O_5 sol yielded inorganic–inorganic composites; after removing V_2O_5, the composite exhibited slit-like mesopores (the cross-sectional dimensions of 2 × 20 nm) and a specific surface area of 207 m^2g^{-1}. Applying weak magnetic field (0.85 T) during preparation yielded large single domain (cm-scale) with unidirectional channel director.

Exfoliated nanosheets can be assembled into porous material with a regulated pore structure by using organic sacrificial template; highly crystalline pore wall is achieved in this case because the crystallinity of the nanosheets is retained. Miyamoto et al. synthesized macroporous solids by casting and drying the colloidal mixtures of niobate nanosheets and monodisperse polystyrene spheres, followed by calcination [28]. When the average lateral size of the nanosheets is small enough (100 nm) compared to the template sphere (330 nm), the porous materials with ordered macroporous solid with interconnecting pores are obtained (Fig. 7.8A). As the average nanosheet size

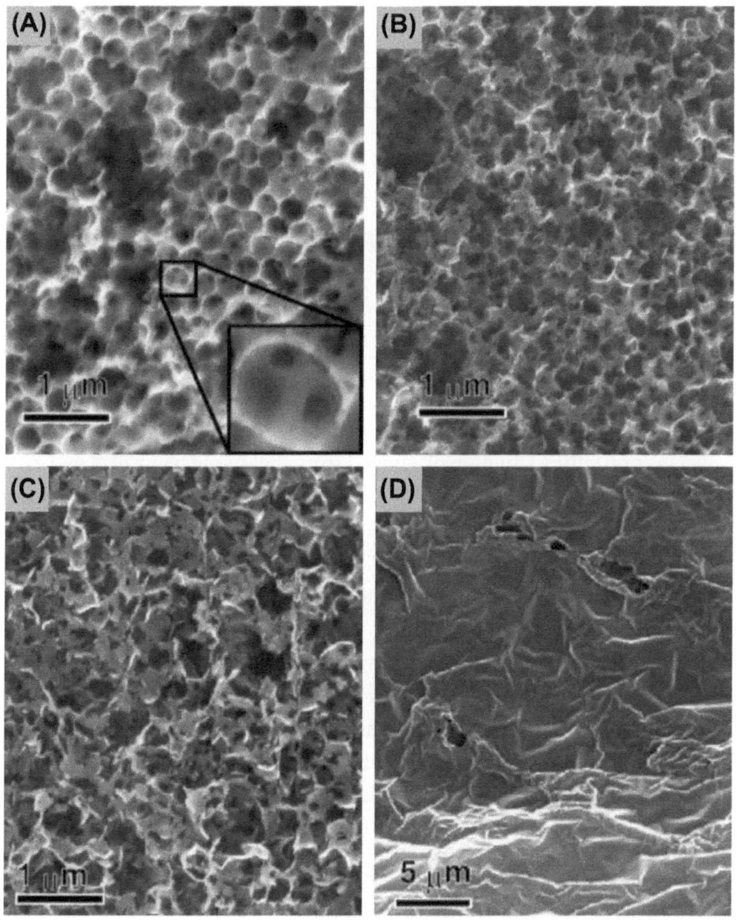

■ **FIGURE 7.8** SEM images of macroporous niobates obtained from the mixtures of polystyrene spheres and niobate nanosheets after calcination for 6 h at 500°C. The mean diameters D of the polystyrene spheres used are (A–D) 330 nm. The mean lateral sizes L of the nanosheets used are (A) 100 nm, (B) 190 nm, (C) 270 nm, and (D) 3 μm. Nb:PS = 25:75 [28].

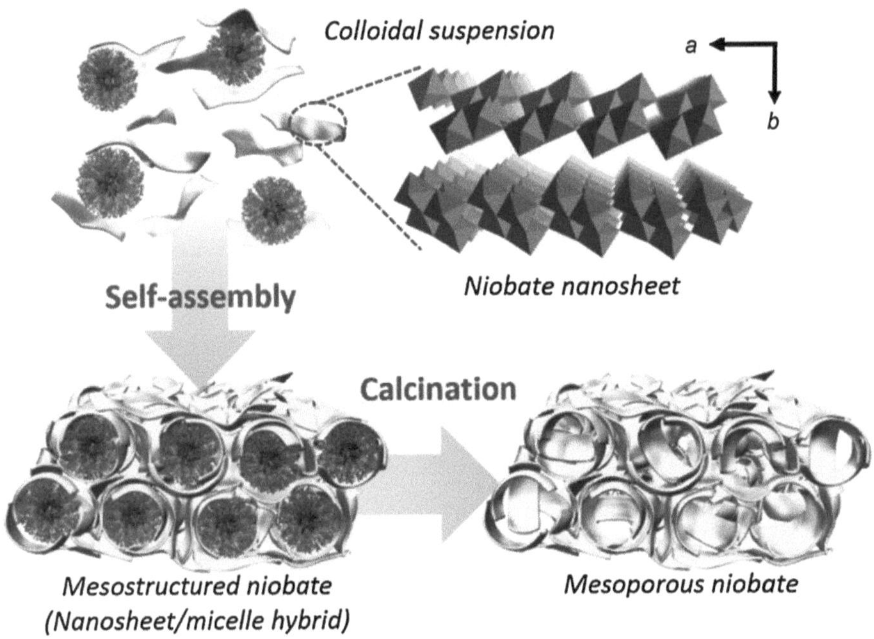

■ **FIGURE 7.9** Synthesis of crystalline mesoporous materials from coassembly of niobate nanosheets and polymeric micelles based on colloidal chemistry [29].

increases, the structure becomes less ordered (Fig. 7.8B–D). The pore wall possesses crystalline structure originated from that of the nanosheet and even the expandable layered structure formed by restacking of the nanosheets. This allows further modification of the porous material by intercalation reaction into the pore wall. Bastakoti et al. used polymeric micelle as a template to assemble niobate and titanate nanosheets into mesoporous solids with crystalline pore wall (Fig. 7.9) [29]. Triblock copolymer, poly(styrene-b-2-vinylpyridine-b-styrene), forms positively charged spherical micelle in a solvent. By adding nanosheet dispersion to the micelle solution, the micelle is coated with negatively charged nanosheets. This mixture was cast and dried on a substrate, followed by calcination, resulting in a disordered mesoporous structure with highly crystalline wall.

7.4 NANOSHEETS FOR SOFT MATERIALS

7.4.1 Nanosheet Liquid Crystal

Due to the huge anisotropy, nanosheets dispersed in a solvent orient themselves spontaneously to form a liquid crystal (LC) phase under certain conditions. In 1938, occurrence of birefringence in aqueous dispersions of

Bentonite clay was first reported by Langmuir [30], followed by establishment of Onsager's theory to roughly explain this phenomenon in 1949 [31]. However, it is only recently that LC nanosheet colloid system was rediscovered and highlighted as new functional materials and a good model system of condensed matter physics.

Gabriel et al. reported that the nanosheet colloid obtained from layered antimonyphosphate $K_3Sb_3P_2O_{14}$ shows LC phase with birefringence [32]. Small-angle X-ray scattering (SAXS) revealed the presence of swollen lamellar structure with very large basal spacing of up to 225 nm (Fig. 7.10). Due to the large spacing and high structural regularity, structural

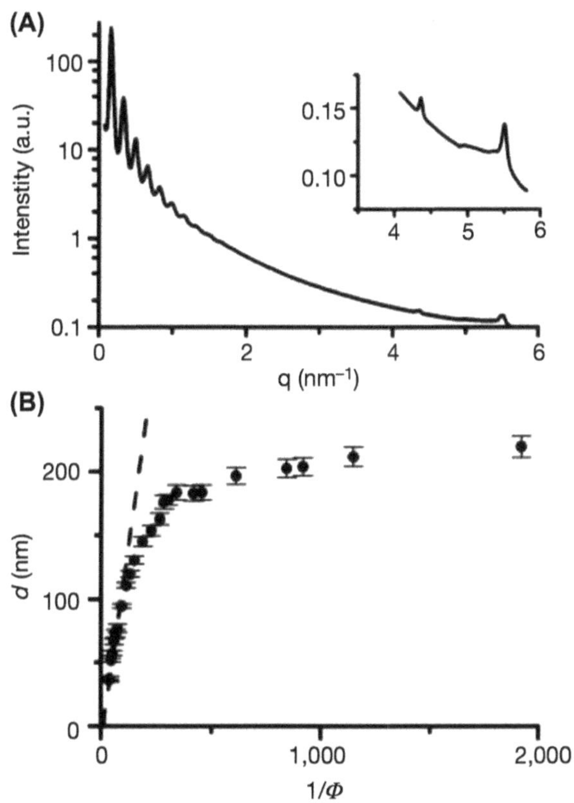

■ **FIGURE 7.10** Small-angle X-ray scattering study of the $H_3Sb_3P_2O_{14}$ suspensions. (A) An example of scattered intensity versus scattering vector modulus q, showing 10 orders of reflection due to the lamellar period. Inset, the thin diffraction lines due to the two-dimensional positional order within the covalent layers ($\varphi = 2.0\%$). (B) Variation of the lamellar period, *d*, with inverse volume fraction. The *dashed straight* line represents the one-dimensional swelling behavior of the lamellar phase [32].

color of blue to red, depending on the nanosheet concentration, is observed. Miyamoto et al. reported the LC phase of the nanosheet colloid obtained from $K_4Nb_6O_{17}$ [33]. In this study, the series of samples with different average lateral size of the nanosheets were prepared by breaking down the as-prepared nanosheets by ultrasonication. It was revealed that the critical concentration for the isotropic-to-LC phase transition decreases as the nanosheet size increases (Fig. 7.11), as predicted by Onsager's theory.

In addition to the materials stated previously, the nanosheet colloids with LC phase have been obtained from layered clay minerals (nontronite, beidellite [34b], fluorohectorite [34c], and fluortetrasilicic mica [34c]), layered titanoniobate $KTiNbO_5$ [35a], layered niobate KNb_3O_8 [34a], layered titanate $H_{1.07}Ti_{1.73}O_4$ [35b], layered perovskite $KCaNb_3O_{10}$ [36], graphene oxide [37], and layered zirconium phosphate [38]. For further application for the synthesis of anisotropic materials and for optical devices, macroscopic alignment of the nanosheets by electric field [39] or magnetic field [27,37a,40] are recently being investigated.

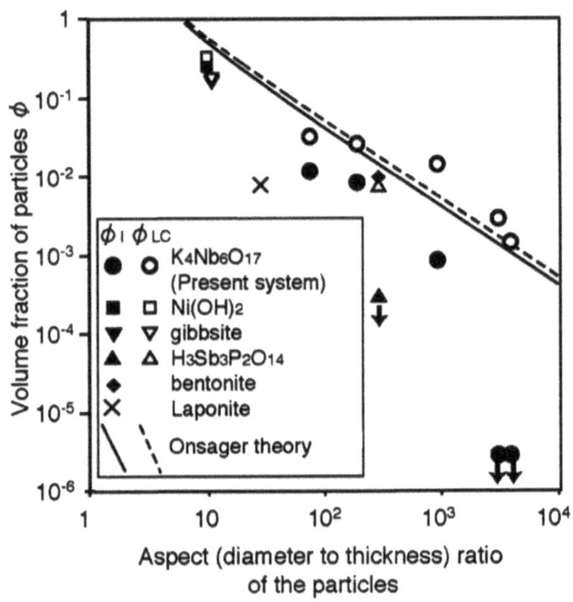

■ **FIGURE 7.11** Relationship between diameter-to-thickness ratio of colloidal disks and the transition concentrations φ_I (isotropic to biphasic) and φ_{LC} (biphasic to liquid crystalline) observed in the present study and in the literature. *Solid and dashed lines* represent the theoretical values numerically calculated based on Onsager's second virial approximation. The *down arrows* indicate that the transition concentration was not determined but is lower than the value indicated by the symbols.

7.4.2 Composite Gel

Nanosheets have been recently utilized to functionalize polymer gel materials. Polymer gels are composed of chemically or physically cross-linked polymer chain network swollen by a solvent. Due to their high biocompatibility, softness, and high response to external stimuli, they are applied for medical materials and soft actuators. While low mechanical strength and uncontrollable microstructures are general problems of gel materials for these applications, the formation of nanosheet/polymer composite gel improves these problems.

Haraguchi et al. reported the clay/poly(*N*-isopropylacrylamide) nanocomposite gel with surprisingly high mechanical property [41]. The gel was synthesized from an aqueous mixture of monomer (NIPAm), redox initiator, and hectorite clay nanosheets. Much higher breaking strain and stress compared with the gel with no added clay nanosheets were observed. It is considered that the very good mechanical property arises from physical cross-linking of the polymer chains on many points on the surface of the clay nanosheets. Initiated by this pioneering work, many composite gels on similar concept have been reported by using other types of nanosheet materials such as grapheme oxide [42]. The composite gels are also synthesized by photopolymerization technique, which allows for easier processing [42b,43]. By combining electric field-induced alignment with the photopolymerization through a photomask, Paineau et al. reported beidelite/acrylamide composite gel with a partly aligned patterned structure with the resolution of 20 μm [44].

While the previously stated systems were synthesized from isotropic nanosheet colloid, Miyamoto et al. took advantage of the hierarchical structure of the liquid crystalline nanosheet colloid to synthesize macroscopically anisotropic composite gels [45]. The anisotropic gel was obtained from the mixture of liquid crystalline fluorohectorite colloid added with monomer and redox initiator in a glass capillary. The nanosheets were aligned along the capillary wall, forming tree-ring-like structure, and this structure was fixed by polymerization. In the SAXS profile during the synthesis process (Fig. 7.12), the lamellar structure with the basal spacing of 35 nm gradually shrinks to 10 nm as the reaction proceeds, while the anisotropic structure is retained. By storing the as-synthesized gel in water, the gel macroscopically swells to equilibrium swelling state, while nanoscopic structure also swells: the basal spacing increased to more than 100 nm. The anisotropic structure causes anisotropic properties in many ways: optical property, thermally induced volume phase transition, elasticity, and diffusion of molecules. By a similar process, the composite gel of poly(acrylamide)- poly(*N*-isopropylacrylamide) copolymer with fixed liquid crystalline structure of

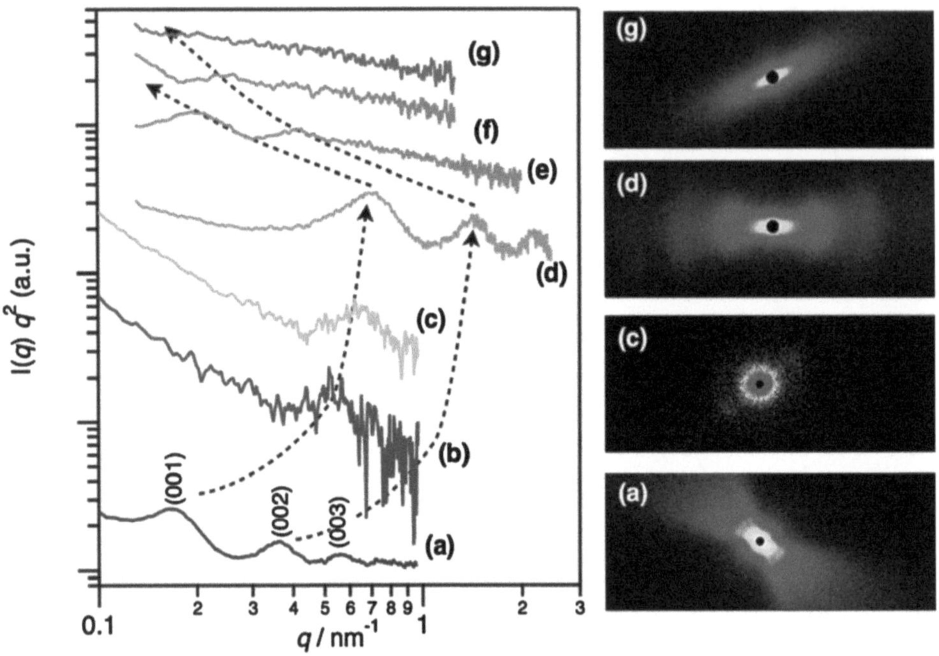

■ **FIGURE 7.12** Small-angle X-ray scattering profiles of the F5-B0-gel system: (a) FHT—NIPA—water mixture (before polymerization), the mixture during polymerization [(b) 1 and (c) 10 min after reaction initiation], (d) as-prepared gel, and the gel swollen for (e) 15, (f) 30, and (g) 60 min in water.

zirconium phosphate nanosheet/polyacrylamide inside was reported by Mejia et al. [46].

Although the previous method is very simple to obtain anisotropic composite gel, the tree-ring-like structure is not suitable for further detailed fundamental study and applications. The sheet-shaped gel with perfectly aligned monodomain structure on cm scale was obtained by prealignment of the liquid crystal nanosheets by electric field as shown in Fig. 7.13 [39a]. If we adsorb a multivalent cationic dye through a holed silicone rubber placed on the gel sheet, the sub-mm resolution pattern is printed and is stably held without diffusion (Fig. 7.14A). Because the dye absorbs light and emits heat, the part of the gel adsorbed with dye shows photothermal response, resulting in reversible deformation into asymmetric shape (Fig. 7.14B). This monodomain anisotropic gel showed not only anisotropic shrinkage but also anomalous temporal swelling in one direction induced by photothermal input.

Recently, Liu et al. also reported monodomain nanosheet/pNIPAm gel with anisotropic optical and rheological properties [47]. The layered titanate or layered perovskite nanosheet was aligned by magnetic field, followed by

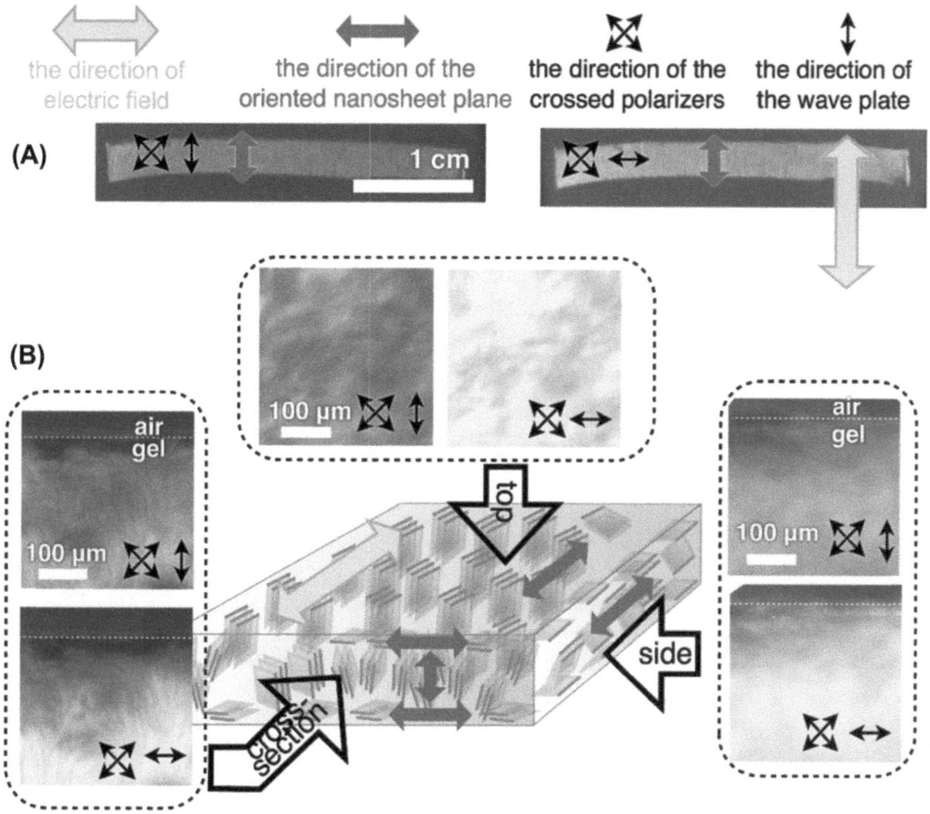

■ **FIGURE 7.13** (A) Photographs and (B) microscopic images and schematic structures of the FHT/PNIPA gels containing 1 wt% of FHT synthesized with in-plane electric field. The images are observed with crossed polarizers and a wave plate [39a].

■ **FIGURE 7.14** Patterned coloration of the FHT/PNIPA gels with TMPyP dye and its partial photoresponsive deformation: (A) photograph (left) and microscopic image (right) of the gel printed with the dye pattern; (B) photoinduced partial deformation of the gel partially adsorbed with the dye as observed by optical microscopy. The *red arrow* indicates the direction of the oriented nanosheet planes [39a].

photopolymerization. Interestingly, the anisotropy of rheological property shows opposite trends, depending on the kind of nanosheet used. In the titanate system, the compressive modulus is higher along the normal of aligned nanosheet plane, whereas it is higher along the parallel direction. It is considered that the titanate nanosheet system has cofacial alignment, or swollen lamellar structure, while the perovskite system shows nematic structure, resulting in the different trend.

REFERENCES

[1] Ogawa M, Kuroda K. Chem Rev 1995;95:399–438.
[2] Furutani Y, Ido K, Sasaki M, Ogawa M, Kandori H. Angew Chem Int Ed Engl 2007; 46:8010–2.
[3] Hata H, Mallouk TE, Kuroda K. Chem Mater 2009;21:985–93.
[4] Suzuki Y, Tenma Y, Nishioka Y, Kawamata J. Chem Asian J 2012;7:1170–9.
[5] [a] Okada T, Matsutomo T, Ogawa M. J Phys Chem C 2010;114:539–45.
[b] Okada T, Watanabe Y, Ogawa M. Chem Commun 2004:320–1.
[6] Shinozaki R, Nakato T. Langmuir 2004;20:7583–8.
[7] Sasai R, Iyi N, Kusumoto H. Bull Chem Soc Jpn 2011;84:562–8.
[8] Lagadic I, Lacroix PG, Clément R. Chem Mater 1999;9:2004–12.
[9] [a] Abe R, Hara M, Kondo JN, Domen K. Chem Mater 1998;10:1647–51.
[b] Abe R, Ikeda S, Kondo JN, Hara M, Domen K. Thin Solid Films 1999;343-344: 156–9.
[10] [a] Abe R, Shinohara K, Tanaka A, Hara M, Kondo JN, Domen K. Chem Mater 1998; 10:329–33.
[b] Sumida T, Abe R, Hara M, Kondo JN, Domen K. Electrochem 1999;67:1224–6.
[c] Sumida T, Takahara Y, Abe R, Hara M, Kondo JN, Domen K, et al. Phys Chem Chem Phys 2001;3:640–4.
[d] Koinuma M, Matsumoto Y, Sumida T, Domen K. Electrochem Solid State Lett 2000;3:481–4.
[e] Sumida T, Abe R, Hara M, Kondo JN, Domen K. J Mater Res 2000;15:2587–9.
[11] [a] Ogawa M, Takahashi M, Kuroda K. Chem Mater 1994;6:715–7.
[b] Ogawa M, Kanaoka N, Kuroda K. Langmuir 1998;14:6969–73.
[c] Sonobe K, Kikuta K, Takagi K. Chem Mater 1999;11:1089–93.
[d] Isayama M, Sakata K, Kunitake T. Chem Lett 1993:1283–6.
[e] Isayama M, Nomiyama K, Kunitake T. Adv Mater 1996;8:641–4.
[12] Osada M, Sasaki T. J Mater Chem 2009;19:2503–11.
[13] [a] Kleinfeld ER, Ferguson GS. Science 1994;265:370–3.
[b] Kleinfeld ER, Ferguson GS. Chem Mater 1996;8:1575–8.
[c] Kleinfeld ER, Ferguson GS. Chem Mater 1995;7:2327–31.
[d] Ferguson GS, Kleinfeld ER. Adv Mater 1995;7:414–6.
[14] [a] Keller SW, Kim H-N, Mallouk TE. J Am Chem Soc 1994;116:8817–8.
[b] Kaschak DM, Mallouk TE. J Am Chem Soc 1996;118:4222–3.
[c] Kaschak DM, Lean JT, Waraksa CC, Saupe GB, Usami H, Mallouk TE. J Am Chem Soc 1999;121:3435–45.
[15] [a] Fang M, Kim CH, Saupe GB, Kim H-N, Waraksa CC, Miwa T, et al. Chem Mater 1999;11:1526–32.
[b] Kim H-N, Keller SW, Mallouk TE. Chem Mater 1997;9:1414–21.

[16] [a] Sasaki T, Ebina Y, Tanaka T, Harada M, Watanabe M, Decher G. Chem Mater 2001;13:4661–7.
[b] Sasaki T, Ebina Y, Fukuda K, Tanaka T, Harada M, Watanabe M. Chem Mater 2002;14:3524–30.
[c] Wang ZS, Sasaki T, Muramatsu M, Ebina Y, Tanaka T, Wang LZ, et al. Chem Mater 2003;15:807–12.

[17] [a] Kotov NA, Haraszti T, Turi L, Zavala G, Geer RE, Dékány I, et al. J Am Chem Soc 1997;119:6821–32.
[b] Lvov Y, Ariga K, Ichinose I, Kunitake T. J Am Chem Soc 1995;117:6117–23.
[c] Lvov Y, Ariga K, Ichinose I, Kunitake T. Langmuir 1996;12:3038–44.
[d] Kotov NA, Magonov S, Tropsha E. Chem Mater 1998;10:886–95.
[e] van Duffel B, Schoonheydt RA. Langmuir 1999;15:7520–9.

[18] [a] Kim DW, Blumstein A, Kumar J, Tripathy SK. Chem Mater 2001;13:243–6.
[b] Kim DW, Blumstein A, Tripathy SK. Chem Mater 2001;13:1916–22.

[19] Sugimoto W, Terabayashi O, Murakami Y, Takatsu Y. J Mater Chem 2002;12:3814–8.

[20] Matsumoto Y, Funatsu A, Matsuo D, Unal U. J Phys Chem B 2001;105:10893–9.

[21] Koinuma M, Seki H, Matsumoto Y. J Electroanal Chem 2002;531:81–5.

[22] Sasaki T, Nakano S, Yamauchi S, Watanabe M. Chem Mater 1997;9:602–8.

[23] Choy J-H, Lee H-C, Jung H, Kim H, Boo H. Chem Mater 2002;14:2468–91.

[24] [a] Abe R, Shinohara K, Tanaka A, Hara M, Kondo JN, Domen K. Chem Mater 1997;9:2179–84.
[b] Abe R, Shinohara K, Tanaka A, Hara M, Kondo JN, Domen K. J Mater Res 1998;13:861–5.
[c] Domen K, Ebina Y, Ikeda S, Tanaka A, Kondo JN, Maruya K. Catal Today 1996;28:167–74.

[25] Takagaki A, Sugisawa M, Lu D, Kondo JN, Hara M, Domen K, et al. J Am Chem Soc 2003;125:5479–85.

[26] Saupe GB, Waraksa CC, Kim H-N, Han YJ, Kaschak DM, Skinner DM, et al. Chem Mater 2000;12:1556–62.

[27] Camerel F, Gabriel J-CP, Batail P. Adv Funct Mater 2003;13:377–81.

[28] Miyamoto N, Kuroda K. J Colloid Interface Sci 2007;313:369–73.

[29] Bastakoti BP, Li Y, Imura M, Miyamoto N, Nakato T, Sasaki T, et al. Angew Chem Int Ed 2015;54:4222–5.

[30] Langmuir I. J Chem Phys 1938;6:873–96.

[31] Onsager L. Ann NY Acad Sci 1949;51:627–59.

[32] Gabriel J-CP, Camerel F, Lemaire BJ, Desvaux H, Davidson P, Batail P. Nature 2001;413:504–8.

[33] [a] Miyamoto N, Nakato T. Adv Mater 2002;14:1267–70.
[b] Miyamoto N, Nakato T. J Phys Chem B 2004;108:6152–9.

[34] [a] Michot LJ, Bihannic I, Maddi S, Funari SS, Baravian C, Levitz P, Davidson P. Proc Nat Acad Sci USA 2006;103:16101–4.
[b] Paineau E, Antonova K, Baravian C, Bihannic I, Davidson P, Dozov I, Impéror-Clerc M, Levitz P, Madsen A, Meneau F, Michot LJ. J Phys Chem B 2009;113:15858–69.
[c] Miyamoto N, Iijima H, Ohkubo H, Yamauchi Y. Chem Commun 2010;46:4166–8.

[35] [a] Nakato T, Miyamoto N, Harada A. Chem Commun 2004:78–9.
[b] Nakato T, Yamashita Y, Kuroda K. Thin Solid Films 2006;495:24–8.

[36] Miyamoto N, Yamamoto S, Shimasaki K, Harada K, Yamauchi Y. Chem Asian J 2011;6:2936–9.
[37] [a] Kim JE, Han TH, Lee SH, Kim JY, Ahn CW, Yun JM, et al. Angew Chem Int Ed 2011;50:3043–7.
[b] Xu Z, Gao C. ACS Nano 2011;5:2908–15.
[38] Mejia AF, Chang Y-W, Ng R, Shuai M, Mannan MS, Cheng Z. Phys Rev E 2012;85: 061708.
[39] [a] Inadomi T, Ikeda S, Okumura Y, Kikuchi H, Miyamoto N. Macromol Rapid Commun 2014;35:1741–6.
[b] Dozov I, Paineau E, Davidson P, Antonova K, Baravian C, Bihannic I, et al. J Phys Chem B 2011;115:7751–65.
[c] Nakato T, Nakamura K, Shimada Y, Shido Y, Houryu T, Iimura Y, et al. J Phys Chem C 2011;115:8934–9.
[40] [a] Commeinhes X, Davidson P, Bourgaux C, Livage J. Adv Mater 1997;9:900–3.
[b] Desvaux H, Gabriel J-CP, Berthault P, Camerel F. Angew Chem Int Ed 2001;40: 373–5.
[c] Michot LJ, Bihannic I, Maddi S, Funari SS, Baravian C, Levitz P, et al. Proc Natl Acad Sci USA 2006;103:16101–4.
[d] Porion P, Faugere AM, Michot LJ, Paineau E, Delville A. J Phys Chem C 2011; 115:14253–63.
[e] de Azevedo EN, Engelsberg M, Fossum JO, de Souza RE. Langmuir 2007;23: 5100–5.
[f] Paineau E, Antonova K, Baravian C, Bihannic I, Davidson P, Dozov I, et al. J Phys Chem B 2009;113:15858–69.
[g] Eguchi M, Angelone MS, Yennawar HP, Mallouk TE. J Phys Chem C 2008;112: 11280–5.
[41] [a] Murata K, Haraguchi K. J Mater Chem 2007;17:3385–8.
[b] Haraguchi K, Li H-J, Song L, Murata K. Macromolecules 2007;40:6973–80.
[42] [a] Ma X, Li Y, Wang W, Ji Q, Xi Y. Eur Polym J 2013;49:389–96.
[b] Liu M, Ishida Y, Ebina Y, Sasaki T, Aida T. Nat Commun 2013;4:2029.
[c] Liu J, Song G, He C, Wang H. Macromol Rapid Commun 2013;34:1002–7.
[d] Fan J, Shi Z, Lian M, Li H, Yin J. J Mater Chem A 2013;1:7433.
[e] Shen J, Yan B, Li T, Long Y, Li N, Ye M. Soft Matter 2012;8:1831–6.
[43] Haraguchi K, Takada T. Macromolecules 2010;43:4294–9.
[44] Paineau E, Dozov I, Bihannic I, Baravian C, Krapf M-EM, Philippe A-M, et al. ACS Appl Mater Interfaces 2012;4:4296–301.
[45] Miyamoto N, Shintate M, Ikeda S, Hoshida Y, Yamauchi Y, Motokawa R, et al. Chem Commun 2013;49:1082–4.
[46] Mejia AF, Ng R, Nguyen P, Shuai M, Acosta HY, Mannan MS, et al. Soft Matter 2013;9:10257.
[47] Liu M, Ishida Y, Ebina Y, Sasaki T, Hikima T, Takata M, et al. Nature 2015;517: 68–72.
[48] Sasai R, Itoh T, Ohmori W, Itoh H, Kusunoki M. J Phys Chem C 2009;113:415–21.
[49] Suzuki Y, Matsunaga R, Sato H, Kogure T, Yamagishi A, Kawamata J. Chem Commun 2009:6964–6.

Nanoarchitectonics of Bio-Materials and for Biomedicals

Engineering DNA Molecules for Morphological Reconfiguration

K. Cervantes-Salguero and S. Murata
Tohoku University, Sendai, Miyagi, Japan

CHAPTER OUTLINE

On the basis of hybridization of DNA molecules, nucleic acid nanotechnology enables us to rationally design and build various nanostructures and nanodevices. Usually these DNA nanostructures are designed as a thermodynamically stable state, therefore they are intrinsically static [1]. These static DNA nanostructures have been utilized as a building block self-assembled into a platform to immobilize diverse molecules and nanoparticles. On the other hand, dynamic reconfiguration of DNA nanostructures is getting more attention because it can change its shape and properties, according to external commands or environmental change. Here, the term "reconfiguration" is simply defined as a change in the conformation of the structure. In this chapter, we review some methods for reconfigurable DNA nanostructures along with their typical examples, and in particular we describe a reconfigurable building block able to self-assemble into higher-order structures.

8.1 ENGINEERING STATIC DNA NANOSTRUCTURES

Although DNA is the most important material, which stores all the genetic information of living organisms, it is also recognized as a useful material for

Supra-materials Nanoarchitectonics. http://dx.doi.org/10.1016/B978-0-323-37829-1.00008-0

building artificial structures at nanoscale. For this purpose synthetic DNA molecules with desired base sequence are widely utilized.

Before the review, we would like to summarize some important properties of DNA. DNA is one of the most well-understood biomolecules. Especially, we can predict its thermodynamic properties such as secondary structures fairly accurately [2]. DNA exists in the form of single-stranded DNA (ssDNA or a strand) or double-stranded DNA (dsDNA or a duplex) in aqueous solution. The backbone of a DNA strand is made of two alternating subunits [sugars (deoxyribose) and phosphate groups] covalently joined each other. At each sugar, one nucleobase is attached, which gives a particular sequence to the strand. Such a sequence is composed of four nucleobases: guanine, adenine, thymine, and cytosine, represented as G, A, T, and C, respectively. The backbone has a directionality defined as $5' \rightarrow 3'$ (from the $5'$ carbon at one sugar to the $3'$ carbon at another sugar). In order to form a DNA duplex, the sequences of both strands must be complementary to each other, namely: G binds to C, and A to T (Fig. 8.1A). This complementarity pairing is called Watson–Crick base pairing. In addition to the base pairing, the two strands must have opposite directions. With such "molecular rules," we are able to design DNA nanostructures.

There are several strategies to construct static DNA nanostructures. The first strategy is called DNA tiling. In this strategy, basic elements called DNA tiles or DNA motifs self-assemble into a larger structure. A tile is an arrangement of several DNA duplexes connected to each other by crossover junctions. The crossover junction is found as Holliday junction in biological cells, which is an intermediate state in the genetic recombination process. The Holliday junction migrates along the strand in a random fashion (Fig. 8.1B). When we utilize the junctions as a structural component, they must be immobilized to make a well-defined structure. This is done by modifying the base sequences around the junction (Fig. 8.1C). In a DNA tile, two immobilized Holliday junctions define the shape of the complex (so it is called a double crossover molecule [3]). Four ends of the helices are decorated with short single-stranded portions called sticky ends, which define connectivity between the tiles. DNA tiles have been widely used as a basic building block for self-assembling 2D crystals [4,5] (Fig. 8.1D). The self-assembly of DNA tiles is sometimes modeled as a process similar to that of crystal growth; namely, monomers (DNA tiles) bind each other to form a nucleate and then grow into a large planar structure. However, more specific interactions among the monomers can be introduced in DNA tiles through base sequence design at the sticky end. DNA tiles are suitable to make periodic 2D patterns with small kinds of DNA strands at low cost.

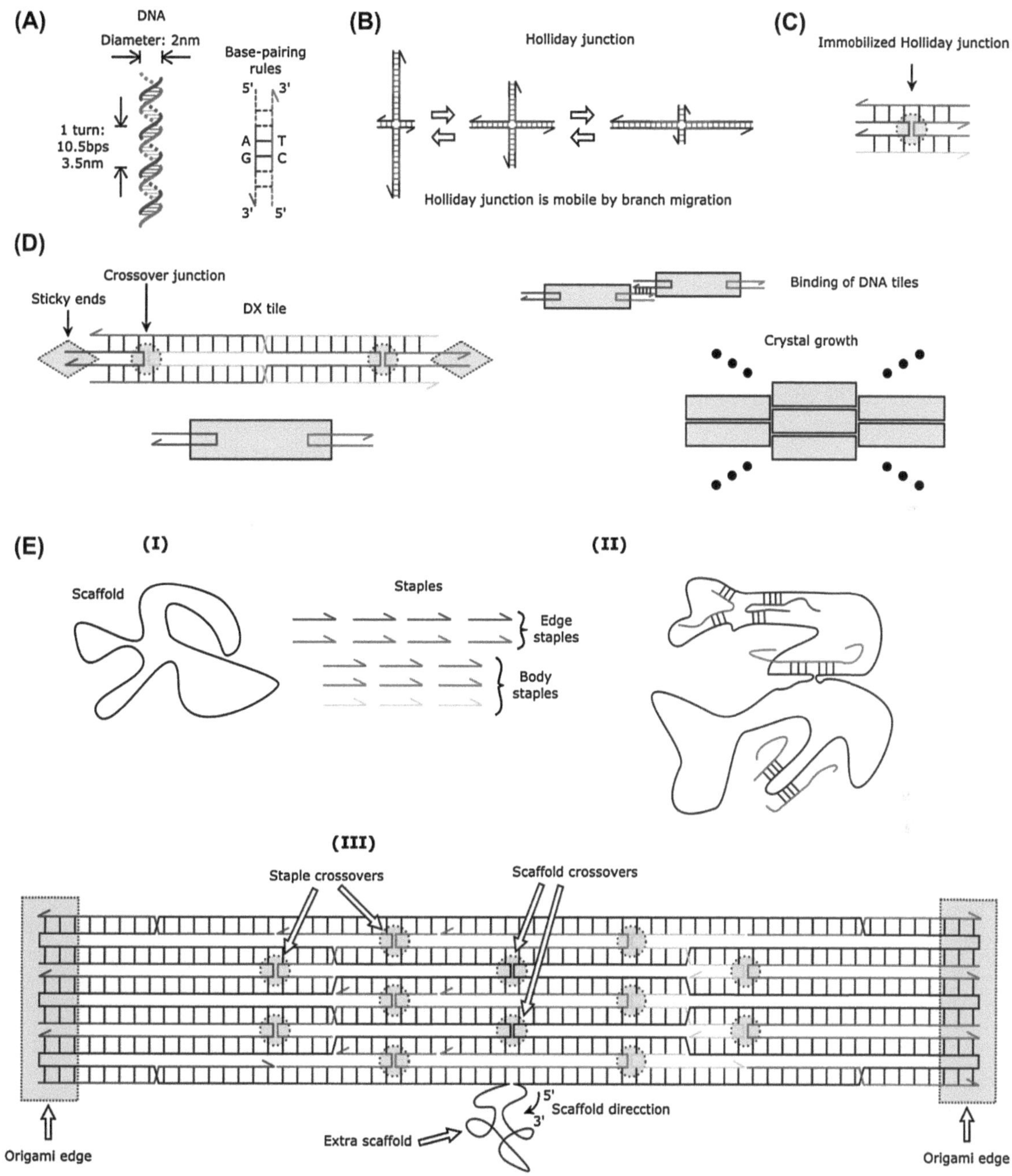

■ **FIGURE 8.1** Various methods of DNA nanofabrication. (A) Left: molecular structure of DNA. The diameter is about 2 nm and a turn of DNA duplex consists of 10.5 base pairs, which makes 3.5 nm. Right: The Watson–Crick base pairing. (B) Holliday junction and its branch migration. (C) Immobilized Holliday junction is the basis for most of the DNA nanostructures. (D) Double crossover tile made of two crossovers and four sticky ends. Self-assembled crystal by binding between sticky ends. (E) DNA origami. (I) A scaffold and a set of staples. (II) By annealing, staples bind to the scaffold and then fix separate sections in the scaffold. (III) Final structure after annealing.

However, depending on the application, we need aperiodic structures and then different strategies should be applied.

The second strategy is called DNA origami. Currently, DNA origami is the most powerful technique to assemble nanostructures with determined sizes and shapes [6–8]. This technique is suitable when structures are required in mega Dalton range. The same as DNA tiles, it exploits the specificity of molecular recognition of DNA. A DNA origami is composed of a long circular single-stranded DNA and hundreds of short single-stranded DNAs called a "scaffold" and "staples," respectively. In principle, staples fold the scaffold into a desired conformation by hybridization to its complementary section, until the whole structure becomes double stranded, which is thermodynamically most stable (Fig. 8.1C). The typical self-assembly protocol for DNA origami requires thermal annealing of a mixture containing the scaffold and the staples. Typical annealing profile usually starts at 95°C, where all the strands exist as single strands, and then decreases gradually to lower temperatures such as 4°C. The scaffold can be as large as ~51 K nucleotides [9], while the most commonly used scaffold, M13mp18 bacteriophage has 7249 nucleotides. The staples, which are synthetic, have different lengths depending on the origami design, usually consisting of less than 50 nucleotides. The methodology of DNA origami is so general that various ways for routing and arranging the scaffold and staples have been demonstrated to make 2D, 3D, and curved nanostructures [10–15]. Furthermore, designing DNA origamis with computer-aided design programs such as caDNAno [16] assists the process of designing DNA nanostructures.

8.2 ENGINEERING DNA NANOSTRUCTURES WITH DYNAMIC MORPHOLOGY

In general, "reconfiguration" is defined as the progressive change in relative position of component parts of the structure during a certain time period. Typically, rigid elements with particular shapes are connected by joints or flexible components to allow the change, and the rigid elements have shapes that can fit specifically between each other in order to achieve mechanical strength. In some cases, components can include some flexibility or compliance. Such properties can be realized at molecular scale by arranging shape and stiffness of DNA nanostructures.

In the design of reconfigurable nanostructure, DNA duplexes and bundles of duplexes are employed as stiff components, while single strands are used as flexible elements. The stiffness of DNA is characterized as the persistence length, and their elastic responses are predictable by the free-jointed chain or the worm-like chain models [17]. Usually, flexible elements

Table 8.1 Driving Strategies to Reconfigure DNA Nanostructures

External forces	
	Strand-exchange (single-stranded short DNA input)[a]
	BZ-transition (salt concentration)
	Photo-responsive bases (light)
	Electric/magnetic fields
Autonomous	
	Enzymatic reactions (DNAzyme, various enzymes)[a]
	Strand-exchange[a]
	Thermal fluctuation (entropic spring)

[a] *Specific to DNA base sequence, otherwise nonspecific.*

can simply be regarded as linear springs following Hookean law if they are not overextended. Arranging desired stiffness over the structure usually requires trial and error [18–21].

The reconfiguration is driven by some forces classified by the source of force (Table 8.1). The external forces include response to additional molecules such as ions, short DNA strands, proteins, and other stimuli. Other stimuli include light, electrical signals, and temperature. Another category is "autonomous," which does not need forces. Among the autonomous methods DNAzymes and various enzymes permit a wide range of operations but usually are sensitive to environments, e.g., temperature and ionic concentrations [22].

In order to drive a specific portion of the structure, driving strategies with some specificity are needed. Toehold-mediated strand displacement [23–25] is often utilized for base-sequence-specific actuation. It is possible to replace a specific strand in the structure with another strand, which makes a desired reconfiguration. In the displacement, an input DNA strand binds to a duplex sequence specifically and, subsequently, an output strand is released into solution (Fig. 8.2A). Driving force of this process is given by free energy of hybridizing base pairs, where complex Y is more stable than complex X. Once the process is completed, it cannot be reversed.

DNA tweezers is an early example of an actuated DNA nanostructure driven by the toehold-mediated strand displacement (Fig. 8.2A, [23]). In DNA tweezers, a short ssDNA segment connecting two dsDNA segments (arms) served as a joint of the tweezers. The closed and open states of the tweezers can be observed by using fluorescence resonance energy transfer (FRET). The tweezers work as a bistate machine where transition between the states is fueled by additional DNA strands that trigger the

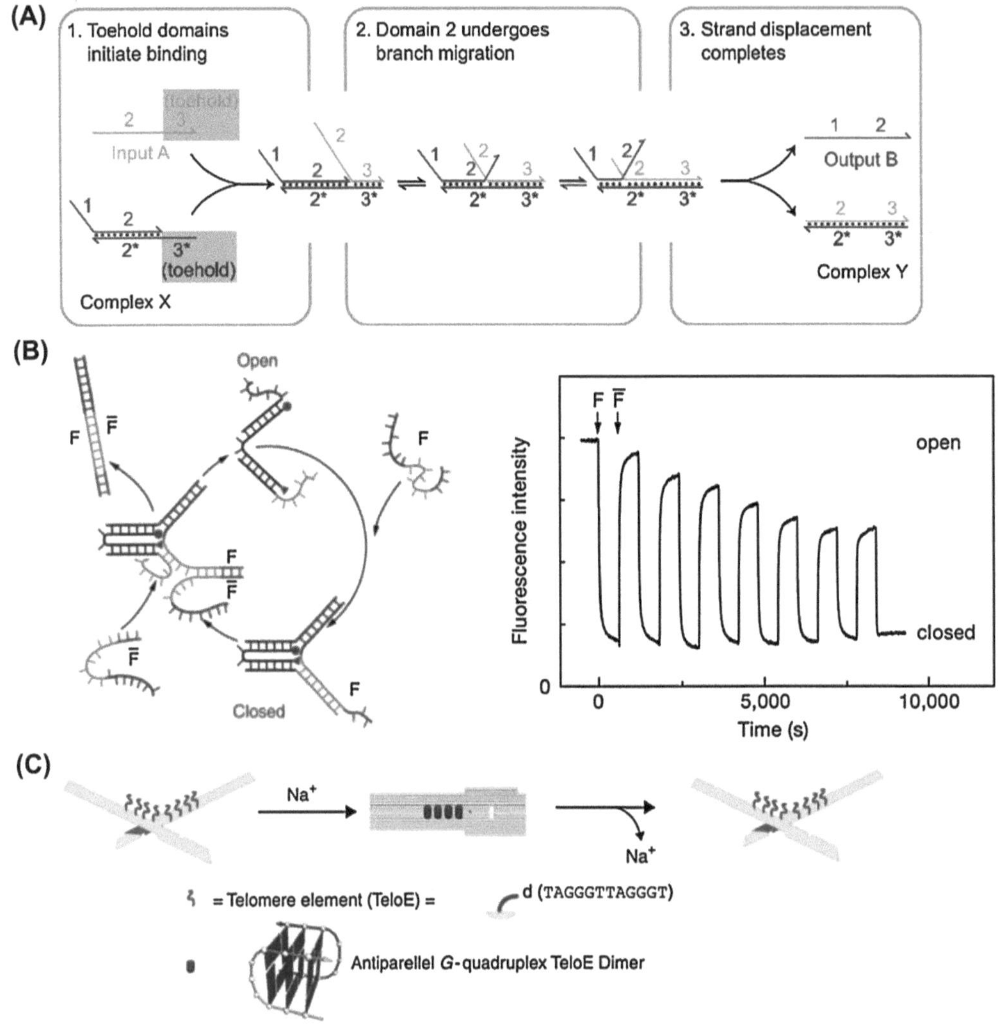

■ **FIGURE 8.2** Methods for driving DNA nanostructures. (A) Toehold-mediated strand displacement. (Reprinted with permission of Nature Publishing Group, copyright 2011 [24].) (B) DNA tweezers. Left: mechanism of DNA tweezers. Reconfiguration is fueled by adding single-stranded DNAs (fuel and antifuel) alternately. The *red circle* is a fluorescent molecule and the red triangle is a quencher to report the reconfiguration by fluorescence resonance energy transfer (FRET). Right: working cycle of tweezers measured by FRET. In an open state fluorescence intensity is high; in a closed state it is low. (Reprinted with permission of Nature Publishing Group, copyright 2000 [23].) (C) "Molecular beacon" made of DNA actuated by addition of ions and other molecules. Addition of Na^+ closes the DNA nanostructure and opens it after their removal. (Reprinted with permission of Nature Publishing Group, copyright 2011 [26].)

toehold-mediated strand displacement. Transformation from open to closed state occurs when an ssDNA called a fuel strand is added. Reversing to the open state is done by adding a complementary strand to the fuel (i.e., antifuel). By applying fuel and antifuel alternatively, it is able to make a cyclic reaction (right in Fig. 8.2B). Other methods for actuating DNA nanostructures have also been explored including DNA-based "single-molecule beacon" driven by Na^+ concentration [26] (Fig. 8.2B).

Embedding functional molecules is another strategy to achieve reconfiguration. For instance, a folding process of DNA origami sheet can be controlled by hydrophobicity of lipid molecules decorated on the surface of the sheet [27]. In situ lipid coating of DNA nanostructures also drives the migration of the nanostructures from hydrophilic to hydrophobic regions [28]. Using artificial photoresponsive nucleobases to control hybridization is also examined [29,30]. It is advantageous because (1) it is free from contamination by waste molecules used in the driving operation, and (2) a fast response can be achieved by designing the machinery as an intramolecular reaction [30]. Another example is the incorporation of methylated nucleobases in a DNA nanomotor for achieving rotation. The methylated bases underwent a B-Z transition when triggered by a change in the Mg^{2+} concentration [31].

Applying electric and magnetic fields are other ways to drive DNA nanostructures. In this category, methods range from harnessing electrically driven diffusion for molecular computation [32], orientating rotational nanomotors by electric fields [33], controlling the reconfiguration of DNA hairpins by applying magnetic fields [34], and magnetic-driven separation of DNA molecules for artificial self-replication [35].

Constrained motion of nanostructures with movable parts by thermal fluctuation have been reported [18–20,36–38]. The motion is not controllable but constrained along programmed paths [35].

8.3 CONSTRAINED MOTION OF DNA NANOSTRUCTURES IN THERMODYNAMIC EQUILIBRIUM

Our group worked on self-assembly of a molecular cluster made of DNA nanostructure [19]. Here, the number of monomers in a cluster can be controlled by the mechanical/elastic properties of a monomer molecule that is made by DNA origami method. The employed strategy is to allow a building block (a monomer) to change its shape under thermal fluctuation. The resultant shapes of the blocks define the number of blocks in the cluster.

1. Cluster formation by reconfiguring blocks
   Clusters self-assembled by the shape-variable blocks can contain a limited number of monomers, and thus their configurations can have multiple forms (hereinafter referred to as polymorphic). We

restrict ourselves to use homogeneous blocks, namely one kind of blocks, to make various cluster configurations.

The configuration of a cluster is determined by the connectivity of a block, i.e., the number of bonds and their orientations. In general, N monomers can take C_N-kinds of configurations where connection among the monomers depends on each monomer's position in the cluster. These monomers must be homogeneous as we assumed and have n bonding arms (A_i, $1 \leq i \leq n$). Relative angles between each pair of arms must be in the defined limit (θ_i, $1 \leq i \leq n$). In some configurations, not all the arms of a monomer in cluster C_N are connected. Thus, the connected arms per monomer may be

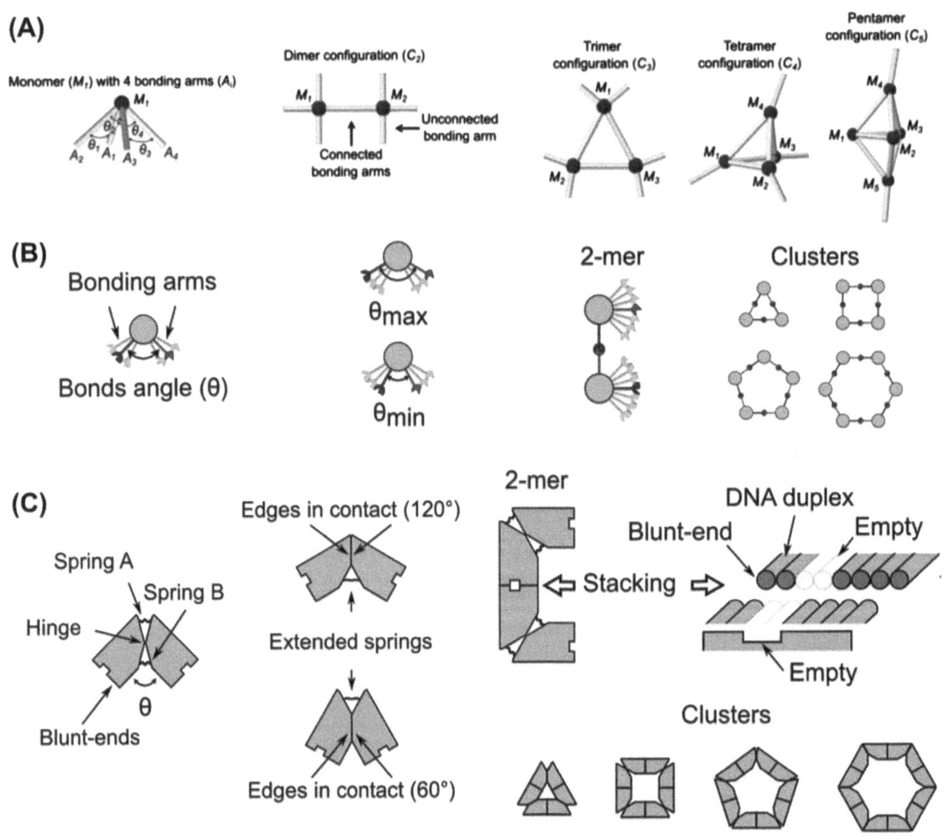

■ **FIGURE 8.3** Concept of shape-variable building blocks. (A) 3D abstract representation of shape-variable monomer (in *purple*) with four bonding arms (in *gray*, A_i, $1 \leq i \leq 4$) and its 3D polymorphic cluster configurations (C_N, $2 \leq N \leq 5$). θ_i indicates the angle between arms A_i and A_{i+1}. M_k ($1 \leq k \leq N$) labels the monomer within the configuration C_N (B) 2D abstract representation of the shape-variable monomer for assembling polymorphic ring-shaped clusters. (Figure is modified from the original article [18].) (C) Monomer geometry that satisfies the angular conditions.

lesser than n; in such a case unused arms should be "inactive" to prevent unspecific bonding (Fig. 8.3A).

For simplicity, we consider 2D case hereafter, where the monomers have only two bonding arms and the angle between bonding arms varies within a certain range. These arms have complementary bonding edges like the sticky end, and make bonds to each other in such a way that a dimer can take a *cis* configuration while preventing a *trans* configuration (Fig. 8.2B).

2. Design of shape-variable block

DNA origami technique [6,8,16] is used to implement the blocks. There are several key properties in the implementation (Fig. 8.3C). The angle between arms must stay between a minimum and a maximum. This constraint is realized by giving a particular shape to the block and a hinge connecting two movable parts in the block. The shape of the parts (oblique cuts) ensures the angle between the connecting arms, 60° and 120°. To keep the angular constraint within the range and to prevent undesired torsional and

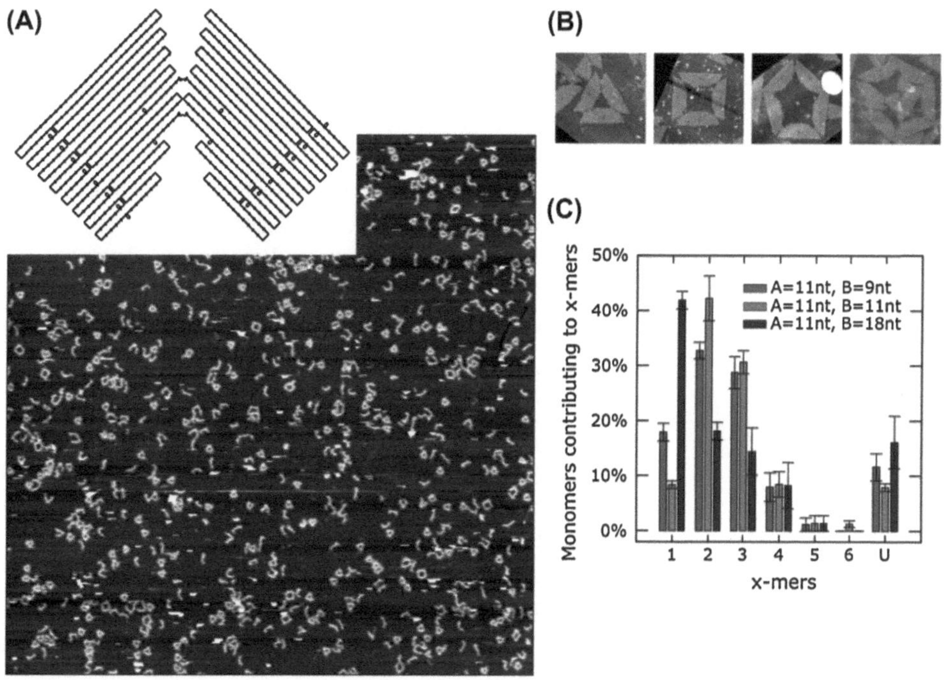

■ **FIGURE 8.4** Polymorphic clusters made of shape-variable building blocks. (A) Atomic force microscopy (AFM) image of the polymorphic clusters deposited on mica substrate. Inset: scaffold arrangement in the origami monomer. (B) Formation of 3-mer, 4-mer, 5-mer, and 6-mer clusters. (C) Size distribution of the clusters. [AFM images in (A) and (B) are 1 × 1 μm and 310 × 300 nm, respectively.] (Figures (B) and (C) are modified from the original article [18].)

rotational motion induced by thermal fluctuation, two single strands A and B (which is a section of origami scaffold) connect two parts. These single strands behave as entropic springs (i.e., tend to shrink to globular state with higher entropy). Stiffness of the springs is tuned by the number of nucleotides in the spring. (Other applications of single-stranded DNA as a spring can be found in Refs. [20,21].) The molecular recognition between bonding arms was programmed by the stacking interaction between blunt ends following a method in [8] (Fig. 8.3C).

3. Experimental demonstration
   Fig. 8.4A and B shows the result of formation experiment by using an ordinary origami annealing protocol. The size of self-assembled clusters depends on a strength of springs. Fig. 8.4C shows the cluster size distribution (number of monomers contributing to each type of cluster) for different lengths of spring B, while keeping length of A constant. The length of spring B affects the numbers of small size clusters ($x \leq 3$), however, larger clusters appear to be constant ($x \geq 4$).

8.4 CONCLUSIONS

Rapidly developing DNA nanotechnology has enabled us to realize nanosystems capable of structural reconfiguration. In this review, after introducing strategies for DNA nanostructure fabrication including DNA tiles and DNA origamis, an approach to realize size-controllable formation of cluster made of DNA origami block is described. The self-assembled cluster size depends on the block shape and mechanical flexibility of the block introduced by entropic springs. It may not be difficult to expand the variety of shape-variable blocks for assembling other clusters and crystals.

Due to space limitation, we could not mention some important issues such as sensing mechanical forces by deformation of DNA nanostructures [39,40]. We envision that driving nanodevices with movable parts by external forces may provide a new way of fuel-free operation. It may be interesting to study how mechanical forces could actuate an assembly process of building blocks. Well-designed nanomechanical structures coupled with molecular computation is one of the future directions toward intelligent molecular machineries.

REFERENCES

[1] Padirac A, Fujii T, Rondelez Y. Nucleic acids for the rational design of reaction circuits. Curr Opin Biotechnol 2013;24:575–80.

[2] Bloomfield VA, Crothers DM, Tinoco Jr I. Nucleic acids: structures, properties and functions. Mill Valley, CA: University Science Press; 1998.

[3] Seeman NC. DNA in a material world. Nature 2003;421:427–31.

[4] Winfree E, Liu F, Wenzler LA, Seeman NC. Design and self-assembly of two-dimensional DNA crystals. Nature 1998;394:539–44.

[5] Evans CG, Hariadi RF, Winfree E. Direct atomic force microscopy observation of DNA tile crystal growth at the single-molecule level. J Am Chem Soc 2012;134: 10485–92.

[6] Rothemund PW. Folding DNA to create nanoscale shapes and patterns. Nature 2006; 440:297.

[7] Castro CE, Kilchherr F, Kim D-N, Shiao EL, Wauer T, Wortmann P, et al. A primer to scaffolded DNA origami. Nat Meth 2011;8:221–9.

[8] Woo S, Rothemund PW. Programmable molecular recognition based on the geometry of DNA nanostructures. Nat Chem 2011;3:620.

[9] Marchi AN, Saaem I, Vogen BN, Brown S, LaBean TH. Toward larger DNA origami. Nano Lett 2014;14:5740–7.

[10] Dietz H, Douglas SM, Shih WM. Folding DNA into Twisted and curved nanoscale shapes. Science August 7, 2009;325(5941):725–30. http://dx.doi.org/10.1126/science.1174251.

[11] Han D, Pal S, Nangreave J, Deng Z, Liu Y, Yan H. DNA origami with complex curvatures in three-dimensional space. Science April 15, 2011;332(6027):342–6. http://dx.doi.org/10.1126/science.1202998.

[12] Ke Y, Douglas SM, Liu M, Sharma J, Cheng A, Leung A, et al. Multilayer DNA origami packed on a square lattice. J Am Chem Soc 2009;131(43):15903–8. http://dx.doi.org/10.1021/ja906381y.

[13] Ke Y, Voigt NV, Gothelf KV, Shih WM. Multilayer DNA origami packed on hexagonal and hybrid lattices. J Am Chem Soc 2012;134(3):1770–4. http://dx.doi.org/10.1021/ja209719k.

[14] Zhang F, Jiang S, Wu S, Li Y, Mao C, Liu Y, et al. Complex wireframe DNA origami nanostructures with multi-arm junction vertices. Nat Nanotechnol 2015;10:779–84. http://dx.doi.org/10.1038/nnano.2015.162.

[15] Benson E, Mohammed A, Gardell J, Masich S, Czeizler E, Orponen P, et al. DNA rendering of polyhedral meshes at the nanoscale. Nature July 23, 2015;523:441–4. http://dx.doi.org/10.1038/nature14586.

[16] Douglas SM, Dietz H, Liedl T, Högberg B, Graf F, Shih WM. Self-assembly of DNA into nanoscale three-dimensional shapes. Nature May 21, 2009;459:414–8. http://dx.doi.org/10.1038/nature08016.

[17] Li J, Wijeratne SS, Qiu X, Kiang CH. DNA under force: mechanics, electrostatics, and hydration. Nanomaterials 2015;5(1):246.

[18] Zhou L, Marras AE, Su H-J, Castro CE. Direct design of an energy landscape with bistable DNA origami mechanisms. Nano Lett 2015;15(3):1815–21. http://dx.doi.org/10.1021/nl5045633.

[19] Cervantes-Salguero K, Hamada S, Nomura SM, Murata S. Polymorphic ring-shaped molecular clusters made of shape-variable building blocks. Nanomaterials 2015;5: 208–17. http://dx.doi.org/10.3390/nano5010208.

[20] Zhou L, Marras AE, Su H-J, Castro CE. DNA origami compliant nanostructures with tunable mechanical properties. ACS Nano 2014;8(1):27–34. http://dx.doi.org/10.1021/nn405408g.

[21] Liedl T, Högberg B, Tytell J, Ingber DE, Shih WM. Self-assembly of three-dimensional prestressed tensegrity structures from DNA. Nat Nanotechnol 2010;5: 520–4.

[22] Baccouche A, Montagne K, Padirac A, Fujii T, Rondelez Y. Dynamic DNA-toolbox reaction circuits: a walkthrough. Methods 67:234–249.

[23] Yurke B, Turberfield AJ, Mills AP, Simmel FC, Neumann JL. A DNA-fuelled molecular machine made of DNA. Nature August 10, 2000;406.

[24] Zhang DY, Winfree E. Control of DNA strand displacement Kinetics using toehold Exchange. J Am Chem Soc 2009;131(47):17303–14.

[25] Zhang DY, Seelig G. Dynamic DNA nanotechnology using strand-displacement reactions. Nat Chem 2011;3:103–13. http://dx.doi.org/10.1038/nchem.957.

[26] Kuzuya A, Sakai Y, Yamazaki T, Xu Y, Komiyama M. Nanomechanical DNA origami 'single-molecule beacons' directly imaged by atomic force microscopy. Nat Commun 2011;2. http://dx.doi.org/10.1038/ncomms1452. Article number: 449.

[27] List J, Weber M, Simmel FC. Hydrophobic actuation of a DNA origami bilayer. Struct Angew Chem Int Ed 2014;53:4236–9.

[28] Yonamine Y, Cervantes-Salguero K, Nakanishi W, Kawamata I, Minami K, Komatsu H, et al. In situ 2D-extraction of DNA wheels by 3D through-solution transport. Phys Chem Chem Phys 2015;17:32122–5.

[29] Liang X, Zhou M, Kato K, Asanuma H. Photoswitch nucleic acid catalytic activity by regulating topological structure with a universal supraphotoswitch. ACS Synth Biol 2013;2(4):194–202.

[30] Kamiya Y, Asanuma H. Light-driven DNA nanomachine with a photoresponsive molecular engine. Acc Chem Res 2014;47(6):1663–72.

[31] Rajendran A, Endo M, Hidaka K, Sugiyama H. Direct and Real-time observation of rotary movement of a DNA nanomechanical device. J Am Chem Soc 2013;135: 1117–23.

[32] Braich RS, Chelyapov N, Johnson C, Rothemund PWK, Adleman L. Solution of a 20-variable 3-SAT problem on a DNA computer. Science 2002;296(5567): 499–502. http://dx.doi.org/10.1126/science.1069528.

[33] Klapper Y, Sinha N, Ng TWS, Lubrich D. A rotational DNA nanomotor driven by an externally controlled electric field. Small 2010;6:44–7.

[34] Hamad-Schifferli K, Schwartz JJ, Santos AT, Zhang S, Jacobson JM. Remote electronic control of DNA hybridization through inductive coupling to an attached metal nanocrystal antenna. Nature January 10, 2002;415:152–5. http://dx.doi.org/10.1038/415152a. Received July 12 2001; Accepted November 16 2001.

[35] Wang T, Sha R, Dreyfus R, Leunissen ME, Maass C, Pine DJ, et al. Self-replication of information-bearing nanoscale patterns. Nature October 13, 2011;478:225–8. http://dx.doi.org/10.1038/nature10500.

[36] Kuzuya A, Yuichi Ohya Y. Nanomechanical molecular devices made of DNA origami. Acc Chem Res 2014;47(6):1742–9.

[37] Castro CE, Su H-J, Marras AE, Zhou L, Johnson J. Mechanical design of DNA nanostructures. Nanoscale 2015;7:5913–21. http://dx.doi.org/10.1039/C4NR07153K.

[38] Marras AE, Zhou L, Su H-J, Castro CE. Programmable motion of DNA origami mechanisms. Proc Natl Acad Sci USA 2015;112:713–8.

[39] Zhang Y, Ge C, Zhu C, Salaita K. DNA-based digital tension probes reveal integrin forces during early cell adhesion. Nat Commun 2014;5. http://dx.doi.org/10.1038/ncomms6167. Article number: 5167.

[40] Blakely BL, Dumelin CE, Trappmann B, McGregor LM, Choi CK, Anthony PC, et al. A DNA-based molecular probe for optically reporting cellular traction forces. Nat Methods 2014;11:1229–32. http://dx.doi.org/10.1038/nmeth.3145.

Smart Polymers With Nanoarchitectonics

M. Ebara
National Institute for Materials Science (NIMS), Tsukuba, Japan

CHAPTER OUTLINE

9.1 INTRODUCTION

There has been a renewal of interest in controlling "nanoarchitectures," and a considerable number of studies have been conducted on designing of novel materials to meet these applications [1–3]. Some special types of polymers, for example, have emerged as a very useful class of polymers and have their own special chemical properties and applications in various areas [4–6]. These polymers are coined with different names, based on their physical or chemical properties, like *stimuli-responsive* or *environmentally sensitive* or *smart* or *intelligent* polymers. We shall use the name *smart* polymers for such polymer systems in this chapter. The characteristic feature that actually makes them "smart" is their ability to respond to very slight

Supra-materials Nanoarchitectonics. http://dx.doi.org/10.1016/B978-0-323-37829-1.00009-2

changes in the surrounding environment. The uniqueness of these materials lies not only in the fast macroscopic changes occurring in their structure but also that these transitions are reversible. The responses are manifested as changes in one or more of the following: shape, surface characteristics, solubility, formation of an intricate molecular assembly, a sol-to-gel transition, and others [7]. The environmental trigger behind these transitions can be either change in temperature or pH shift, increase in ionic strength, presence of certain chemicals, etc. More recently, changes in electric and magnetic field, light or radiation forces have also been reported as stimuli for these polymers. In this chapter, I describe some biomedical applications of smart polymers.

9.2 CLASSIFICATION ON THE BASIS OF STIMULI

9.2.1 Physical Stimuli

Smart polymers can be classified as either physical- or chemical-stimuli-responsive polymers. The physical stimuli, such as temperature, electric or magnetic fields, and mechanical stress, will affect the level of various energy sources and alter molecular interactions at critical onset points. Chemical stimuli, such as pH, ionic factors, and chemical agents, will change the interactions between polymer chains or between polymer chains and solvent at the molecular level. Recently, biochemical stimuli have been considered as another category that involves the responses to antigen, enzyme, ligand, and other biochemical agents. Some systems have been developed to combine two stimuli-responsive mechanisms into one polymer system, the so-called dual-responsive polymer systems.

Temperature-responsive polymers are probably the most commonly studied class of environmentally sensitive polymer systems. Of the many temperature-responsive polymers, poly(*N*-isopropylacrylamide) (PNIPAAm) is probably the most extensively used because its lower critical solution temperature (LCST) is close to body temperature. Copolymers of NIPAAm can also be made using other monomers to alter the LCST. PNIPAAm can take many forms: dissolved in aqueous solution, adsorbed or grafted on aqueous−solid interfaces, or cross-linked in the form of hydrogels. Grafting of PNIPAAm onto solid surfaces makes them "smart" surfaces where the surface energy can be controlled with temperature [8−10]. PNIPAAm also can be functionalized by chemical conjugation to biomolecules such as a targeting ligand to yield a large and diverse family of polymer-biomolecule hybrid systems that can respond to biological as well as to physical and chemical stimuli [11,12]. Although these grafting or conjugation processes can produce "smart" surfaces or "smart" conjugations, derivatizations of PNIPAAm side chains or end groups to form reactive groups have resulted in disappearance of the "smartness" from PNIPAAm. For example, if NIPAAm monomer is copolymerized with a reactive monomer such as acrylic acid (AAc), the LCST increases and may even disappear [13]. Based on the hypothesis that an appropriate proportion of hydrophobicity and hydrophilicity in the molecular structure of PNIPAAm is required for the sharp phase transition to occur, the authors have newly designed a series of monomers that have a parallel structure with NIPAAm monomer to deceive water molecules into behaving like those around PNIPAAm chains even though they have very hydrophilic reactive groups such as carboxyl, amino, and hydroxyl groups, i.e., 2-carboxy-isopropylacrylamide (CIPAAm) [14–16], 2-amino-isopropylacrylamide (AIPAAm) [17], and 2-hydroxy-isopropylacrylamide (HIPAAm) [18] (Fig. 9.1). Use of these monomers for copolymerization with NIPAAm can synergistically combine the individual properties of the two or more components to yield new and desirable properties. One could say that these polymers are "double or multismart."

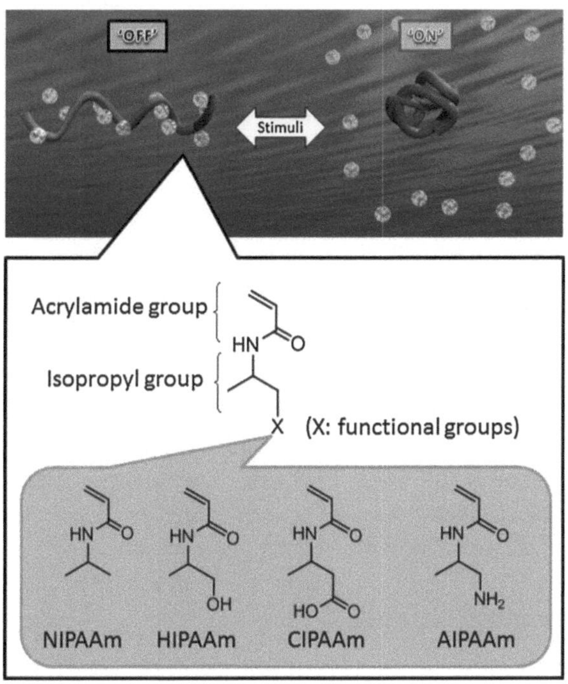

■ **FIGURE 9.1** Schematic illustration of temperature-responsive phase transition behavior of poly (*N*-isopropylacrylamide) (PNIPAAm) in aqueous solution and chemical structures of newly designed *N*-isopropylacrylamide-based functional monomers.

Photoresponsive polymers also have been used in a number of biotechnology applications, such as photocontrolled enzymatic bioprocessing [19], phototriggered targeted drug delivery systems [20], and photocontrolled separation/recovery systems in bioMEMs formats. Since the light stimulus can be imposed instantly and delivered in specific amounts with high accuracy, light-sensitive hydrogels may possess special advantages over others [21]. For example, the sensitivity of temperature-sensitive hydrogels is rate limited by thermal diffusion, while pH-sensitive hydrogels can be limited by hydrogen ion diffusion. The capacity for instantaneous delivery of the sol−gel stimulus makes the development of light-sensitive hydrogels important for various applications in both engineering and biochemical fields. Typical photoreactive guests in polymers are azobenzene [22], triphenylmethane [23], and spiropyran [24] groups, which have been entrapped, cross-linked, and introduced as side chains or part of the main chain in polymer matrices. The authors have integrated a photoinitiated proton-releasing reaction of *o*-nitrobenzaldehyde (NBA) into pH-responsive hydrogels [25]. NBA-integrated hydrogels demonstrated quick release of protons upon UV irradiation, allowing the pH inside the gel to decrease below the pK_a of the polymer within 1 min. Spatial control of gel shrinkage was also made possible by irradiating UV light to a limited region of the gel through a photomask.

Electric current can also be used as an environmental signal to induce responses of polymers. Electrosensitive hydrogels undergo shrinking or swelling in the presence of an applied electric field. Osada and Hasebe [26] reported an electrically activated artificial muscle system that is contracted by an electrical stimulus under isothermal conditions. They reported that the addition of NaCl increased the rate of water release, whereas the addition of organic solvents such as ethanol, acetone, or water reduced the rate of water release, and the contraction that resulted from the electrostatic interaction between charged macromolecules and the electrodes led to extensive dehydration of the gel. Tanaka et al. [27] studied the effect of electric current on the contraction behavior of partially hydrolyzed polyacrylamide gels in a mixture of 50% acetone and water. They observed that the contraction was most significant and rapid in water, whereas with increasing acetone percentage, the rate of contraction decreased gradually.

9.2.2 Chemical Stimuli

While physical stimuli are advantageous because they allow local and remote control, they result in a discontinuous response when the stimulus is turned "off." In other words, only the illuminated region is active, and continuous illumination is necessary. In the human body, the appearance of numerous bioactive molecules is tightly controlled to maintain a normal metabolic balance via the feedback system called homeostasis. For example, hormones or cytokines not only act locally (local signals) but also travel to other locations in the body via blood circulation (remote signals) [28]. A representative process is blood coagulation, which is a complex sequence involving numerous clotting factors. The concentration gradients of protons (pH), ions, and oxidizing or reducing agents are also important characteristics observed in living systems. The human body exhibits variations in pH along the gastrointestinal tract, tumoral areas, inflamed or infected tissues, and the endosomal lumen. From these regards, pH-responsive polymers are made of polymeric backbones with ionic pendant groups that can accept or donate protons in response to an environmental pH change. As the environmental pH changes, the degree of ionization in a pH-responsive polymer is dramatically changed at a specific pH known as pK_a or pK_b. This rapid change in the net charge of ionized pendant groups causes abrupt volume transition by generating electrostatic repulsive forces between ionized groups, which creates a large osmotic swelling force.

Glucose-sensitive hydrogels have also been developed by many researchers because one of the most challenging problems in controlled drug delivery is the development of self-regulated (modulated) insulin delivery systems. The delivery of insulin is different from the delivery of other drugs, since insulin must be delivered in an exact amount at the exact time of need. Thus, self-regulated insulin delivery systems require glucose-sensing ability and an automatic shut-off mechanism. Many hydrogel systems have been developed for modulating insulin delivery, and all of them have a glucose sensor built into the system. Kataoka and coworkers have developed a self-regulated insulin delivery system using phenylboronic acid (PBA), a synthetic molecule capable of reversibly binding with 1,2- or 1,3-*cis*-diols including glucose [29–31]. PBA and its derivatives are known to form covalent complexes with polyol compounds including glucose. The glucose-dependent shift in the equilibrium of PBA between the uncharged and anionically charged forms, when coupled with a properly amphiphilic three-dimensional backbone (or gel), could induce a reversible change in the volume of the gel. The resultant rapid change in hydration under certain conditions could cause a localized dehydration of the gel surface, that is, the so-called skin layer, thus offering a method for instantly controlling the permeation of gel-loaded insulin.

Yoshida and coworkers have demonstrated an autonomic swelling-shrinking oscillation by integrating the chemical oscillation of the Belousov–Zhabotinsky (BZ) reaction into the hydrogel [32,33]. The BZ reaction is often analogically compared with the TCA cycle, which is a key metabolic process taking place in the living body. The overall process of the BZ reaction is the oxidation of an organic substrate, such as malonic acid (MA) or citric acid, by an oxidizing agent (bromate ion) in the presence of a strong acid and a metal catalyst. A copolymer gel that consists of NIPAAm and ruthenium tris(2,2′-bipyridine) (Ru(bpy)32+) was prepared. The Ru(bpy)32+, acting as a catalyst for the BZ reaction, was appended to the polymer chains of NIPAAm. The poly(NIPAAm-*co*-Ru(bpy)32+) gels have a phase transition temperature owing to the thermosensitive constituent NIPAAm. They have further demonstrated a novel biomimetic "self-walking" gel actuator that can walk spontaneously with a worm-like motion without switching of external stimuli [23].

9.3 APPLICATIONS OF SMART POLYMERS

9.3.1 Smart Hydrogels

Smart hydrogels are promising materials for controlling drug delivery, since they change their properties in response to specific stimuli. Temperature-responsive hydrogels have been studied most extensively to obtain an "on-off" drug release profile in response to a stepwise temperature change. Okano and coworkers have achieved complete and rapid on-off regulation of drug permeation in response to stepwise temperature changes by using temperature-responsive hydrogels comprising PNIPAAm [34]. A dense gel surface layer (skin layer) is established immediately after temperature increase above the hydrogel's collapse temperature. The formed skin layer is dense enough to stop or retard the flux of water inside the gel to the outside of the gel. Gel surface skin layer formation can be controlled by changing the nanoarchitectures of polymer network, namely, the lengths of alkyl side chains on comonomers used during gel copolymerization. As mentioned before, glucose-responsive insulin release was also performed using PBA gels. Matsumoto and coworkers have developed this glucose-responsive hydrogel system to operate under physiological conditions (pH 7.4, 37°C), aiming for future use in a self-regulated insulin delivery system to treat diabetes mellitus. The release of insulin from the gel has been continuously controlled by the skin layer with close correspondence to each additional pattern of glucose (Fig. 9.2) [31].

Smart hydrogels in the form of microgels and nanogels have also been developed by many researchers because they display many advantages when they are used in biological applications. PNIPAAm gel particles are typically prepared by precipitation polymerization or emulsion polymerization. In the precipitation polymerization approach, the polymerization of monomers and cross-linkers is initiated in water by a free radical initiator

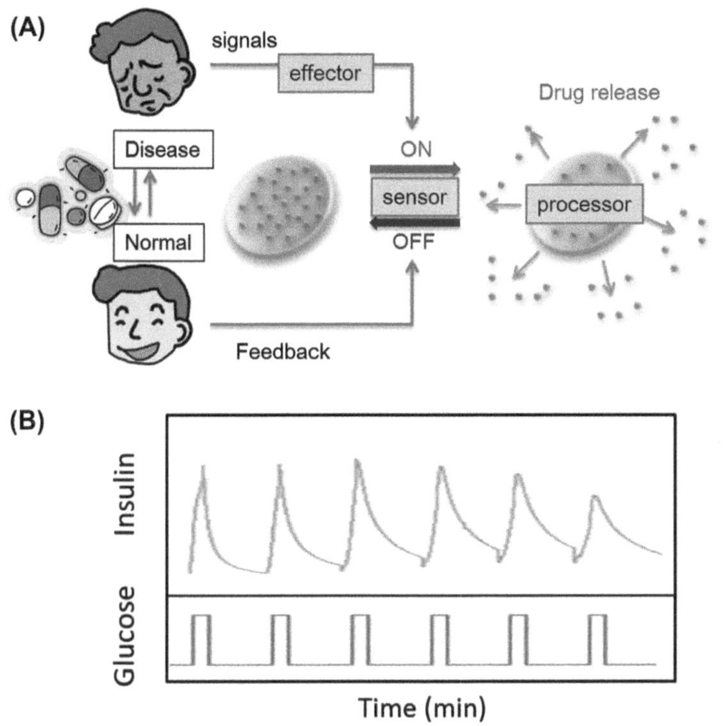

■ **FIGURE 9.2** (A) Design concept for "smart" drug delivery systems with sensor, processor, and effector functions, which can respond directly to a person's individual needs. (B) Time course of changes in fluorescence intensity of FITC-labeled insulin released from glucose-sensitive gels under physiological conditions.

at temperatures above the LCST. Once the growing polymer chains reach a critical length, they collapse and phase separate to form colloidal particles. Omura et al. [35] developed a smart nanogel system that fully swells at normal physiological pH values but undergoes a sharp volume phase transition at moderately acidic pH values. The nanogels were prepared by precipitation polymerization of NIPAAm and propylacrylic acid (PAAc) in the presence of the cross-linker. At room temperature, the NIPAAm-PAAc nanogels were discrete, spherical structures with diameters ranging from 200 to 250 nm. The hydrodynamic diameter of the nanogels decreased to ca. 100–150 nm when the solution temperature was increased to 37°C. At 37°C, when the pKa was below that of the NIPAAm-PAAc, the gels collapsed and aggregated. However, at 37°C and a physiological pH of 7.4, the nanogels did not fully collapse owing to the charge–charge repulsion derived from the ionized carboxyl groups of the PAAc. Thus, such nanogel particles could be useful for releasing drugs in regions of local acidosis, including sites of infection, tumors, ischemia, and intracellular endosomes.

9.3.2 Smart Surfaces

The phase transition behavior and conformational switching of smart polymer-grafted surfaces in response to external stimuli play an important role in changes in biomolecular interactions as well as in determining the permeability, mechanical, and optical properties of surfaces. Stimuli-responsive features are attractive for the remote and timely control of surface functions depending on circumstances. A novel suite of cell culture dished with nanometer-sized surface-grafted PNIPAAm layers was developed by Okano's group for the cultivation of anchor-dependent mammalian cells at 37°C. This technology allows to collect intact cells or intact cell sheets only by reducing temperature. When directly transplanted to human patients for corneal regeneration or heart failure, cell sheets adhered rapidly, avoiding the need for sutures [36,37]. The advantages of cell sheet engineering over more conventional strategies are that (1) cell sheets can adhere to the host tissues without the need for sutures, (2) it avoids the use of scaffold materials that can potentially be inflammatory, (3) it avoids the use of deleterious enzymes that are used to remove cells from culture dishes, and (4) control over the spatial distribution of cells by layering cell sheets, etc. The authors have successfully immobilized synthetic cell-adhesive peptides Arg-Gly-Asp-Ser (RGDS) onto temperature-responsive polymer-grafted surfaces, and demonstrated cell adhesion and detachment under serum-free conditions [38–40]. The PNIPAAm was first functionalized by copolymerization with a reactive comonomer having both a carboxyl group. RGDS were then immobilized onto the copolymer. Bovine aortic endothelial cells both adhered and spread on these surfaces even under serum-free conditions at 37°C, similar to those in 10% serum-supplemented culture. Spread cells promptly detached from the surfaces on lowering culture temperatures below the lower critical solution temperature of the polymer, 32°C. These surfaces would be useful for serum-free culture for tissue-engineering applications.

9.3.3 Smart Nanofibers

Recently, as fiber demand has been increasing continuously, diverse techniques have been developed to produce fibers of macromolecules or polymers. Various types of fiber can be fabricated using different techniques, including single fiber, nonwoven sheet fibers, core–shell or hollow fibers, small-fragment fibers, and hydrogel-like fibers, and these techniques are also versatile. These techniques have both unique merits and demerits. Smart polymer-based fibers are expected to show rapid and sharp responses compared with other types of material; in particular, nanosized fibers show a very high response rate. This high response rate offers a possibility for manipulation of smart fibers for advanced applications. Furthermore, reversibly alterable properties can be used as on-off systems that are triggered by applied stimuli on-demand remotely [41]. We have recently developed a smart hyperthermia nanofiber mesh that is capable of simultaneously realizing hyperthermia and chemotherapy of tumors (Fig. 9.3). The nanofiber is described with simultaneous heat generation and drug release in response to on-off switching of alternating magnetic field (AMF) for

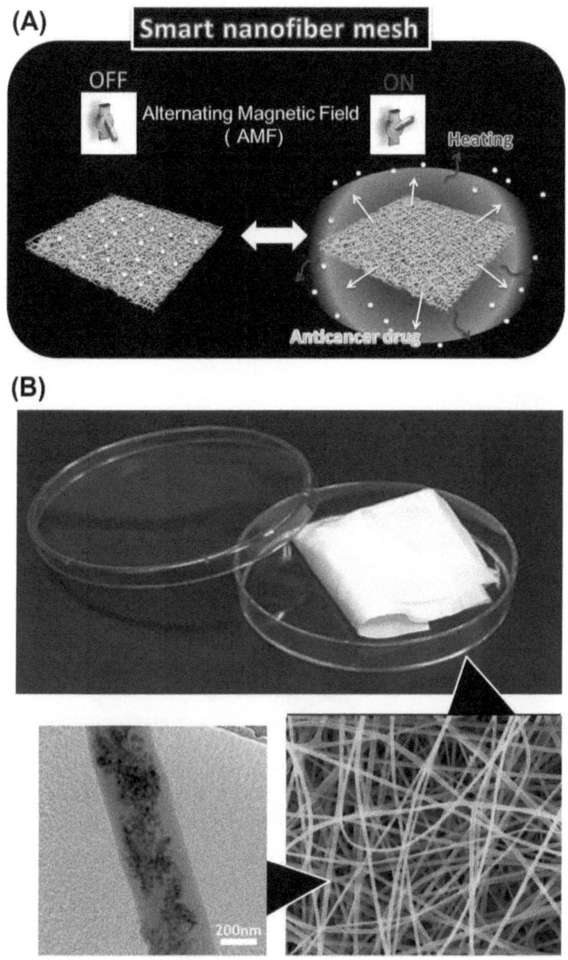

■ **FIGURE 9.3** (A) Schematic illustrations of a smart anticancer nanofiber mesh. (B) Images of smart nanofiber mesh produced by electrospinning method from a solution of P(NIPAAm-*co*-HMAAm) mixed with a solution of magnetic nanoparticles and anticancer drug.

induction of cancer apoptosis [42]. The nanofiber is composed of a chemically cross-linkable temperature-responsive polymer with an anticancer drug and magnetic nanoparticles, which serve as a trigger of drug release and a source of heat, respectively. Seventy percent of cancer cells died in only 5 min application of AMF by the double effects of heat and drug in vitro. No animal died with the implantation of nanofibers and the animal body weight displayed a growth trend in vivo. These results indicate that nanofibers not only improved the double effect of chemotherapy and hyperthermia but also alleviated the side effects associated with the combination therapy.

9.3.4 Smart Shape Memory

Shape-memory polymers (SMPs) are a new class of smart materials, which have the capability to change from a temporary shape "A" to a memorized permanent shape "B" upon application of an external stimulus. In recent years, SMPs have attracted much attention from basic and fundamental research to industrial and practical applications due to the cheap and efficient alternative to well-known metallic shape-memory alloys. Fig. 9.4 shows the schematic images of thermally induced crystal/amorphous transition in SMPs. SMPs currently cover a broad range of applications in

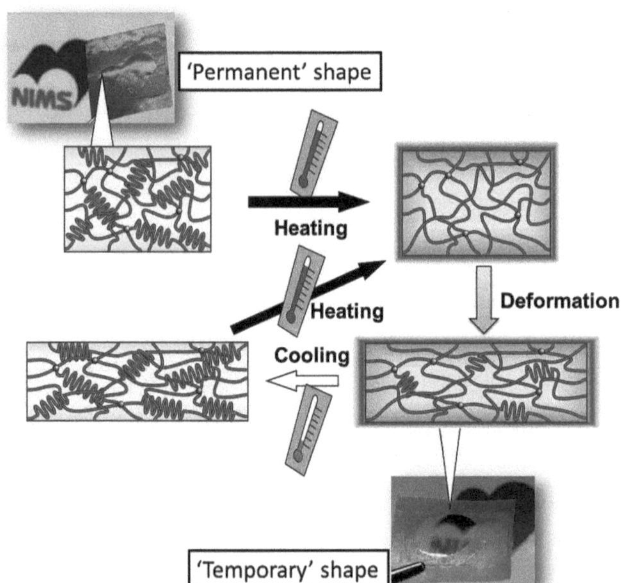

■ **FIGURE 9.4** Schematic illustrations of thermally induced shape-memory effect in shape-memory polymers.

biomedical fields. Especially, new developments of SMPs have been promoted as implantable medical devices in minimally invasive surgery, where a compacted device could be passed through a smaller incision and deployed to its full shape once inside the body. The first study and first clinical trial of an SMP stent was the Igaki–Tamai stent, which is a balloon-expandable poly(L-lactide) stent [43].

SMPs have also emerged as powerful tools for basic cell studies as well as promising biomedical applications. Neuss et al. [44] were the first to demonstrate the bulk shape-memory effect on adhered cells [44]. They characterized the behaviors of L929 mouse fibroblasts, human MSCs, human mesothelial cells, and rat mesothelial cells on a poly(ε-caprolactone) PCL-based SMP network. The first example of the shape-memory surface (SMS) applied to a dynamic cell culture substrate was shown by Davis et al. [45]. To enable surface shape memory during cell culture, they prepared a glassy SMS from Norland Optical Adhesive 63 (NOA-63) because the recovery temperature can be tuned by manipulating the cross-linking condition in this system. After plating the cells on the substrate with temporal grooves, they triggered shape-memory activation using a change in temperature tailored to be compatible with mammalian cells, such as mouse embryonic fibroblasts, thereby causing a topographic transformation back to the original flat surface. Although cell viability remained higher than 95% before and after topography change and temperature increase, the large, irregular dimensions of the surface patterns limited the degree of control over fibroblast cell morphology.

The author's group has also been developing an SMS system with dynamically tunable nanopatterns to direct cell fate [46]. The shape-memory nanopatterns were prepared by chemically cross-linking semicrystalline PCL in a mold to show shape-memory effects over its melting temperature. Permanent surface patterns were first generated by cross-linking the PCL macromonomers in a mold, and temporary surface patterns were then embossed onto the permanent patterns. The temporary surface patterns could be easily triggered to transition quickly to the permanent surface patterns by a 37°C heat treatment. Using this substrate, we have successfully demonstrated time-dependent cell orientation changes by inducing nanotopographical transition from grooves with a height of 300 nm to flat. We have also succeeded in inducing a 90° rotation of the cell orientation by using shape-memory nanogrooves [47].

9.4 CONCLUSIONS

Although this section focuses solely on the roles of smart polymer-based nanoarchitectonics within the fields of medicine, they have an extensive range of applications. They are expected to help address many problems facing today's society. These technologies, for example, may make it possible to manufacture programmable materials that require less energy to produce less waste than conventional materials. Recent progress in polymeric synthesis has led to intriguing new "smart" polymer systems, resulting from novel approaches for synthesizing random, block, or graft copolymers as well as efficient control of molecular weight and its distributions. Living radical polymerization, such as atom transfer radical polymerization, reversible addition fragmentation, and nitroxide-mediated free-radical polymerization, have been introduced to provide polymers with very narrow polydispersities and a living nature to the polymerization. Furthermore, smart materials focus on many types of novel materials, including not only polymers but also ceramics, hybrid composites, shape-memory alloys, magnetorheological fluids, electrorheological materials, nanotubes etc. These materials will enable us to develop better, faster, cheaper, and more powerful technology that has the potential for improvements in health, safety, quality of life, particularly in the resource-limited environments, such as aerospace, or in low-infrastructure sites, such as in the developing world.

REFERENCES

[1] Pancrazio JJ. Nanomed 2008;3:823–30.
[2] Lafuma A, Quere D. Nat Mater 2003;2:457–60.
[3] Zheng Y, Gao X. Soft Matter 2007;3:178–82.
[4] Hoffman AS. In: Ratner BD, Hoffman AS, Schoen FJ, Lemons JE, editors. Biomaterials science: an introduction to materials in medicine. 2nd ed. Elsevier; 2004. p. 107–15.
[5] Jilie K, Li M. In: Galaev I, Mattiasson B, editors. Applications in biotechnology and biomedicine. 2nd ed. CRC Press; 2008. p. 247–68.
[6] Lowe AB, McCormick CL. Stimuli-responsive water soluble and amphiphilic polymers. In: McCormick CL, editor. Acs symposium series, vol. 780. American Chemical Society; 2001. p. 1–13.
[7] Ebara M, Kikuchi A, Sakai K, Okano T. In: Yui N, Mrsny RJ, Park K, editors. Understanding and designing fast responsive polymeric systems. CRC Press; 2004. p. 219–44.
[8] Ebara M, Yamato M, Hirose M, Aoyagi T, Kikuchi A, Sakai K, et al. Biomacromolecules 2003;4:344–9.
[9] Uenoyama S, Hoffman AS. Radia Phys Chem 1988;32:605–8.
[10] Lahann J, Langer R. MRS Bull 2005;30:185–8.
[11] Stayton PS, Shimoboji T, Long C, Chilkoti A, Chen G, Harris JM, et al. Nature 1995; 378:472–4.
[12] Ebara M, Yamato M, Aoyagi T, Kikuchi A, Sakai K, Okano T. Biomacromolecules 2004;5:505–10.
[13] Chen G, Hoffman AS. Nature 1995;373:49–52.

[14] Aoyagi T, Ebara M, Sakai K, Sakurai Y, Okano T. J Biomater Sci Polym Ed 2000;11: 101–10.
[15] Ebara M, Aoyagi T, Sakai K, Okano T. J Polym Sci Part A Polym Chem 2001;39: 335–42.
[16] Ebara M, Aoyagi T, Sakai K, Okano T. Macromolecules 2000;33:8312–6.
[17] Yoshida T, Aoyagi T, Kokufuta E, Okano T, Polym J. Sci Part A Polym Chem 2003; 41:779–87.
[18] Maeda T, Yamamoto K, Aoyagi T. J Colloid Interface Sci 2006;302:467–74.
[19] Shimoboji T, Larenas E, Fowler T, Kulkarni S, Hoffman AS. PS Stay Proc Natl Acad Sci 2002;99:16592–6.
[20] Techawanitchai P, Idota N, Uto K, Ebara M, Aoyagi T. Sci Technol Adv Mat 2012; 13:064202.
[21] Tomatsu I, Peng K, Kros A. Adv Drug Del Rev 2011;63:1257–66.
[22] Shinkai S, Kinda H, Manabe O. J Am Chem Soc 1982;104:2933–4.
[23] Mamada A, Tanaka T, Kungwatchakun D, Irie M. Macromolecules 1990;23: 1517–9.
[24] Sumaru K, Ohi K, Takagi T, Kanamori T, Shinbo T. Langmuir 2006;22:4353–6.
[25] Techawanitchai P, Ebara M, Idota N, Aoyagi T. Colloids Surf B Biointerfaces 2012; 99:53–9.
[26] Osada Y, Hasebe M. Chem Lett 1985;14:1285–8.
[27] Tanaka T, Nishio I, Sun S-T, Ueno-Nishio S. Science 1982;218:467–9.
[28] Nichols TC, Fischer TH, Deliargyris EN, Baldwin AS. Ann Periodontol 2001;6: 20–9.
[29] Kataoka K, Miyazaki H, Bunya M, Okano T, Sakurai Y. J Am Chem Soc 1998;120: 12694–5.
[30] Matsumoto A, Ikeda S, Harada A, Kataoka K. Biomacromolecules 2003;4:1410–6.
[31] Matsumoto A, Yoshida R, Kataoka K. Biomacromolecules 2004;5:1038–45.
[32] Suzuki D, Sakai T, Yoshida R. Angew Chem Int Ed 2008;47:917–20.
[33] Yoshida R, Takahashi T, Yamaguchi T, Ichijo H. Adv Mater 1997;9:175–8.
[34] Okano T, Bae YH, Jacobs H, Kim SW. J Control Release 1990;11:255–65.
[35] Omura T, Ebara M, Lai JJ, Yin X, Hoffman AS, Stayton PS. J Nanosci Nanotechnol 2014;14:2557–62.
[36] Nishida K, Yamato M, Hayashida Y, Watanabe K, Yamamoto K, Adachi E, et al. N Engl J Med 2004;351:1187–96.
[37] Nishida K, Yamato M, Hayashida Y, Watanabe K, Maeda N, Watanabe H, et al. Transplantation 2004;77:379–85.
[38] Ebara M, Yamato M, Aoyagi T, Kikuchi A, Sakai K, Okano T. Tissue Eng 2004;10: 1125–35.
[39] Ebara M, Yamato M, Aoyagi T, Kikuchi A, Sakai K, Okano T. Adv Mater 2008;20: 3034–8.
[40] Ebara M, Yamato M, Aoyagi T, Kikuchi A, Sakai K, Okano T. Biomaterials 2008;29: 3650–5.
[41] Kim Y-J, Ebara M, Aoyagi T. Angew Chem Intl Ed 2012;51:10537–41.
[42] Kim Y-J, Ebara M, Aoyagi T. Adv Func Mater 2013;23:5753–61.
[43] Tamai H, Igaki K, Kyo E, Kosuga K, Kawashima A, Matsui S, et al. Circulation 2000;102:399.

[44] Neuss S, Blomenkamp I, Stainforth R, Boltersdorf D, Jansen M, Butz N, et al. Biomaterials 2009;30:1697.
[45] Davis KA, Burke KA, Mather PT, Henderson JH. Biomaterials 2011;32:2285.
[46] Ebara M, Uto K, Idota N, Hoffman JM, Aoyagi T. Adv Mater 2012;24:273.
[47] Ebara M. Sci Technol Adv Mater 2015;16:014804.

Crystal-Induced Nanoarchitectonics of Organic Polymer Materials

Y. Oaki
Keio University, Kohoku-ku, Yokohama, Japan

CHAPTER OUTLINE

10.1 INTRODUCTION

Nanoarchitectonics directs development of functional materials with unprecedented properties that are not conceivable from just aggregation of unit nanomaterials [1,2] and is one of the most effective approaches for hierarchical morphology control of materials. Morphologies of materials from nanoscopic to macroscopic scales involve their properties [3,4]. Living organisms have performed nanoarchitectonics in nature earlier than researchers [5–12]. Hierarchically organized structures from nanoscopic to macroscopic scales are observed in a variety of biominerals, such as seashells, bone, and teeth. Biological nanoarchitectonics contribute the emergence of unprecedented properties that are not conceivable from each component. If the similar hierarchical morphology design is achieved in synthetic systems, the emergent functions and enhanced properties can be

Supra-materials Nanoarchitectonics. http://dx.doi.org/10.1016/B978-0-323-37829-1.00010-9

extracted from materials. Bioinspired approaches have attracted much interest in recent years [13–24]. Inspired by biomineralization, organic molecules and macromolecules facilitate morphogenesis of crystals through controlled growth. A variety of interaction schemes between the crystals and organic molecules are proposed in previous works. For example, adsorption of organic molecules on the specific crystal surface induces morphology changes of crystals through habit modification [25,26]. When diffusion-controlled condition of solutes is achieved around the crystals in the dense polymer matrices [27–30], the morphologies are changed with an increase in the polymer density. Organic macromolecules can act as the templates to control specific crystal polymorph and orientation [31–35]. Hierarchical morphologies from nanoscopic to macroscopic scales are generated from combination of these effects [36–38]. A variety of techniques for control of reaction and crystal growth have been developed in crystalline inorganic materials. Our group also has studied biological and biomimetic nanoarchitectonics for morphology control of crystalline inorganic materials [39–50]. One of the characteristic findings is the nanostructures in calcium carbonate ($CaCO_3$) biominerals, such as nacreous layer and sea urchin spine [43,49,50]. A variety of $CaCO_3$ biominerals are regarded as single-crystalline objects. In contrast, we found that a variety of complex morphologies can be generated from the hierarchical organization of nanocrystals as the units with incorporation of biological macromolecules. Furthermore, the similar hierarchical morphologies have been reproduced in synthetic systems.

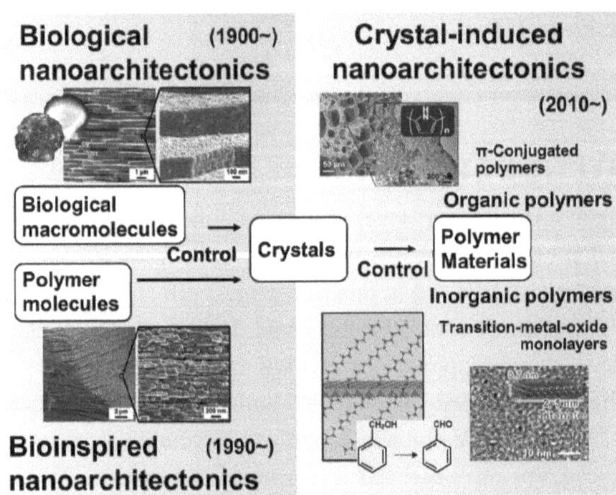

■ **FIGURE 10.1** Schematic illustrations of the relationship between biological, bioinspired, and crystal-induced nanoarchitectonics. Crystal-induced nanoarchitectonics is a new concept for hierarchical morphology control of organic and inorganic polymer materials by using crystals with controlled morphology and orientation. *Reproduced with permission from Oaki Y, Imai H. Angew Chem Int Ed 2005;44:6571, Copyright 2015 Wiley-VCH; Oaki Y, Kijima M, Imai H. J Am Chem Soc 2011;133:8594; Nakamura K, Oaki Y, Imai H. J Am Chem Soc 2013;135:4501; Honda M, Oaki Y, Imai H. Chem Mater 2014;26:3579, Copyright 2015 American Chemical Society.*

Our recent challenges are focused on applications of these basic findings to nanoarchitectonics of organic and inorganic polymer materials (Fig. 10.1) [51–64]. In previous works, crystal morphologies are controlled by polymers. Based on these works, hierarchical morphologies of organic polymer materials can be controlled by crystals. We have recently developed nanoarchitectonics of organic and inorganic polymer materials by using crystalline materials. In the present chapter, three new concepts for nanoarchitectonics of organic polymer materials are introduced. Interspace, surface, and structure of crystals are used for synthesis and hierarchical morphogenesis of organic polymer materials, such as conductive polymers [51–61]. Nanoarchitectonics of inorganic polymer materials, such as transition metal-oxide monolayers with the controlled lateral size and surface chemistry, are shown in reports elsewhere [62–64]. The synthetic approaches, resultant structures, and properties of the organic polymer materials are summarized in the present chapter.

10.2 ORGANIC POLYMER MATERIALS FOR NANOARCHITECTONICS

Morphologies of organic materials in a range of length scales involve their properties. Molecular design and synthesis are an important approach for preparation of organic materials with specific properties [65–71]. Self-organization of designed molecules facilitates formation of functional soft materials [71,72]. In general, morphologies of organic materials in molecular and nanometer scales can be tuned on the basis of molecular design. However, it is not easy to achieve hierarchical morphology control of organic materials in multiple scales from nanoscopic to macroscopic scales by typical methods. The hierarchical morphologies in multiple length scales have potentials for improvement of properties. Fig. 10.2 shows the potential applications of organic polymer materials with controlled morphologies in

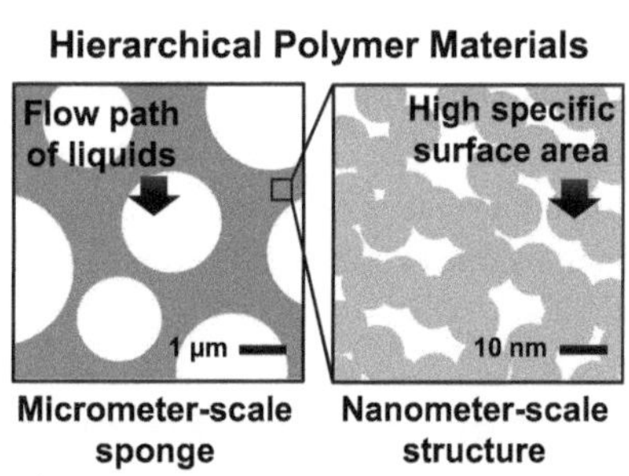

■ **FIGURE 10.2** A schematic illustration of hierarchically organized polymer materials and their structure characteristics. The structures in different scales have effects on emergence of the different functions. When the resultant polymer materials have both micrometer-scale spongy shapes and nanoscale particles, flow path of liquids and high specific surface area for reactions can be ensured in the hierarchical morphologies.

different length scales. The nanostructures, smaller than 100 nm in scales, provide the high specific surface area and smaller pore that are suitable for reaction and adsorption of molecules [52,53,55,73]. The microstructures, larger than 100 nm in scales, contribute formation of microscale pores and easy handling of materials. The microscale pores serve as flow path of liquids and solutions. These hierarchical structures can be applied to electrochemical devices, catalysts, adsorbents, and separation materials [52,53,55,73]. Although a variety of inorganic materials show the enhanced properties based on these strategies, the nanoarchitectonics are not fully applied to organic materials, such as polymers. Synthesized polymers can form their organized structures by a number of methods, such as phase separation, layer-by-layer assembly, and electrospinning [74–80]. In contrast, it is not easy to achieve simultaneous synthesis and hierarchical morphology control of polymer materials. A wide variety of templates, such as porous materials and self-assembled molecules, are used for morphogenesis of polymer materials [81–95]. Generalizable approaches for hierarchical morphogenesis of polymer materials are required for advances of nanoarchitectonics. The hierarchical morphology control of organic materials has potential for enhancement and emergence of the properties.

Our group has reported on bioinspired approaches for control of crystal growth [39–48]. A variety of approaches have been developed for hierarchical organization of inorganic crystals and inorganic/organic composites. Our recent challenge is focused on applications of these approaches to simultaneous synthesis and hierarchical morphogenesis of polymer materials [51–64]. The challenges can be regarded as crystal-induced nanoarchitectonics of organic polymer materials. The surface of crystals, interspace between the unit crystals, and structure itself are used for synthesis and hierarchical morphology control of organic polymer materials. In the present chapter, these three new approaches are introduced to develop crystal-induced nanoarchitectonics of organic polymer materials. The interspace of the unit nanocrystals is used for simultaneous synthesis and hierarchical morphogenesis of polymer materials. The surface of crystals serves as the oxidants and reaction sites for oxidative polymerization and subsequent morphogenesis. The crystal structure itself is used for morphology control, topochemical polymerization, and application. These concepts can be generalizable as nanoarchitectonics of a variety of organic polymer materials.

10.2.1 Polymerization and Morphogenesis in Interspace of Nanocrystals

A new nanoarchitectonics for organic polymer materials is synthesis and simultaneous morphogenesis of organic polymer materials by using hierarchically organized structures consisting of the oriented nanocrystals [51−55]. Our group found that hierarchical structures of biominerals, such as nacreous layer and sea urchin spine, consisted of the oriented nanocrystals 20−100 nm in size (Fig. 10.3) [43,49,50]. The finding implies that the oriented nanocrystals form the composites with incorporation of biological macromolecules in the interspace of the unit nanocrystals. Since the building block is a nanocrystal, a variety of the macroscopic morphologies are generated from the organization of the unit crystals with preservation of the single-crystalline structure. Recent reports suggest that the similar hierarchical structures consisting of the unit nanocrystals are formed through nonclassical crystallization routes [96−102]. We found that the oriented nanocrystals have the interspace between the units with removal of the originally incorporated biological macromolecules [43,50]. In our previous work, a variety of organic molecules as a guest were introduced in the host nanospace from the solutions (Fig. 10.4) [41,43,103]. Based on the results, we

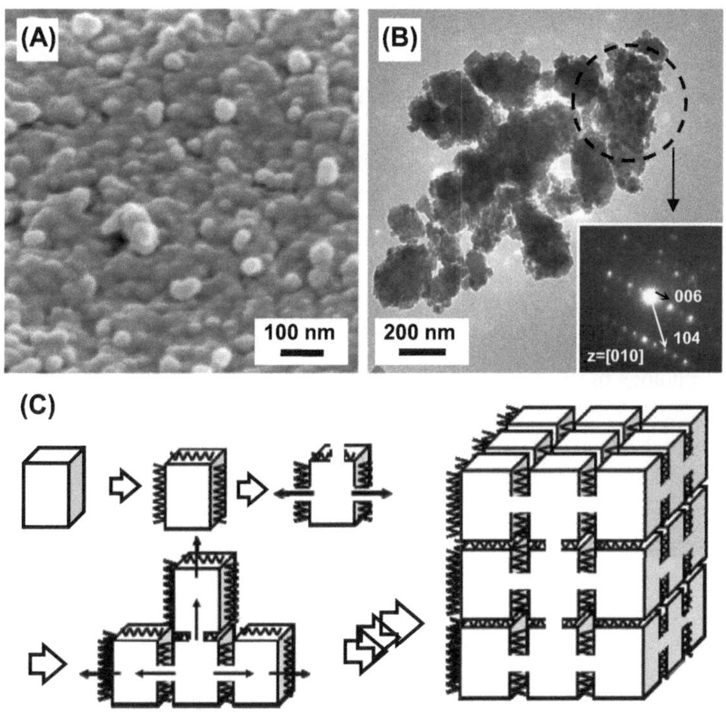

■ **FIGURE 10.3** Nanostructures in a biomineral, namely a sea urchin spine. (A) FESEM image. (B) FETEM image and its selected-area electron diffraction pattern (the inset). (C) Schematic illustration of the oriented nanocrystals and their proposed growth mechanisms. *Reproduced with permission from Oaki Y, Kotachi A, Miura T, Imai H. Adv Funct Mater 2006;16:1633, Copyright 2015 Wiley-VCH.*

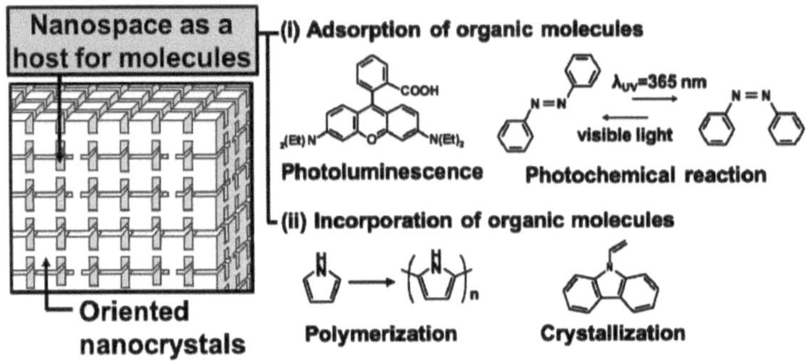

■ **FIGURE 10.4** Schematic illustrations of the interspace in the oriented nanocrystals. The interspace can adsorb and incorporate organic molecules [43,49–55].

developed the methods for hierarchical morphology replication from the oriented nanocrystals to polymer materials [51–55]. The monomers are introduced in the nanoscale interspace with immersion in the neat liquids. When incorporation and polymerization of the monomers are achieved in the nanoscale interspace, the original crystals and polymers form composites. The replicated polymer materials are obtained after dissolution of the original materials. In this way, the hierarchical morphologies can be replicated from the original materials to polymer materials through simultaneous polymerization and morphogenesis. Although templates for morphology replication are well studied in previous work, it is not easy to achieve hierarchical morphogenesis in multiple length scales.

Here we introduce a model case for hierarchical morphology replication from sea urchin spine consisting of the oriented calcite $CaCO_3$ nanocrystals to polypyrrole (PPy) [51,52]. Fig. 10.5 shows the original hierarchical morphology of sea urchin spine, a $CaCO_3$ biomineral, and replicated PPy architecture. The original sample of sea urchin spine showed the macroscopic sponge shapes 10–50 μm in the pore size and nanocrystals 20–80 nm in size (Fig. 10.5A–C) [49,50]. The sample was immersed in neat liquid of the monomer, namely pyrrole (Py). The excess amount of Py monomer was absorbed by a paper towel. The oriented nanocrystals with inclusion of the monomer were then immersed in the solution containing an oxidative agent, such as chloroform solution containing trivalent iron chloride. The oxidative polymerization proceeded in the nanospace. The composites of $CaCO_3$ and PPy were obtained after the polymerization. Since the polymers were formed in the interspace of the oriented nanocrystals, the unit nanocrystals were not clearly observed on the field-emission scanning electron microscopy (FESEM) images. The PPy-replicated architecture was obtained after dissolution of the original $CaCO_3$ (Fig. 10.5D–F). The hierarchical morphology

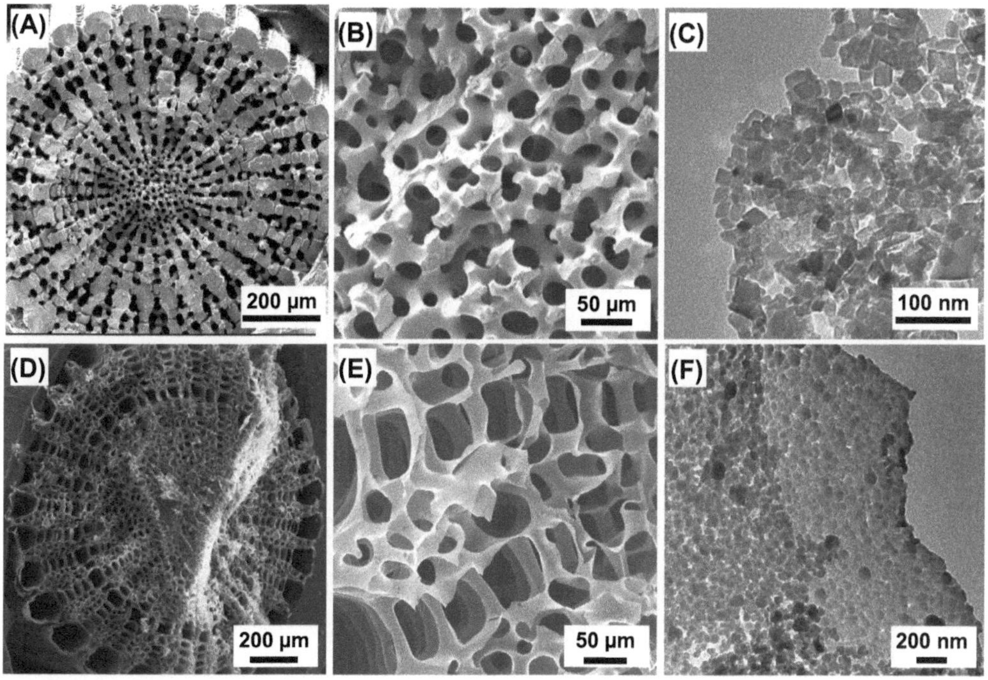

■ **FIGURE 10.5** Hierarchical morphology replication from a biomineral to an organic polymer material. (A—C) FESEM (A and B) and FETEM (C) images of an original $CaCO_3$ biomineral, namely a sea urchin spine. (D—F) FESEM (D and E) and FETEM (F) images of the replicated PPy. *Reproduced with permission from Oaki Y, Kijima M, Imai H. J Am Chem Soc 2011;133:8594, Copyright 2015 American Chemical Society.*

consisting of the nanoparticles and microscopic sponges was preserved after the replication processes (Fig. 10.5). The PPy architectures showed the improved electrochemical properties as an electrode active material of supercapacitor rather than a commercial PPy microfiber. Whereas a commercial PPy microfiber 500 nm in diameter showed the specific capacity around $100\ F\ g^{-1}$ at the scan rate of $1\ mV\ s^{-1}$, the improved capacity, namely $250\ F\ g^{-1}$, was observed on the PPy-replicated architecture. The improved electrochemical performance is ascribed to the hierarchical structure of PPy including the micrometer-scale pores for smooth diffusion of the electrolytes and nanoparticles with the high specific surface area. The present approach for the hierarchical morphology replication is applied to morphogenesis of the other polymer materials through oxidative polymerization, such as poly(3-hexylthiophene) and poly(3,4-ethylenedioxythiophene) [52,53].

The approach was applied to the schemes other than oxidative polymerization [54,55]. The poly(vinyl carbazole) (PVCz) and polystyrene (PSt) with the hierarchical morphologies were obtained by using the same original hierarchical structures, such as a sea urchin spine. The hierarchical morphologies in nanometer and micrometer scales contribute improvement of their properties. For example, the resultant PVCz, as a donor, formed a 6.6 times higher amount of charge–transfer complex with 2,4,7-trinitro-9-fluorenone, as an acceptor, than commercial PVCz powder [54]. The hierarchical structure showed the more remarkable effects in the PSt-replicated architecture as a hydrophobic adsorbent (Fig. 10.6). The adsorption properties as a

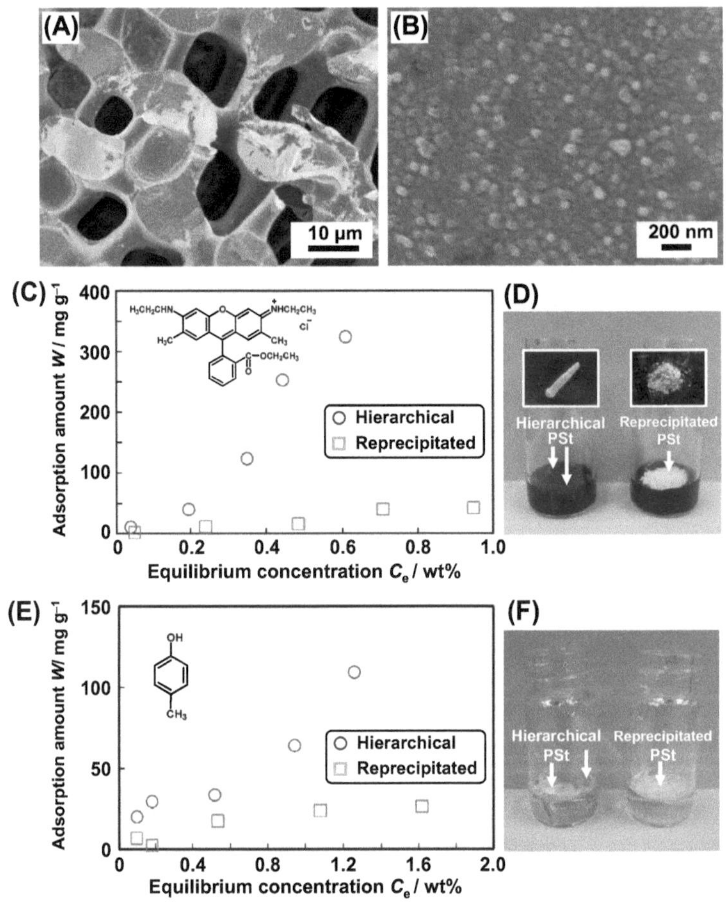

■ **FIGURE 10.6** Hierarchically organized PSt through the morphology replication from a biomineral. (A and B) FESEM images, (C) adsorption isotherms of the hierarchical PSt (*circles*) and reprecipitated PSt in methanol from the toluene solution (*squares*) in the rhodamine 6G aqueous solution; (D) photographs of the hierarchical (left) and reprecipitated (right) PSt put in the dye aqueous solutions; (E and F) adsorption isotherms and photographs of the same samples in *p*-cresol aqueous solutions. *Reproduced with permission from Sato K, Oaki Y, Imai H. Chem Commun 2015;51:7919, published by The Royal Chemical Society.*

hydrophobic adsorbent were studied in the aqueous solution containing rhodamine 6G, a water-soluble dye, and *p*-cresol, a small organic molecule, as models for industrial wastewaters [55]. The reprecipitated and commercial PSt (Amberlite XAD200) samples were used as the reference. Fig. 10.6 shows the adsorption isotherms of the organic molecules on the PSt hierarchical architectures. The hierarchical PSt showed improved adsorption properties than the reprecipitated one in all the initial concentration ranges of the organic molecules. The adsorption properties of the hierarchical PSt were comparable to those of the commercial one. The pores originating from the sponge structure in the original material contribute the penetration of the aqueous solutions and the smooth diffusion of the target organic molecules (Fig. 10.6D and F) [55]. In addition, the PSt nanoparticles contribute the high specific surface area, approximately 82.4 $m^2\ g^{-1}$, for the adsorption. The hierarchical morphogenesis of polymer materials has potential for improving their properties. These results suggest that the hierarchical morphologies can be replicated from the original inorganic crystals, such as biominerals consisting of water-soluble compounds, to a variety of organic polymer materials. Recently, a variety of hierarchical morphologies consisting of the oriented nanocrystals were prepared through bioinspired approaches [13–24,96–102]. The present new nanoarchitectonics provides organic polymers with the tailored hierarchical morphologies.

10.2.2 Polymerization and Morphogenesis Using Surface of Oxidant Crystals

In the previous section, polymerization was performed in the interspace of the oriented nanocrystals. Since the hierarchical morphologies are replicated from the original materials to the polymers, the approach is regarded as a family of template methods. The resultant morphologies of resultant polymers are limited to those of the original materials. In the present section, the surface of crystals is used for nanoarchitectonics of conductive polymer materials [56–59]. Our intention here is oxidative polymerization of heteroaromatic monomers on the crystal surface of inorganic oxidative agents (Fig. 10.7) [57,58]. The oxidant crystals can provide the oxidative agents with high density and rate on the solid surface. Therefore, formation of the nanostructures is expected on the surface through simultaneous reaction and polymerization at the interface. The neat liquid of monomers and powder of oxidant crystals were separately put in the same sample bottle. The evaporated monomer vapor was adsorbed and polymerized on the surface of oxidant crystals. In addition, we found that the conductive polymers were deposited on the inner wall of the sample bottle (Fig. 10.8) [59]. The results indicate the presence of two deposition schemes in this system.

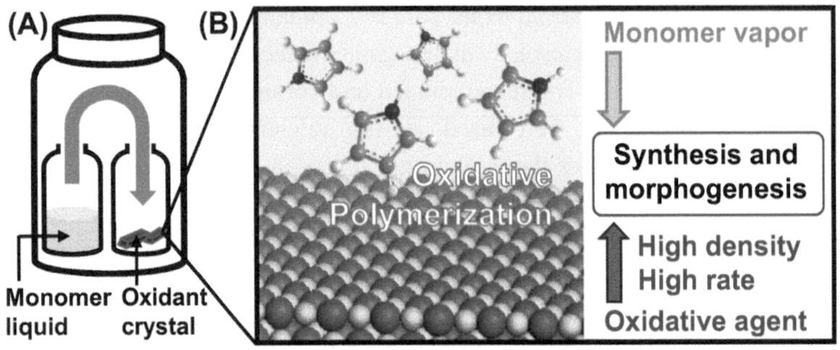

■ **FIGURE 10.7** Schematic illustrations of the experimental setup (A) and polymerization reaction (B) at the interface of solid oxidant crystal and monomer vapor. *Reproduced with permission from Kuwabara K, Oaki Y, Muramatsu R, Imai H. Chem Commun 2015;51:9698, published by The Royal Chemical Society.*

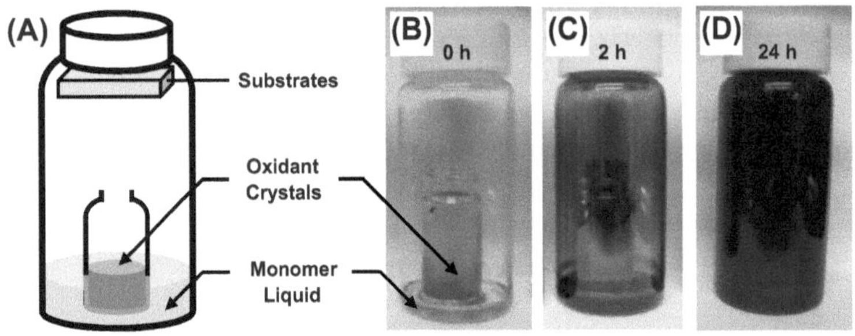

■ **FIGURE 10.8** Schematic illustrations of the experimental setup (A) and time course photographs of the sample bottle containing separately placed oxidant crystals and monomer liquids at 0 h (B), 2 h (C), and 24 h (D). *Reproduced with permission from Muramatsu R, Oaki Y, Kuwabara K, Hayashi K, Imai H. Chem Commun 2014;50:11840, published by The Royal Chemical Society.*

When monomer vapor is supplied on the surface of oxidant crystals, the polymerization proceeds on the solid surface [57]. The color of the oxidant crystals was changed after the polymerization. The polymerization experiments were performed by a variety of combinations of heteroaromatic monomers and inorganic oxidative agents [57]. Formation of the conductive polymers was observed in a number of the combinations. Whether or not the conductive polymers were formed is ascribed to the natures of the monomers and oxidant crystals, such as resonance stability and electron density of the monomers and reactivity of the oxidant crystals. Fig. 10.9 shows the morphologies of polyaniline (PAni) formed on the surface of copper nitrate trihydrate [$Cu(NO_3)_2 \cdot 3H_2O$] crystals. The resultant conductive polymers showed the nanoparticles around 200 nm in size on the surface of the oxidant crystals in the early stage (Fig. 10.9C). After the reaction proceeded, the films several micrometer in thickness consisting of the primary nanoparticles were obtained on the crystal surface (Fig. 10.9F and G). The similar

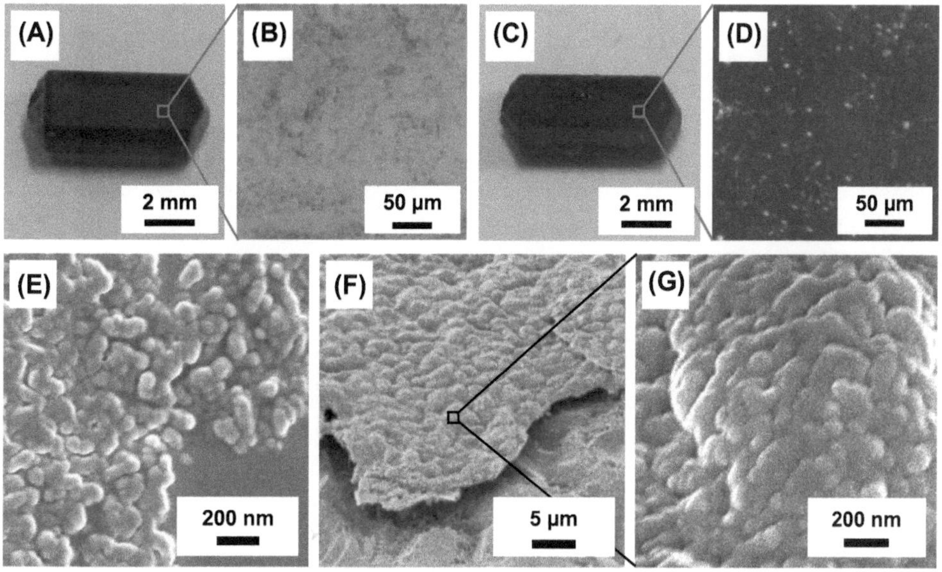

■ **FIGURE 10.9** Formation of PAni on the surface of an oxidant inorganic crystal, namely $Cu(NO_3)_2 \cdot 3H_2O$. (A and B) photograph (A) and optical microscopy image (B) of $Cu(NO_3)_2 \cdot 3H_2O$ single crystal; (C and D) photograph (C) and optical microscopy image (D) of PAni on the surface of $Cu(NO_3)_2 \cdot 3H_2O$ single crystal; (E—G) FESEM images of the PAni nanoparticles (E and F) and films (G) consisting of the nanoparticles on the surface in the early and later stages of the growth, respectively. *Reproduced with permission from Kuwabara K, Oaki Y, Muramatsu R, Imai H. Chem Commun 2015;51:9698, published by The Royal Chemical Society.*

morphogenesis was observed on the other combinations of monomers and oxidant crystals. The present approach induced formation of the conductive polymer nanoparticles and their accumulated films on the crystal surface of oxidative agents. If the macroscopic morphology of the original crystals is tuned, the hierarchical structures can be obtained by the present approach. As a simple model, the original oxidant crystal was freeze-dried to obtain the porous morphology (Fig. 10.10A). When Py monomer was polymerized on the surface of the porous copper sulfate pentahydrate crystals, the porous spongy PPy consisting of the nanoparticles was obtained with dissolution of the inorganic species (Fig. 10.10B and C). The electrochemical properties were measured on a commercial PPy, the aggregates of the PPy film consisting of the nanoparticles, and the sponge PPy (Fig. 10.10D—F). The hierarchical PPy architecture contributed the improvement of the specific capacity of supercapacitor [57]. The sponge PPy consisting of the nanoparticles showed the improved specific capacity rather than that of commercial microfibers and film consisting of the nanoparticles. The micrometer-scale and nanoscale morphologies, respectively, facilitate the smooth diffusion of electrolytes solutions and smooth reaction originating from the high specific surface area. The hierarchical morphology has effects on the

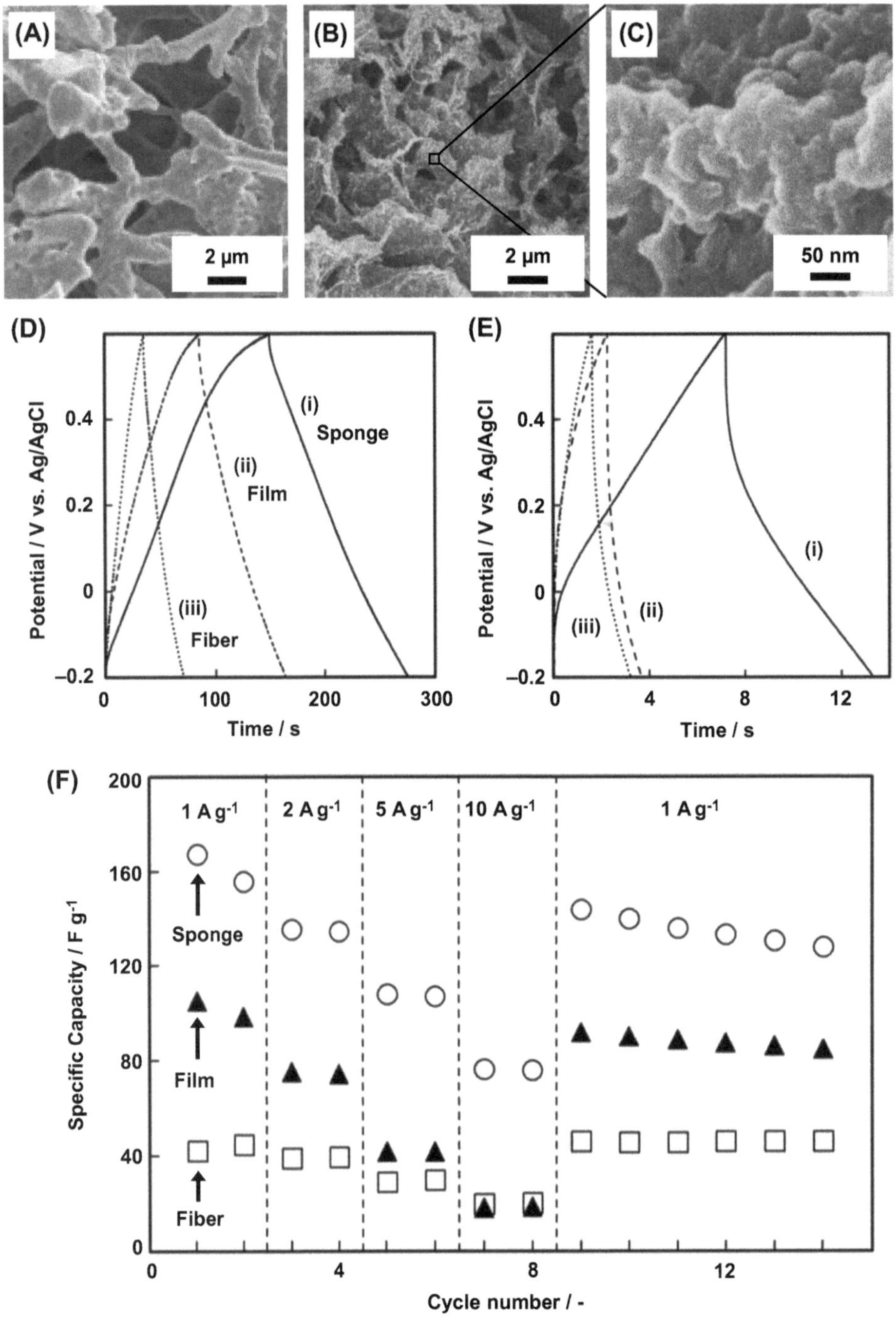

■ **FIGURE 10.10** Sponge PPy formed on the freeze-dried oxidant crystal and its electrochemical application. (A) FESEM image of the original $CuSO_4 \cdot 5H_2O$ after freeze-drying; (B and C) FESEM images of the PPy formed on the freeze-dried $CuSO_4 \cdot 5H_2O$; (D and E) chronopotentiograms of the sponge (i), film (ii), and fibrous (iii) PPy samples at the current density of 1 A g^{-1} (D) and 10 A g^{-1}; (F) relationship between the specific capacity and current density of the samples. *Reproduced with permission from Kuwabara K, Oaki Y, Muramatsu R, Imai H. Chem Commun 2015;51:9698, published by The Royal Chemical Society.*

electrochemical properties. Furthermore, when the polymerization was performed on the oxidant crystals with the macroscopic dendritic morphologies, the shapes were replicated to the conductive polymers.

In a couple of the combinations we found deposition of the conductive polymers on the inner wall of the sample bottle in addition to the surface of the oxidant crystals (Fig. 10.8) [59]. When substrates and substances were put in the sample bottle, conductive polymers, such as PPy and PEDOT, were obtained on the substrates (Figs. 10.8 and 10.11). The nanoparticles were formed on the substrates, such as glass and potassium bromide single crystal, in the early stage of the deposition. As the reaction time increased, the growth of the nanoparticles formed the film on the substrates (Fig. 10.11A). The wrinkled shapes of the films were observed on the surface with the further increase in the reaction time (Fig. 10.11B). In the present system, the deposition of PPy on the separately placed substrates was initiated by generation of the activated monomer species. For example, the combinations of Py and iron nitrate nonahydrate ($Fe(NO_3)_3 \cdot 9H_2O$) spontaneously generate the nitrated Py monomer ($Py-NO_2$) on the surface of the inorganic crystal [59]. As well as the Py monomer, vapor of the activated monomer is adsorbed on the surface of the substrates. The polymerization is initiated by the activated $Py-NO_2$ monomer with supply of the Py monomer on the substrates. Therefore, conductive polymers, such as PPy and PEDOT, were formed in the sample bottles through spontaneous generation of the activated monomers. The present method can be used as the coating and casting of the conductive polymers with a variety of substrates and substances under low temperature and ambient pressure. When a sea urchin spine as a model of complex hierarchical morphology is used as the substance, the

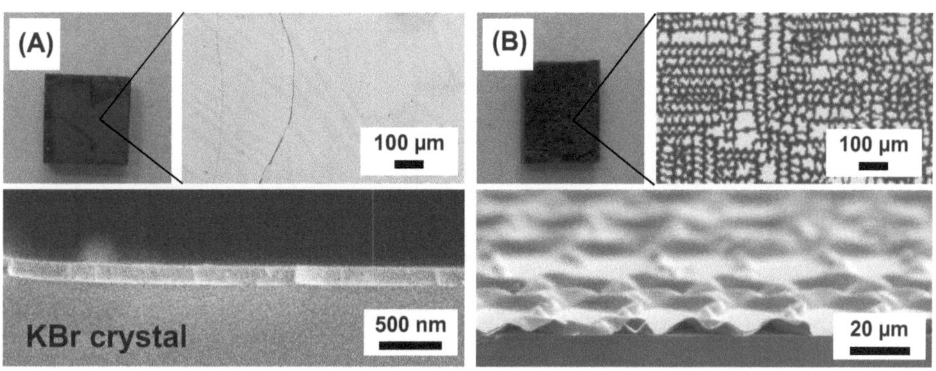

■ **FIGURE 10.11** PPy formed on the substrates placed in the sample bottle containing the oxidant crystals and monomer liquids, as shown in Fig. 10.8. Optical microscopy and FESEM images of the PPy on a glass substrate in the early (A) and later (B) stages of the deposition. *Reproduced with permission from Muramatsu R, Oaki Y, Kuwabara K, Hayashi K, Imai H. Chem Commun 2014;50:11840, published by The Royal Chemical Society.*

hierarchical morphology can be replicated to PPy by the casting. Coating of conductive polymers contributes improvement of electron conductivity on electrode active materials [104]. Lithium titanate ($Li_4Ti_5O_{12}$) is an anode active material of lithium ion battery. The PPy coating was performed on a commercial $Li_4Ti_5O_{12}$ nanocrystal around 100 nm in size [59]. The thin PPy layer, approximately 1.7 wt%, was partially coated on the surface of the $Li_4Ti_5O_{12}$ nanocrystals (Fig. 10.12A). The charge–discharge reactions

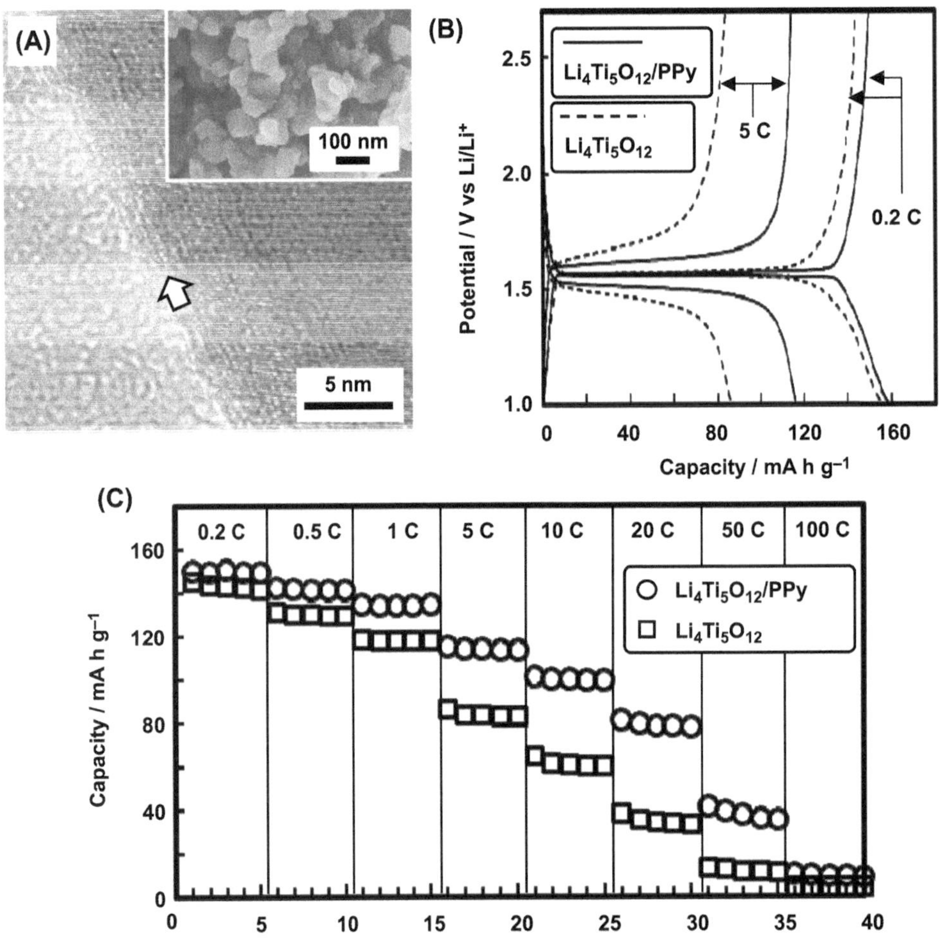

■ **FIGURE 10.12** PPy coating on the $Li_4Ti_5O_{12}$ nanocrystals and its electrochemical properties. (A) FETEM and FESEM (the inset) images of the PPy-coated $Li_4Ti_5O_{12}$ nanocrystals around 100 nm in size; (B) charge–discharge curves of the $Li_4Ti_5O_{12}$ nanocrystals and PPy-coated ones at 0.2 C and 5 C; (C) Relationship between the specific capacity and current density of the samples (1 C = 0.175 mA g^{-1}). *Reproduced with permission from Muramatsu R, Oaki Y, Kuwabara K, Hayashi K, Imai H. Chem Commun 2014;50:11840, published by The Royal Chemical Society.*

were performed on commercial $Li_4Ti_5O_{12}$ nanocrystals and PPy-coated $Li_4Ti_5O_{12}$ nanocrystals in the potential range of 1.0–2.7 V (vs Li/Li^+) (Fig. 10.12B). In general, the charge–discharge curves of $Li_4Ti_5O_{12}$ show the plateaus at 1.5 V vs. Li/Li^+ [105]. The overpotential on the charge and discharge reactions, corresponding to the deviations of the potentials from 1.5 V vs. Li/Li+, was decreased on the PPy-coated samples. The improved specific capacity was observed on the PPy-coated $Li_4Ti_5O_{12}$ nanocrystals (Fig. 10.12C), when the current density of the charge and discharge reactions was increased. The PPy coating contributes the improvement of the electron conductivity. The present technique can be applied to a variety of conductive coating on electrode active materials for improvement of electrochemical performances.

In previous works, conductive polymers were synthesized from the monomer vapors, such as vapor phase polymerization (VPP) and oxidative chemical vapor deposition (oCVD) methods [106–114]. The VPP method provides conductive polymers in polymer matrices containing oxidative agents by diffusion of vapor of heteroaromatic monomers. In the oCVD method, both vapors of monomers and sublimated oxidative agents produce conductive polymers on substrates under a vacuum condition at high temperature. The coating of conductive polymers is achieved on substrates by these methods. However, it is not easy to achieve spontaneous generation of complex hierarchical morphologies from nanoscopic to macroscopic scales. In our present methods, the conductive polymers are directly obtained on the solid surface under ambient pressure and low temperature below 60°C. Furthermore, the complex hierarchical morphologies can be obtained by the present approaches.

10.2.3 Polymerization and Morphogenesis Using Monomer Solid Crystals

In the previous Sections 10.2.1 and 10.2.2, the interspace and surface of solid crystals are used for morphology control of organic polymer materials. In the present section, crystal structure itself is used for polymerization and morphogenesis. Growth control of a monomer crystal and its subsequent topochemical polymerization facilitate morphology and orientation control of the resultant polymer materials. Biomineralization is a model for growth control of inorganic crystals with association of organic polymers in aqueous media. It is not so easy to apply the biomimetic approaches directly to growth control of organic crystals in organic media. In previous works, tuning of the solvents during the recrystallization process facilitated morphological control of organic crystals [115–123]. Biomimetic approaches were applied to morphological control of water-soluble organic crystals

[124–128]. If biomimetic approaches are applied to morphology and orientation control of organic crystals in nonpolar organic media, organic materials with the controlled morphology and orientation can be obtained without complex molecular design and synthesis. Here we focused on a fish scale as a model for growth control of organic crystals [129–134]. The bright silver luster of a fish skin is ascribed to the laminated structure of platy guanine crystals and cytoplasm (Fig. 10.13A and C). The morphology and orientation of guanine crystals are controlled through biological routes. Biomimetic approaches were applied to morphology and

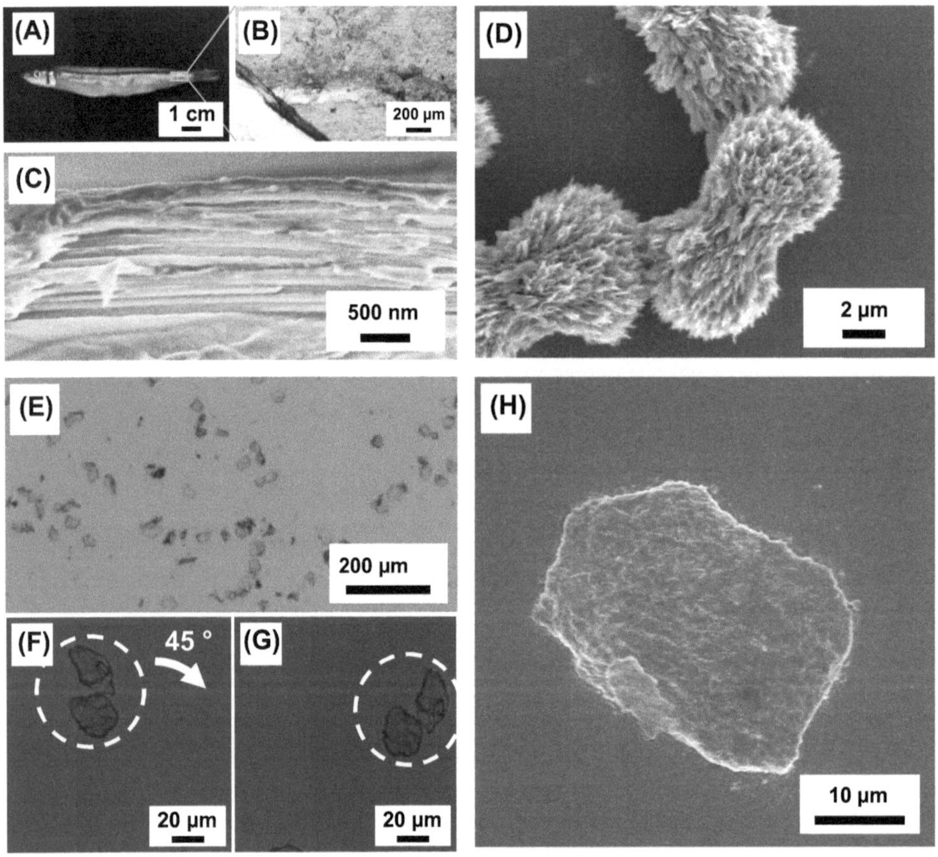

■ **FIGURE 10.13** Morphology and orientation of guanine crystals in a fish scale and synthetic systems. (A) photograph of *Spratelloides gracilis*, a member of the round herring family; (B and C) optical microscopy (B) and FESEM (C) images of the scale consisting of the platy guanine crystals; (D) guanine crystals grown from the ammonia solution on a bare glass substrate with gradual evaporation of water; (E) optical microscopy image of guanine crystals on the amino-group-modified substrate with addition of an additive organic polymer; (F and G) the optical microscopy images observed under the crossed polarizers with each rotation of 45 degree; (H) FESEM image of the guanine crystal grown on the modified surface with an additive organic polymer. *Reproduced from Oaki Y, Kaneko S, Imai H. J Mater Chem 2012;22:22686, with permission from The Royal Society of Chemistry.*

orientation control of guanine crystals [135]. The guanine crystals were grown from the ammonia solution with gradual evaporation of water. The surface-modified substrates and additive organic molecules facilitate formation of the guanine crystals with the similar morphology and orientation to those of the biogenic structure (Fig. 10.13D–H) [135]. Our next challenge is application of biomimetic approaches for morphology and orientation control of functional organic crystals grown in nonpolar organic media [60,61]. We adopted 10,12-pentacosadiynoic acid (PCDA) as an organic crystal for morphology and orientation control [60]. PCDA, insoluble in aqueous media, is a diacetylene (DA) derivative with layered crystal structure [136,137]. Polydiacetylenes (PDAs) are synthesized from the DA monomers in the organized states through topochemical polymerization. If the distance between the diacetylene moieties is less than 0.5 nm, topochemical polymerization proceeds in the organized states by irradiation of UV light, γ-ray, and electron beam [138–141]. The facts indicate that morphology and orientation control of the PCDA monomer crystals direct those of PDAs.

The butanol solution of PCDA was used as the precursor for the crystal growth (Fig. 10.14). The crystal growth proceeded on the substrates with gradual evaporation of the solvent. The ribbon-like PCDA crystals of several-tens micrometer in width and around 100 nm in thickness were obtained throughout the substrates with surface modification by amino groups (Fig. 10.14A) [60]. A decrease in the width and an increase in thickness were observed with addition of polyvinylpyrrolidone (PVP). Although the PCDA crystals were formed on bare glass substrates, homogeneous deposition throughout the substrates was not achieved. When the ribbons were observed by optical microscopy under crossed polarizers, the image turned from bright to dark with each rotation of 45 degree (Fig. 10.14B and C). The XRD pattern showed the intensified peaks corresponding to the interlayer distance of the layered crystal structures at $2\theta = 1.92$, 5.66, 9.39, 13.5, and 16.9 degree, as indicated by the circles in Fig. 10.15A. In contrast, the diffraction peaks assignable to the distance between the alkyl chains was not observed around $2\theta = 22$ degree, as indicated by the white arrow in Fig. 10.15A. The optical microscopy and XRD analyses suggest that the ribbon-like PCDA crystals formed the single-crystalline oriented structure (Fig. 10.15B and C). The layered structure with the interlayer distance of 4.61 nm was arranged parallel to the substrate (Fig. 10.15B and C). In the present study, the solvents, surface functional groups on the substrates, and additive organic polymers play important roles for morphology and orientation control of PCDA crystals. Among a number of solvents, butanol is suitable for morphology and orientation control of PCDA crystals. The slower crystal

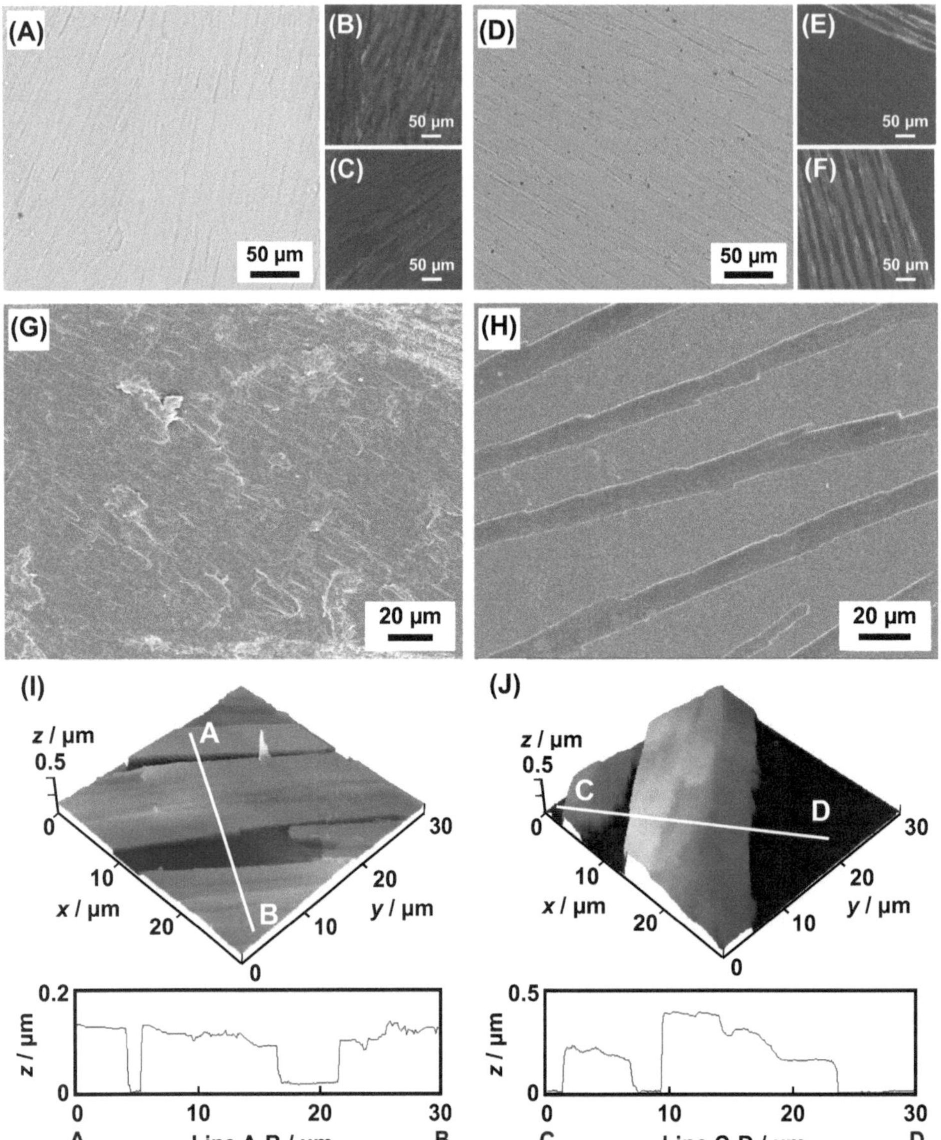

■ **FIGURE 10.14** PCDA ribbons prepared on the NH_2-modified substrates from the 1-butanol solution (the left panels A—C, G, I) and those prepared under the same conditions in the presence of PVP (the right panels D—F, H, J). (A—F) optical microscopy images (A and D) and those observed under crossed polarizers (B and E) with the rotation of 45 degree (C and F); (G and H) FESEM images, (I and J) AFM images and their height profiles of the lines A—B and C—D. *Reproduced with permission from Okaniwa M, Oaki Y, Kaneko S, Ishida K, Maki H, Imai H. Chem Mater 2015;27:2677. Copyright 2015 American Chemical Society.*

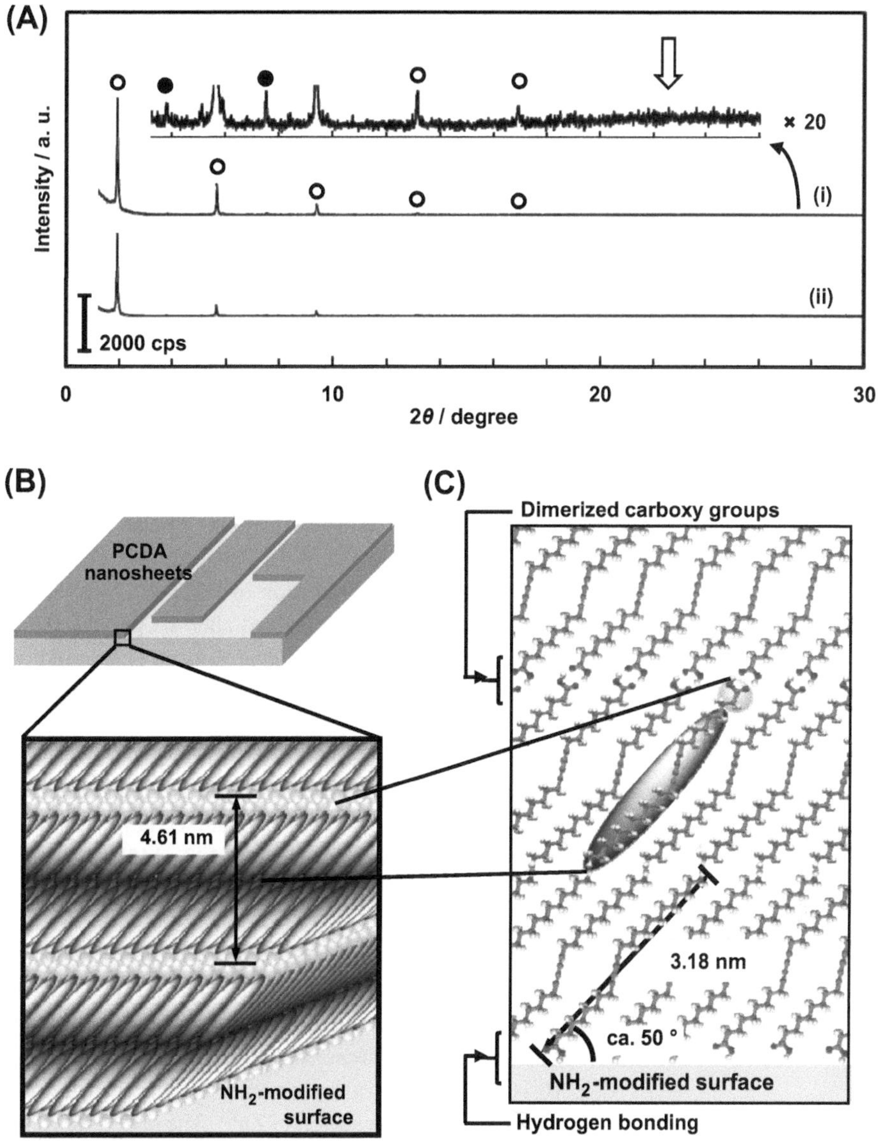

■ **FIGURE 10.15** Orientation of the PCDA molecules and crystals on the substrates. (A) XRD patterns of the PCDA ribbons prepared from the 1-butanol solution (i) and those prepared under the same conditions in the presence of PVP (ii); (B and C) the schematic illustrations of the crystal and molecular orientations on the substrates. *Reproduced with permission from Okaniwa M, Oaki Y, Kaneko S, Ishida K, Maki H, Imai H. Chem Mater 2015;27:2677. Copyright 2015 American Chemical Society.*

growth is achieved by lower vapor pressure, appropriate polarity, and interaction to PCDA. The surface modification by amino groups facilitates the growth of PCDA crystals throughout the substrates. The surface amino groups contribute an increase in wettability, interaction with PCDA leading to the specific orientation, and promotion of heterogeneous nucleation. The additive molecules, namely PVP, change the width and thickness of PCDA ribbons through adsorption to the specific crystal face (Fig. 10.14D−F,H,J). These mechanisms for morphology and orientation control are similar to those of inorganic crystals through biomimetic approaches. The results suggest that the biomimetic approach can be applied to controlled crystal growth of organic compounds in nonpolar organic media.

Intercalation chemistry of guest ions and molecules has been widely studied in inorganic layered compounds. However, intercalation chemistry in organic materials was not fully studied in previous works. We found that metal cations, such as zinc cation (Zn^{2+}), were intercalated in the interlayer space consisting of the dimerized carboxy groups (Fig. 10.15C) [60]. The topochemical polymerization of Zn^{2+}-intercalated PCDA crystal (PCDA-Zn^{2+}) was performed by irradiation of UV light. The PDA-Zn^{2+} with the layered crystal structure was obtained after the polymerization. In general, PDA shows the irreversible color change from blue to red with heating up to around 75°C. The color change is ascribed to the slight torsion of the conjugated backbone in PDA originating from the thermal motion of alkyl chains. The color change is generally irreversible because the distorted conformation is not rearranged after the cooling. The reversible color changes, namely thermochromism, were achieved on PDA derivatives with the designed molecular structures [142−145]. In a couple of previous reports [146−149], thermochromism was achieved on the PDA crystals grown on zinc oxide crystals. However, the relationship between the thermochromism and intercalation of metal cations was not studied in previous reports. Here we studied the color change behavior of the PDA-Zn^{2+} composite with heating and cooling processes. The resultant PDA-Zn^{2+} composites showed the thermochromism with heating and cooling processes (Fig. 10.16A−C), whereas the irreversible color change was only observed on the PDA without intercalation of Zn^{2+}. The reversible color changes were achieved at least 10 times (Fig. 10.16D). The intercalated Zn^{2+} contributes the anchoring of the original conformation of PCDA crystals. Therefore, the anchoring effects induce the returning of the color from red to blue with the cooling. Moreover, the crystal growth, intercalation of Zn^{2+}, and topochemical polymerization were achieved on a paper. The thermochromic paper was obtained through the same processes.

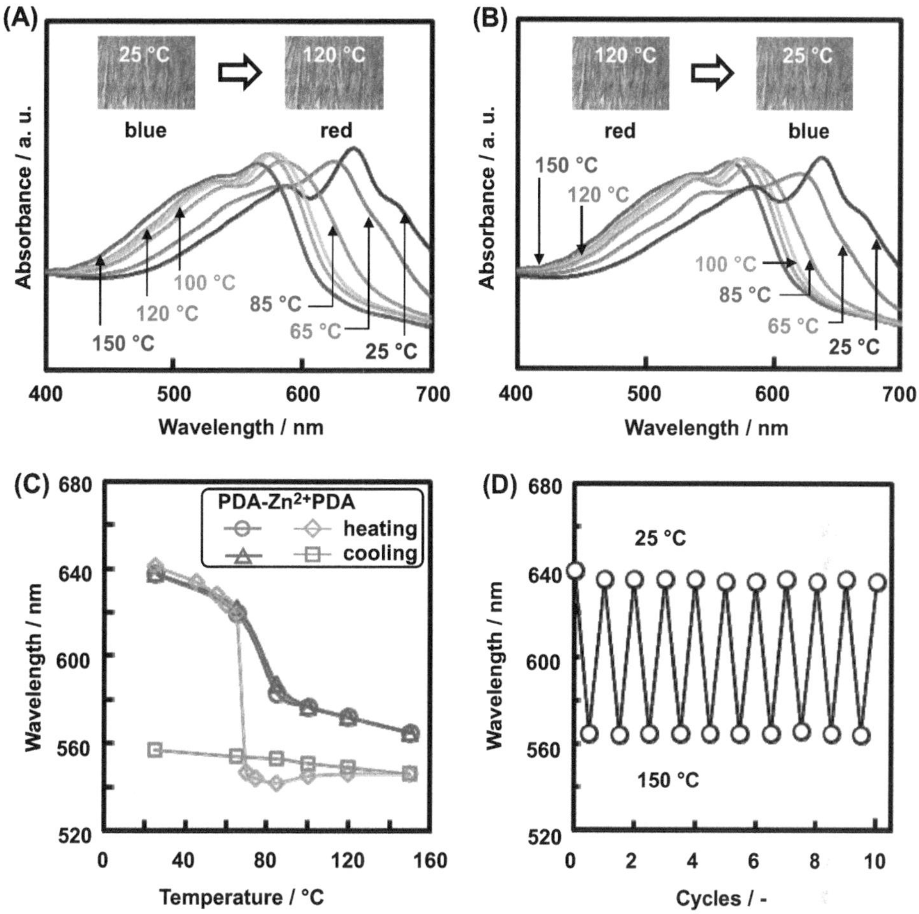

■ **FIGURE 10.16** Thermochromism of PDA after the intercalation of Zn^{2+} (PDA-Zn^{2+}) from the aqueous solution. (A and B) UV—Vis spectra and photographs of the PDA-Zn^{2+} in the heating (A) and cooling processes (B); (C and D) the peak positions of the absorption maximum with the changes of the temperature (C) and their cycle performance (D). *Reproduced with permission from Okaniwa M, Oaki Y, Kaneko S, Ishida K, Maki H, Imai H. Chem Mater 2015;27:2677. Copyright 2015 American Chemical Society.*

In the present section, we can control the morphology and orientation of an organic crystal, namely PCDA, on the substrates with gradual evaporation of the solvents through advanced biomimetic approach. The intercalation of an inorganic cation generated the composite with thermochromic properties based on the reversible conformation changes. Biomimetic approaches and intercalation chemistry in inorganic materials can be applied to development functional organic materials.

10.3 CONCLUSIONS

The present chapter shows crystal-induced nanoarchitectonics of organic polymer materials. Molecular design, synthesis, and self-organization have been well studied in previous works. However, hierarchical morphologies of organic polymer materials from nanoscopic to macroscopic scales are not fully controlled by conventional approaches. We showed that crystals play important roles for hierarchical morphology control of polymers. The interspace, surface, and structure of crystals facilitate morphogenesis of polymer materials, such as conductive polymers. Size, morphology, and orientation of crystals can be tuned by a variety of approaches. The results indicate that crystals with tuned size, morphology, and orientation can be used for syntheses and morphogenesis of organic polymer materials with controlled structures and properties. The approaches are distinguished from those based on molecular design and synthesis. Crystal-induced nanoarchitectonics of organic polymer materials, as a new concept, have versatile potential for generation of functional polymer materials without complex molecular design and synthesis.

REFERENCES

[1] Ariga K, Ji Q, Nakanishi W, hill JP, Aono M. Mater Horiz 2015;2:406.
[2] Ariga K, Li M, Richards G, Hill JP. J Nanosci Nanotech 2011;13:1.
[3] Mann S. Angew Chem Int Ed 2001;39:3392.
[4] Antonietti M, Ozin GA. Chem Eur J 2004;10:28.
[5] Addadi L, Weiner S. Angew Chem Int Ed Engl 1992;31:153.
[6] Weiner S, Addadi L. J Mater Chem 1997;7:689.
[7] Bäuerlein E. Angew Chem Int Ed 2003;42:614.
[8] Sumper M, Brunner E. Adv Funct Mater 2006;16:17.
[9] Mann S. In: Compton RG, Davies SG, Evans J, editors. Biomineralization principles and concepts in bioinorganic materials chemistry. Oxford: Oxford University Press; 2001.
[10] Bäuerlein E, Behrens P, Epple M, editors. Handbook of biomineralization. Weinheim: Wiley-VCH; 2007.
[11] Arakaki A, Nakazawa H, Nemoto M, Mori T, Matsunaga T. J R Soc Interface 2008;5:977.
[12] Arakaki A, Shimizu K, Oda M, Sakamoto T, Nishimura T, Kato T. Org Biomol Chem 2015;13:974.
[13] Mann S. Nature 1993;365:499.
[14] Kato T. Adv Mater 2000;12:1543.
[15] Yu S-H, Cölfen H. J Mater Chem 2004;14:2124.
[16] Imai H, Oaki Y, Kotachi A. Bull Chem Soc Jpn 2006;79:1834.
[17] Xu A-W, Ma Y, Cölfen H. J Mater Chem 2007;17:415.
[18] Meldrum FC, Cölfen H. Chem Rev 2008;108:4332.
[19] Sommerdijk NAJM, de With G. Chem Rev 2008;108:4499.

[20] Mann S. Nat Mater 2009;8:781.
[21] Kato T, Sakamoto T, Nishimura T. MRS Bull 2010;35:127.
[22] Kröger N, Sandhage KH. MRS Bull 2010;35:122.
[23] Imai H, Oaki Y. MRS Bull 2010;35:138.
[24] Yao HB, Fang HY, Wang XH, Yu SH. Chem Soc Rev 2011;40:3764.
[25] Qi L, Cölfen H, Antonietti M, Li M, Hopwood JD, Ashley AJ, et al. Chem Eur J 2001;7:3526.
[26] Yu SH, Antonietti M, Cölfen H, Hartmann J. Nano Lett 2003;3:379.
[27] Henisch HK. Crystals in gels and liesegang rings. Cambridge: Cambridge University Press; 1988.
[28] Sugawara A, Ishii T, Kato T. Angew Chem Int Ed 2003;42:5299.
[29] Sakamoto T, Oichi A, Oaki Y, Nishimura T, Sugawara A, Kato T. Cryst Growth Des 2009;9:622.
[30] Oaki Y, Imai H. Cryst Growth Des 2003;3:711.
[31] Kato T, Suzuki T, Irie T. Chem Lett 2001:186.
[32] Nishimura T, Ito T, Yamamoto Y, Yoshio M, Kato T. Angew Chem Int Ed 2008; 47:2800.
[33] Koumoto K, Saito N, Gao Y, Masuda Y, Zhu P. Bull Chem Soc Jpn 2008;81:1337.
[34] Masuda Y. J Cerm Soc Jpn 2007;115:101.
[35] Dey A, Bomans PHH, Müller FA, Will J, Frederik PM, de With G, et al. Nat Mater 2010;9:1010.
[36] Xu AW, Antonietti M, Yu SH, Cölfen H. Adv Mater 2008;20:1333.
[37] Oaki Y, Imai H. Adv Funct Mater 2005;15:1407.
[38] Oaki Y, Adachi R, Imai H. Polym J 2012;44:612.
[39] Oaki Y, Imai H. J Am Chem Soc 2004;126:9271.
[40] Oaki Y, Imai H. Chem Commun 2005:6011.
[41] Oaki Y, Imai H. Adv Mater 2006;18:1807.
[42] Oaki Y, Hayashi S, Imai H. Chem Commun 2007:2841.
[43] Oaki Y, Imai H. Angew Chem Int Ed 2007;46:4951.
[44] Oaki Y, Imai H. Chem Eur J 2007;13:8564.
[45] Oba M, Oaki Y, Imai H. Adv Funct Mater 2010;20:4279.
[46] Oaki Y, Yagita N, Imai H. Chem Eur J 2012;18:110.
[47] Oaki Y, Nakamura K, Imai H. Chem Eur J 2012;18:2825.
[48] Yagita N, Oaki Y, Imai H. Chem Eur J 2013;19:4419.
[49] Oaki Y, Imai H. Small 2006;2:66.
[50] Oaki Y, Kotachi A, Miura T, Imai H. Adv Funct Mater 2006;16:1633.
[51] Oaki Y, Kijima M, Imai H. J Am Chem Soc 2011;133:8594.
[52] Kijima M, Oaki Y, Munekawa Y, Imai H. Chem Eur J 2013;19:2284.
[53] Munekawa Y, Oaki Y, Imai H. Langmuir 2014;30:3236.
[54] Munekawa Y, Oaki Y, Sato K, Imai H. Nanoscale 2015;7:3466.
[55] Sato K, Oaki Y, Imai H. Chem Commun 2015;51:7919.
[56] Oaki Y, Oki T, Imai H. J Mater Chem 2012;22:21195.
[57] Kuwabara K, Oaki Y, Muramatsu R, Imai H. Chem Commun 2015;51:9698.
[58] Oaki Y, Muramatsu R, Imai H. Polym J 2015;47:183.
[59] Muramatsu R, Oaki Y, Kuwabara K, Hayashi K, Imai H. Chem Commun 2014;50: 11840.

[60] Okaniwa M, Oaki Y, Kaneko S, Ishida K, Maki H, Imai H. Chem Mater 2015;27:2677.
[61] Okaniwa M, Oaki Y, Imai H. Bull Chem Soc Jpn 2015;88:1459.
[62] Nakamura K, Oaki Y, Imai H. J Am Chem Soc 2013;135:4501.
[63] Honda M, Oaki Y, Imai H. Chem Mater 2014;26:3579.
[64] Honda M, Oaki Y, Imai H. Chem Commun 2015;51:10046.
[65] Kato T, Mizoshita N, Kishimoto K. Angew Chem Int Ed 2006;45:38.
[66] Murphy AR, Fréchet JMJ. Chem Rev 2007;107:1066.
[67] Aida T, Meijer EW, Stupp SI. Science 2012;335:813.
[68] Comelissen JJLM, Rowan AE, Nolte RJM, Sommerdijk NAJM. Chem Rev 2001;101:4039.
[69] Dankers PYW, Meijer EW. Bull Chem Soc Jpn 2007;80:2047.
[70] Li M, Ishihara S, Ji Q, Akada M, Hill JP, Ariga K. Sci Technol Adv Mater 2012;13:053001.
[71] Shrestha LK, Ji Q, Mori T, Miyazawa K, Yamauchi Y, Hill JP, et al. Chem Asian J 2013;8:1662.
[72] Nakanishi T, Naito M, Takeoka Y, Matsuura K. Curr Opin Colloid Interface Sci 2011;16:482.
[73] Sato K, Oaki Y, Takahashi D, Toshima K, Imai H. Chem Eur J 2015;21:5034.
[74] Ariga K, Yamauchi Y, Rydzek G, Ji Q, Yonamine Y, Wu KCW, et al. Chem Lett 2014;43:36.
[75] Ariga K, Yamauchi Y, Mori T, Hill JP. Adv Mater 2013;25:6477.
[76] Zander N. Polymers 2013;5:19.
[77] Agarwal S, Greiner A, Wendorff J. Prog Polym Sci 2013;38:963.
[78] Frenot A, Chronakis I. Curr Opin Colloid Interface Sci 2003;8:64.
[79] Lin J, Tian F, Shang Y, Wang F, Ding B, Yu J, et al. Nanoscale 2013;5:2745.
[80] Li X, Ding B, Lin J, Yu J, Sun G. J Phys Chem C 2009;113:20452.
[81] Kelly TL, Wolf MO. Chem Soc Rev 2010;39:1526.
[82] Sada K, Takeuchi M, Fujita N, Numata M, Shinkai S. Chem Soc Rev 2007;36:415.
[83] Bein T, Enzel P. Angew Chem Int Ed Engl 1989;28:1692.
[84] Cardin DJ. Adv Mater 2002;14:553.
[85] Tajima K, Aida T. Chem Commun 2000:2399.
[86] Caruso RA, Antonietti M. Chem Mater 2001;13:3272.
[87] Uemura T, Horike S, Kitagawa K, Mizuno M, Endo K, Bracco S, et al. J Am Chem Soc 2008;130:6781.
[88] Uemura T, Kaseda T, Kitagawa S. Chem Mater 2013;25:3772.
[89] Horike S, Shimomura S, Kitagawa S. Nat Chem 2009;1:695.
[90] Uemura T, Yanai N, Kitagawa S. Chem Soc Rev 2009;38:1228.
[91] Thomas A, Goettmann F, Antonietti M. Chem Mater 2008;20:738.
[92] Moon S, McCarthy T. Macromolecules 2003;36:4253.
[93] Chakraborty S, Colón Y, Snurr R, Nguyen S. Chem Sci 2015;6:384.
[94] Cepak V, Martin C. Chem Mater 1999;11:1363.
[95] Sozzani P, Bracco S, Comotti A, Simonutti R, Valsesia P, Sakamoto Y, et al. Nat Mater 2006;5:545.
[96] Cölfen H, Mann S. Angew Chem Int Ed 2003;42:2350.
[97] Cölfen H, Antonietti M. Angew Chem Int Ed 2005;44:5576.
[98] Niederberger M, Cölfen H. Phys Chem Phys Chem 2006;8:3271.

[99] Zhou L, O'Brien P. Small 2008;4:1566.
[100] Song RQ, Cölfen H. Adv Mater 2010;22:1301.
[101] Zhou L, O'Brien P. J Phys Chem Lett 2012;3:620.
[102] Oaki Y, Imai H. Spr Nanosci 2013;1:1.
[103] Oaki Y, Imai H. Bull Chem Soc Jpn 2009;82:613.
[104] Liu X, Li H, Li D, Ishida M, Zhou H. J Power Sources 2013;243:373.
[105] Wang Y, Li H, He P, Hosono E, Zhou H. Nanoscale 2010;2:1294.
[106] Bhattacharyya D, Howden RM, Borrelli DC, Gleason KK. J Polym Sci B Polym Phys 2012;50:1329.
[107] Coclite AM, Howden RM, Borrelli DC, Petruczok CD, Yang R, Yagüe JL, et al. Adv Mater 2013;25:5392.
[108] Ojio T, Miyata S. Polym J 1986;18:95.
[109] Mohammadi A, Hasan MA, Liedberg B, Lundstrom I, Salaneck WR. Syn Met 1986;14:189.
[110] Kim J, Kim E, Won Y, Lee H, Suh K. Syn Met 2003;139:485.
[111] Winther-Jensen B, West K. Macromolecules 2004;37:4538.
[112] Kim J, Kwon M, Min Y, Kwon S, Ihm DW. Adv Mater 2007;19:3501.
[113] Lock JP, Im SG, Gleason KK. Macromolecules 2006;39:5326.
[114] Bai H, Li C, Chen F, Shi G. Polymer 2007;48:5259.
[115] Nalwa HS, Kasai H, Okada S, Oikawa H, Matsuda H, Kakuta A, et al. Adv Mater 1993;5:758.
[116] Horn D, Rieger J. Angew Chem Int Ed 2001;40:4330.
[117] Zhao YS, Fu H, Peng A, Ma Y, Xiao D, Yao J. Adv Mater 2008;20:2859.
[118] Li R, Hu W, Liu Y, Zhu D. Acc Chem Res 2010;43:529.
[119] Zhao L, Yang W, Luo Y, Zhai T, Zhang G, Yao J. Chem Eur J 2005;11:3771.
[120] Tachikawa T, Chung HR, Masuhara A, Kasai H, Oikawa H, Nakanishi H, et al. J Am Chem Soc 2006;128:15944.
[121] Zhao YS, Xu J, Peng A, Fu H, Ma Y, Jiang L, et al. Angew Chem Int Ed 2008;47: 7301.
[122] Lei Y, Liao Q, Fu H, Yao J. J Phys Chem C 2009;113:10038.
[123] Jiang L, Hu W, Wei Z, Xu W, Meng H. Adv Mater 2009;21:3649.
[124] Cong HP, Yu SH. Chem Eur J 2007;13:1533.
[125] Cong HP, Yu SH. Adv Funct Mater 2008;18:195.
[126] Ma Y, Mehltretter G, Plüg C, Rademacher N, Schmidt MU, Cölfen H. Adv Funct Mater 2009;19:2095.
[127] Huang M, Shilde U, Kumke M, Antonietti M, Cölfen H. J Am Chem Soc 2010;132: 3700.
[128] Yu H, Qi L. Langmuir 2009;25:6781.
[129] Cunningham JT, MacMunn CA. Phil Trans R Soc Lond B 1893;184:765.
[130] Denton EJ. Phil Trans R Soc Lond B 1970;258:285.
[131] Oshima N, Fujii R. Cell Motil Cytoskelt 1987;8:85.
[132] Levy-Lior A, Pokroy B, Levari-Sivan B, Leiserowitz L, Weiner S, Addadi L. Cryst Growth Des 2008;8:507.
[133] Levy-Lior A, Shimoni E, Schwartz O, Gravish-Regev E, Oron D, Oxford G, et al. Adv Funct Mater 2010;20:320.
[134] Yoshioka S, Matsuhana B, Tanaka S, Inouye Y, Oshima N, Kinoshita S. J R Soc Interface 2011;8:56.

[135] Oaki Y, Kaneko S, Imai H. J Mater Chem 2012;22:22686.
[136] Wegner G. Makromol Chem 1972;154:35.
[137] Day D, Ringsdorf H. J Polym Sci Polym Lett Ed 1978;16:205.
[138] Lauher JW, Fowler FW, Goroff NS. Acc Chem Res 2008;41:1215.
[139] Matsumoto A, Odani T. Macromol Rapid Commun 2001;22:1195.
[140] Baughman RH. J Polym Sci Polym Phys Ed 1974;12:1511.
[141] Coe S, Kane JJ, Nguyen TL, Toledo LM, Wininger E, Fowler FW, et al. J Am Chem Soc 1997;119:86.
[142] Wenzel M, Atkinson GH. J Am Chem Soc 1989;111:6123.
[143] Carpick RW, Sasaki DY, Burns AR. Langmuir 2000;16:1270.
[144] Shimogaki T, Matsumoto A. Macromolecules 2011;44:3323.
[145] Tanioku C, Matsukawa K, Matsumoto A. ACS Appl Mater Interfaces 2013;5:940.
[146] Hatada M, Katanosaka A, Minami N. J Photopolym Sci Tech 2006;19:63.
[147] Triphol N, Rungruangviriya N, Potal R, Triphol R. J Colloid Interface Sci 2011; 356:481.
[148] Patlolla A, Zunino J, Frenkel AI, Iqbal Z. J Mater Chem 2012;22:7028.
[149] Chanakul A, Traiphol N, Traiphol R. J Colloid Interface Sci 2013;389:106.

Nanoarchitectonics for Cyclodextrin-Mediated Solubilization and Nanoassembly of Therapeutic Agents

K. Nobusawa[1] and M. Naito[2]

[1]*Osaka University, Osaka, Japan;* [2]*National Institute for Materials Science, Tsukuba, Japan*

CHAPTER OUTLINE

11.1 SUPRAMOLECULAR MANIPULATION: NANOMEDICINE WITH HOST–GUEST NANOARCHITECTONICS

Supramolecular manipulation of nanostructured materials has made significant contributions in current biomedical applications. For curing diseases or repairing organ, nanomachines deliver tailor-made-designed drugs targeted to affected areas. Such improvement in quality of life with nanotechnology is nearing reality. These offshoots of nanotechnology are referred to as "nanomedicine" [1]. Along with this initiative, a variety of concepts in nanomedicine have been proposed with organic/inorganic/metal/polymeric materials, as well as nanofabrication. Among them, we

Supra-materials Nanoarchitectonics. http://dx.doi.org/10.1016/B978-0-323-37829-1.00011-0

specifically focused on the potential application of cyclodextrin (CD)-based host–guest system as a successful example of supramolecular manipulation in nanomedicine. Utilizing their unique ability to solubilize and stabilize small molecules, the CDs and their derivatives have been used as an excipient in pharmaceutical applications for decades [2–5]. Recently, on the basis of unique characteristics of CD derivatives such as their responsiveness to stimuli, rational delivery and targeting of pharmaceutical, therapeutic, and diagnostic agents have been possible. In this chapter, we briefly introduce the fundamental techniques for solubilization of drugs using CDs, followed by discussion of designs through supramolecular nanoassembly for enhancement of drug solubility and therapeutic efficiency. We also illustrate CD-based supramolecular manipulations for biological applications, mainly focusing on phototherapeutic applications.

11.2 CYCLODEXTRIN INCLUSION COMPLEX FORMATION TOWARD NANOASSEMBLED SYSTEMS

A number of techniques have been exploited for equipping therapeutically prospective materials with pharmaceutical formulations to increase drug solubility, bioavailability, and targeted drug delivery. CDs have been used as a potential candidate to solubilize poorly water-soluble drugs and control the pharmacokinetics in clinical environments in which they have low toxicity, encapsulation capability of functional molecules through host–guest interactions, and excellent biocompatibility. By 1953, Freudenberg, Cramer, and Plieninger had received the first patent of CDs in drug formations [6]. Although this patent was only one-and-half pages, it involved the most important concepts on drug formations that are still as valid. Over 30 drugs with CD inclusion complexes are in today's market worldwide [7]. Three CDs are well known as host molecules: the α-, β-, and γ-CD of cyclic-oligosaccharides are composed of 6, 7, and 8 D-glucose units that are covalently linked by an α-1,4 linkage, respectively. These truncated cone structures have primary and secondary hydroxyl groups positioned in the narrower and wider rims of CDs (Fig. 11.1). A specified number of glucose units can encapsulate different classes of guest molecules. CDs have been used for the retention and kinetic release of drugs, delivery control by chemical modification, and preventing degradation of unstable materials in oral and other deliveries [8]. Host:guest inclusion complexes are generally categorized into 1:1, 1:2, 2:1, and 2:2 stoichiometry (Fig. 11.1) in which driving forces of the complexation are such as van der Waals or hydrophobic interactions, the release of contained water molecules in CD cavity, and dimensional fitting.

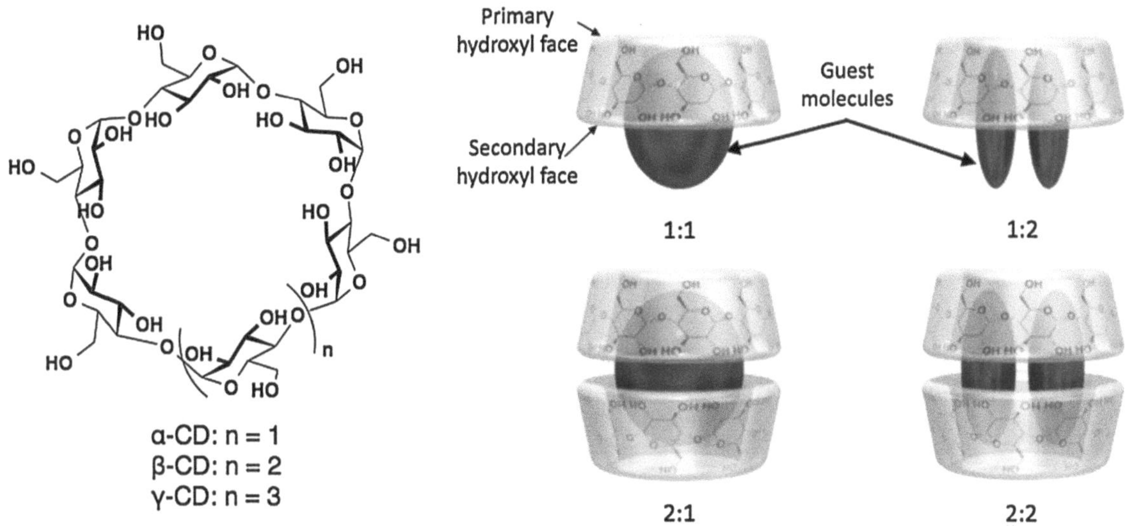

■ **FIGURE 11.1** Chemical structures of three major CDs and schematic illustrations of CD/guest inclusion complexes.

The CD/drug complexes are usually supposed to display the 1:1 stoichiometry. In effective designs for practical use, in order to achieve complex stability and drug solubility, chemical modification of CD can be applied. The chemical features of CD enable various functionalities of the primary and secondary hydroxyl groups. For more applicability, proper designs on each hydroxyl side are applied, which result in certain properties, such as amphiphilic molecules that form self-assembled nanostructures. One of the manners is methylation of hydroxyl groups of CD to improve the water solubility of host molecules themselves, expand the hydrophobic space of the cavities, and distort CD ring structures, improving their ability to associate with guest molecules. For example, paclitaxel (PTX), which is one of the most widely used medications in the treatment of cancers, is subjected to the 1:1 stoichiometry with β-CD, and the stability constants and the antitumor activity can be modulated by methylation of hydroxyl groups of β-CD [9]. Albendazole (ABZ), which has potent antiparasitic activity, is encapsulated into randomly methylated β-CD in the 1:1 complex formulation [10]. In this case, the apparent aqueous solubility of ABZ increased noticeably with CD, and the corresponding solubility was enhanced by a factor of 7600 compared to that without CD. Shape complementarity and hydrogen bonding formation between host and guest molecules can together enhance solubility in the host–guest system.

From the above examples of CD/ABZ inclusion complex, it is suggested that aggregates composed of the inclusion complex and/or CD itself can afford to enhance the hydrosolubility of drugs (Fig. 11.2). According to other reports of hydrosoluble drugs in the host–guest complexation manner, the enhancement of apparent drug solubility can be designed based on self-association, aggregation of the inclusion or noninclusion complexes [11–13]. Considering that CD/drug inclusion complexes are probable components of building blocks in nanoarchitectures, it is important to explore their assembling behavior when developing novel nanoassemble systems. One example can be seen in a CD/curcumin inclusion complex [13]. Curcumin, which is a kind of lipophilic polyphenol, is known to possess anticancer properties. To explore the delivery properties of CD as a carrier of curcumin, a complex of β-CD and curcumin was prepared, and the authors illustrated the self-assembley or nanoassembley process for the improvement of drug solubility and delivery characteristics of curcumin using β-CD. The different loading capacities of curcumin could be controlled with up to 26% of curcumin. The self-assembled complex enhanced the intracellular uptake of curcumin and improved its therapeutic efficacy against the cancer cells compared to free curcumin. Sun et al. reported that a class of aza-arm modified β-CDs acts as building blocks to provide a CD/drug-assembled architecture. The CDs not only hold PTX but also act as supramolecular amphiphiles for further self-assembling into vesicular structures in aqueous solution [14].

As a versatile approach for drug loading and delivery, supramolecular conjugates of CD-branched polymer and adamantine-flanking gold nanoparticle have been reported [15]. In this study, high affinity of host–guest interaction between adamantine and CD adducts was employed; as a result, the conjugate of cyclodextrin-modified hyaluronic acid (HACD) and gold nanoparticle provided a three-dimensional porous structure. This biocompatible platform allowed the encapsulation and delivery of various anticancer drugs, such as doxorubicin hydrochloride, PTX, camptothecin, irinotecan

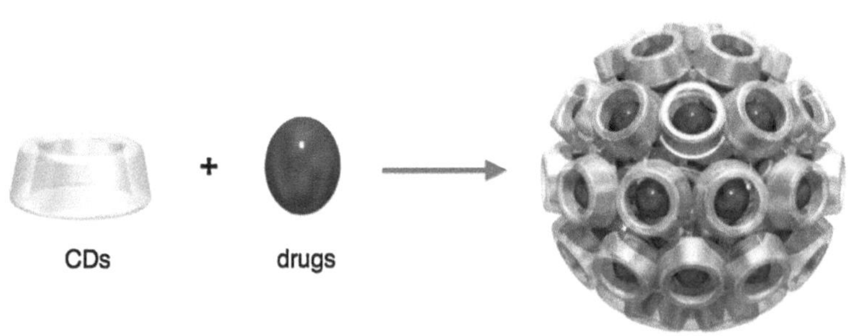

■ **FIGURE 11.2** Schematic illustration of the assembly of the inclusion complex via host–guest interaction of CD and drugs.

hydrochloride, and topotecan hydrochloride. The conjugates exhibited pH-responsive drug release under mild acidic conditions. Additionally, due to the high efficiency of their cellular uptake ability by HA-receptor-mediated endocytosis, the conjugates with anticancer drugs efficiently inhibited the growth of human breast cancer MCF-7 cells and drug toxicity to normal cells was decreased.

Fullerenes are one of the promising therapeutic applicants. The excellent properties originating from water-soluble fullerene have opened up new opportunities for applications in life science and clinical fields, such as antioxidant agents, diagnostic imaging/sensing, photodynamic therapy (PDT) and even as drug carriers [16]. PDT has received much attention in modern phototherapeutic applications because of its therapeutic efficacy and treatment modality for the malignant and diseased cells. In PDT, photosensitizers (PSs) accumulate in desired cells or tissues, and convert tissue oxygen into highly reactive oxygen species (ROS), such as singlet oxygen ($^1O_2$) or free radicals with the help of external energy from light [17]. In the case of fullerenes, they show high ROS generation efficiencies, e.g., a quantum yield of $^1O_2$ generation by [60]fullerene (C_{60}) is 0.96 (532 nm) [18]. However, these intact fullerenes are inherently insoluble in water. Nowadays, a variety of attempts to develop facile and efficient methods for water solubilization of fullerenes with chemical treatment exist, such as polyhydroxylated fullerenes (fullerenols) [19], water-soluble fullerene polymers [20] and a type of amphiphilic fullerenes [21]. Physical treatments can also be used using bio or bio-related materials [22,23]. CD and its derivatives have been employed as effective candidates for C_{60} hydrosolubilization without chemical treatments to C_{60}. Nakanishi et al. investigated the biological activity of the γ-CD/C_{60} inclusion complex, and the complex exhibited an efficient photodriven DNA-cleaving ability via visible light irradiation with the presence of NADH (β-nicotinamide adenine dinucleotide, reduced form) under an O_2-saturated condition [24]. Zhang et al. illustrated that γ-CD polymer (γ-CDP) formed by polymerization of γ-CD with epichlorohydrin incorporates C_{60} with high solubility, and the γ-CDP/C_{60} conjugate shows phototherapeutic properties [25]. The γ-CDP/C_{60} conjugate was active generating $^1O_2$ under ultraviolet A irradiation, and exhibited a higher phototoxicity against mouse melanoma cell lines B16-F10 compared to pristine C_{60}.

In most cases for the complexation of monomeric γ-CD and C_{60}, however, to achieve sufficient binding affinity between C_{60} and γ-CD and retain the complex stability, excess amounts of γ-CD are required in the solution. For example, removing excess γ-CD can cause the slow decomposition of the complex [26]. On the other hand, a precipitate was observed when a high concentration of γ-CD was used to soluble C_{60}. Although the precipitate containing γ-CD and C_{60} can be soluble in boiling water, the solution color was changed to yellow, which is attributed to the formation of C_{60} aggregates. Therefore, maintaining its solubility and stability in aqueous media is still an issue.

The formulation of the CD/C_{60} inclusion complex is commonly thought to display the 2:1 stoichiometry as shown in Fig. 11.3A. Indeed, the 2:1 complex formulation of γ-CD was directly substantiated by means of X-ray crystal structure analysis using a malonic acid derivative of C_{60} encapsulated in γ-CDs (Fig. 11.3B) [27]. Furthermore, this experiment highlights the importance of the hydrogen bonding interactions between the CD's wider rims facing each of the CDs. The result suggests that the multiple interface interactions are attributed to the stabilization of the inclusion complex. Thus, the regioselective substitutions of CD in a 2:1 complex should be investigated further for establishing supramolecular architectures.

Liu et al. showed that when dimer β-CD linked with metallobridged spacer, C_{60} exhibited a higher water solubility [28] compared to neutral β-CD, which had previously been reported [29]. From STM observation, this host–guest interaction system provided self-assembled supramolecular polymers with regular linear arrangement of the C_{60} array, and this assembly had the ability to act as the photodriven DNA cleaver (Fig. 11.4). The same group also reported the construction of a well preorganized nanosupramolecular architecture with a CD-dimer with Tb^{3+} and C_{60} units [30]. The complexation of CD-Tb polyads and C_{60}, characterized by means of NMR, UV–vis adsorption spectral, XRD, and TG-DTA results suggested a 1:3 mole ratio between CD-Tb polyads and C_{60} units. Moreover, STM

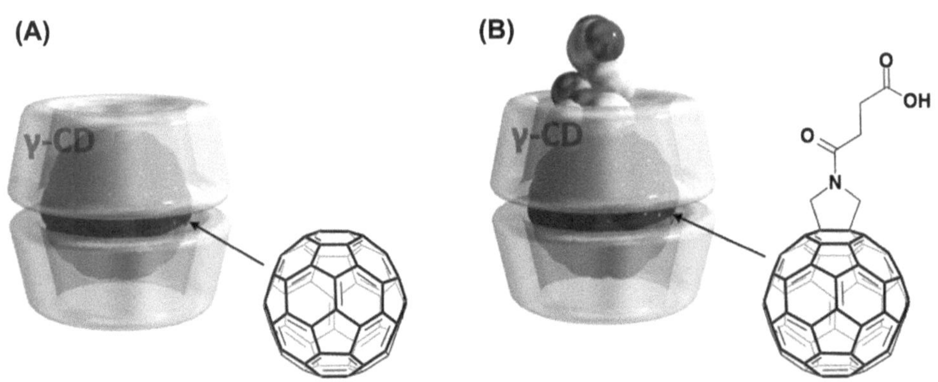

■ **FIGURE 11.3** Schematic illustration of 2:1 inclusion complex of γ-CD and (A) C_{60} and (B) malonic acid derivative. The structural integrity of host–guest molecular stoichiometry of the complex in (B) was clarified as 2:1 by means of X-ray diffraction analysis.

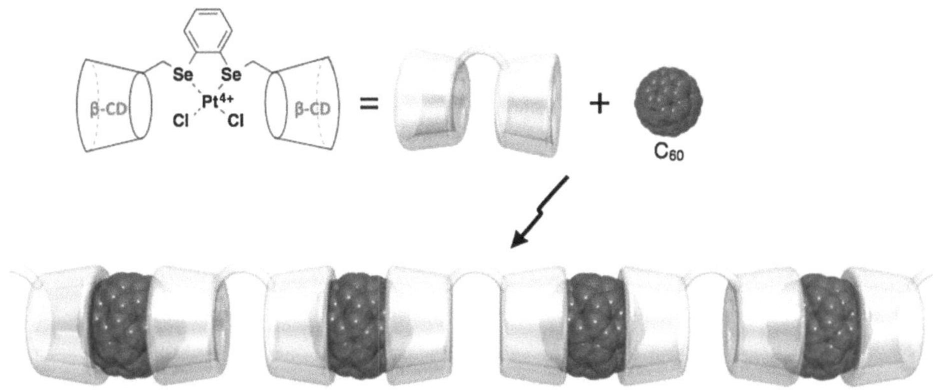

■ **FIGURE 11.4** Schematic illustration of C_{60} self-assembled polymer with metallobridged β-CD dimer.

observation secured the structural evidence for the assembly resulting in a trefoil bundle-shaped nanoarchitecture. The assembly with C_{60} caused luminescence quenching of the polyads through energy transfer to C_{60}.

While covalently linked multiple CDs have previously been reported for constructing supramolecular assembly with C_{60}, we recently proposed a novel hierarchically assembled nanoarchitecture via noncovalent bonding of the inclusion complex for hydrosolubilization of C_{60} (Fig. 11.5) [31]. In new material design, all 6-hydroxyl groups of γ-CD are substituted into pyridyl groups with triazole by employing click chemistry. This pyridyl terminated γ-CD (PCD) exhibited 90 times higher hydrosolubilizing capability for C_{60} than nonsubstituted γ-CD. Because nonsubstituted CD or a modified CD without pyridyl groups (HCD in Fig. 11.5A) did not show such high-solubilizing abilities, appended pyridine in CD is essential in this system. From UV−vis absorption spectral analysis, the mole ratio between C_{60} and PCD in the complex solution was evaluated to be 1:1.8−2.0. We also estimated an abundance ratio of CDs with or without forming the inclusion complex by ^{1}H NMR studies. As a result, it was evaluated that 94% PCD forms the inclusion complex, namely, there are extremely fewer noninclusion complexes in the aqueous media. On the other hand, in γ-CD solution, 58% γ-CD remained as the noninclusion form. Furthermore, the NMR analysis suggested that PCD/C_{60} units are associated with high-ordered supramolecular structures when associated with C_{60}, in which the pyridyl groups may act as noncovalent interaction moieties to assemble PCD/C_{60} units. Morphological investigation of AFM and molecular mechanics simulation supported that the PCD/C_{60} inclusion complex spontaneously formed fibril structures. Consequently, PCD/C_{60} complexes are thought to be stabilized in the formed fibrous architecture.

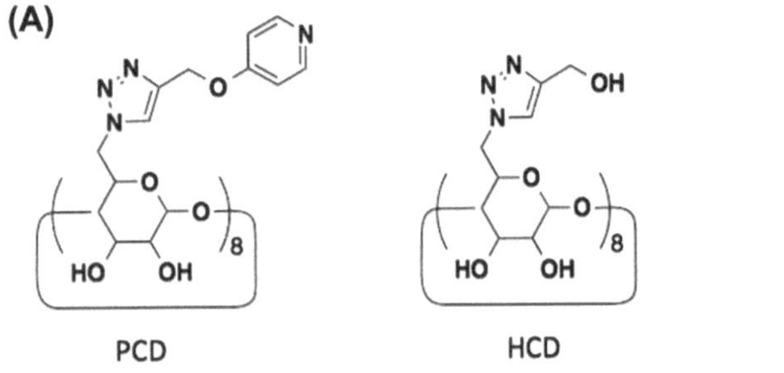

	Maximum concentration of C_{60}
PCD	72.7 mM
HCD	1.1 mM
γCD	0.8 mM

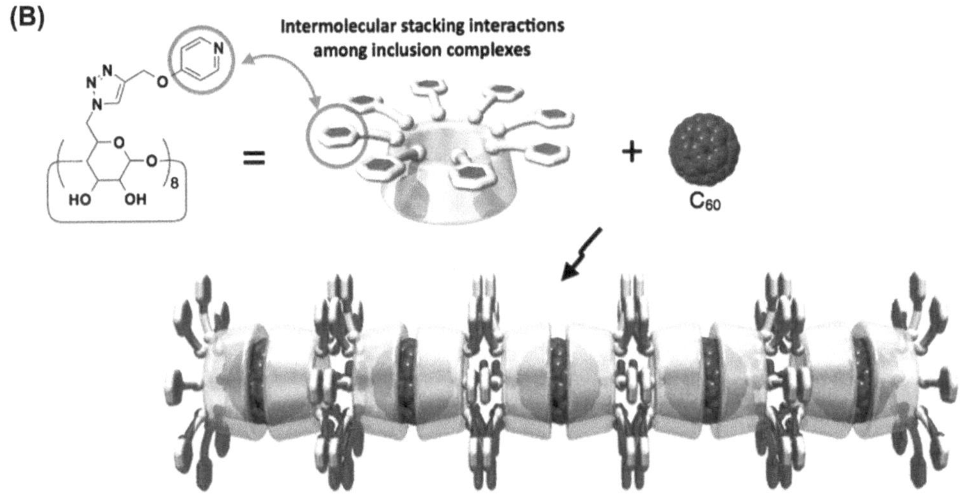

■ **FIGURE 11.5** (A) Chemical structures of cyclodextrin derivatives and C_{60} concentrations in the inclusion complex solution. (B) Schematic representation of hierarchical assembly of inclusion complex of C_{60} through host–guest complexation.

The PCD/C_{60} inclusion complex was prepared by a solid-state mechanochemical process [32]. To briefly explain, this method is called high-speed vibration milling (HSVM), in which powders of C_{60} and CDs are vigorously shaken with a mixing ball. The resulting mixture is dissolved in water and purified by filtration (or centrifugation in our study) to remove the insoluble segments. In our study, the γ-CD/C_{60} inclusion complex prepared by HSVM had relatively high solubility up to 0.8 mM. In contrast, a concentration of extracted C_{60} in aqueous CD solution under reflux was 0.08 mM [26]. Although the PCD/C_{60} inclusion complex was fabricated by HSVM method, a maximum concentration of C_{60} reached 72.7 mM, which is much higher than when using γ-CD. This mechanochemical technique can also be a useful, facile but efficient method in the nanofabrication of carbon materials, such as blending nanoscale functional materials toward the development of a manipulation system of carbon nanotube [33], and selective edge activation of graphene [34].

11.3 NANOASSEMBLY CONTROL USING A CD INCLUSION COMPLEX FOR PDT APPLICATION

Liposome, which is formed through self-assembly of amphiphilic lipids in aqueous media, is one of the excellent nanocapsules that has a major practical use for drug loading and targeted drug delivery. Drugs in liposomes commonly represent two types of formations; hydrophilic drugs are introduced in an interior aqueous core, and hydrophobic drugs are embedded into lipid membranes (Fig. 11.6). One of the challenges is to achieve enough drug-loading efficiency for therapeutic use. CD inclusion complexes are promising drug loaders because poorly water-soluble therapeutic agents can retain hydrosolubility by being encapsulated in the CD cavity during loading processes. One method is to confine drugs to the liposome interior during liposome formation by lipid film hydration with CD/drug complex solution. This kind of system is called "drugs-in-cyclodextrin-in-liposome" [35]. For example, Piel et al. showed that betamethasone as a model drug and its inclusion complexes with various kinds of CDs can be encapsulated in the liposome aqueous core [36]. It was revealed that drug encapsulation efficiencies are higher in all CDs tested compared to those without CD. CDs showed a lower affinity for a lipid component, and allowed liposome to entrap high concentrations of the drug with or without minimal interruption of the lipid components, conserving the drug-holding liposome from destruction.

CD-based inclusion complexes also raise the prospect of modulating self-assembly in which molecular assembly can be regulated in the liposomal

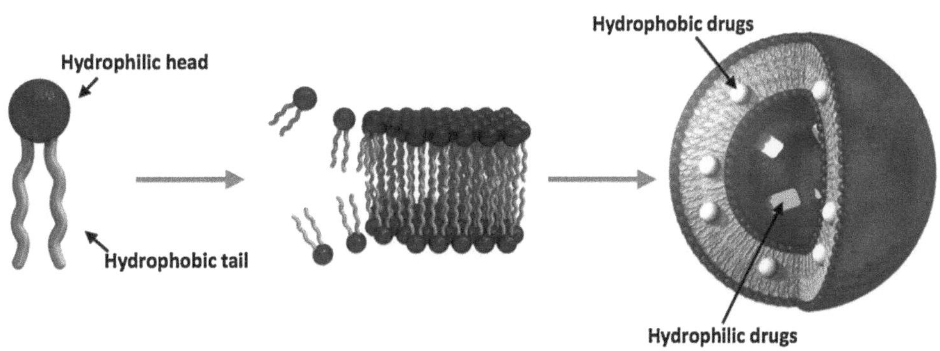

■ **FIGURE 11.6** Schematic representation of liposome formation through the self-assembly of amphiphilic lipids and drug-loaded liposome.

membranes. Hydrophobic fullerenes are inherently entrapped within the liposomal membranes, and these fullerene-liposome conjugates exhibit photoactive properties as therapeutic agents [23]. A conventional way for introducing fullerenes to lipid membranes is to mix fullerenes and lipids in organic solution, and prepare dry thin films before the liposome formation via hydration in aqueous solution (Fig. 11.7A) [37,38]. In contrast, Ikeda et al. have provided a new strategy using host−guest complexes to efficiently induce fullerenes into liposomal membranes; it is called the exchange method (Fig. 11.7B) [23,39,40]. In this method, the loading efficiency of 10 mol% C_{60} incorporation against lipid concentration was achieved. The value was considerably higher than that in a previous study, which was less than 3% [38]. The C_{60} loading was achieved by heat assist at 80°C, and the exchange method of C_{70} took place at room temperature within a

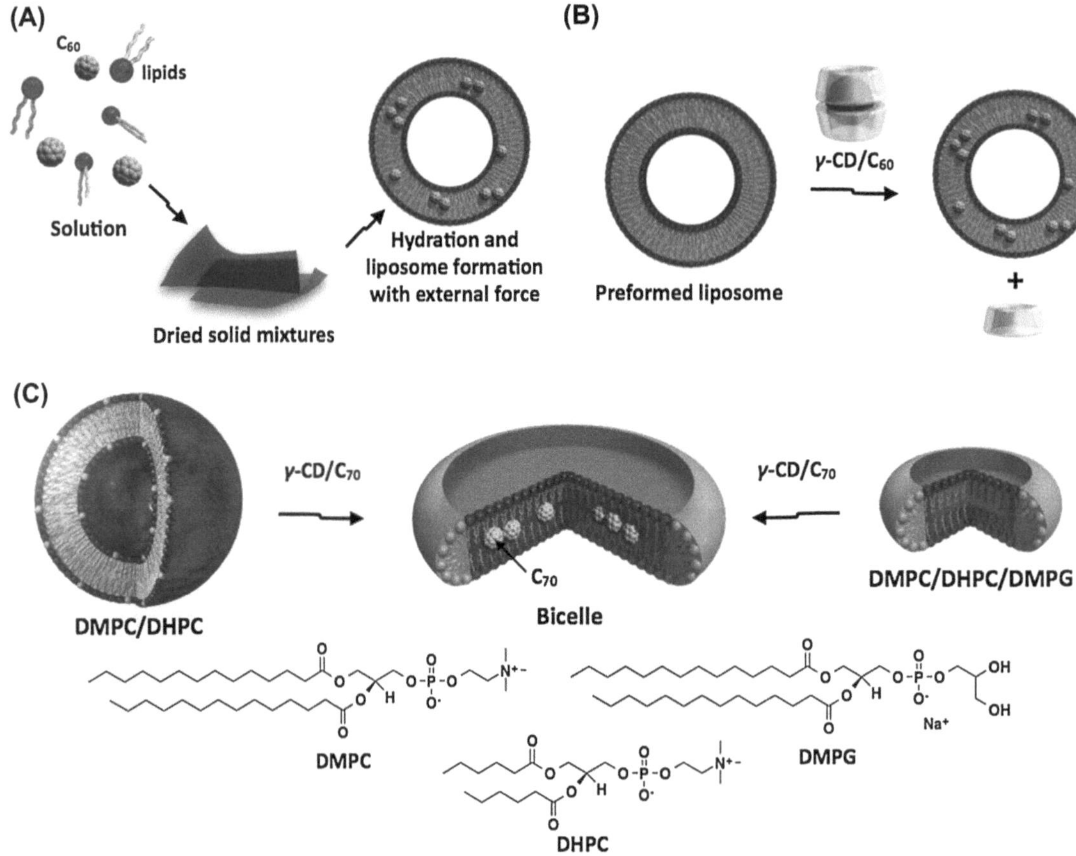

■ **FIGURE 11.7** Schematic representation of fullerene-loaded liposome prepared by (A) hydration of premixed components of fullerenes and lipids and (B) fullerene exchange method. (C) C_{70} induced nanoassembly conversion on lipid bilayers by utilizing the exchange method.

few minutes using γ-CD/C_{70}. These fullerene-loaded liposomes were investigated based on their biological activities. As a result, fullerenes within cationic liposomes exhibited noticeably efficient DNA photocleavage activities compared to γ-CD inclusion complexes, and the C_{70}-loaded liposome showed 3.5 times higher ability than the C_{60}-liposome. The influences of preparation methods have been investigated in terms of changes in the morphology and stability of liposomes and photodynamic studies of fullerenes, C_{60} and C_{70} [41]. As a result, the exchange reaction showed advantageous properties with easier control of the vesicle size, and higher stability of the liposome solution. Moreover, from the assay of cell viability against HeLa cells, the C_{60} or C_{70}-loaded liposomes in the exchange method enhanced photodynamic activities compared to the corresponding liposomes prepared by the conventional method, despite collecting identical fullerene concentrations in both cases. These clear advantages in photoresponse action are considered to be caused by the degree of aggregates of fullerenes in lipid bilayers. Further experimental and computational-simulation studies proved the location of C_{60} in liposomal membrane when taking advantage of high concentrated C_{60}-loaded liposome. The results showed that C_{60} is incorporated in the hydrophobic core of the lipid bilayers [42]. Furthermore, this exchange method provides an interesting aspect of the morphological conversion of the liposome to a disk-like assembled structure, which is commonly known as a bicelle, when inducing fullerene C_{70} using such methods (Fig. 11.7C) [43]. Bicelles are generally prepared using a mixture of lipids with long- and short-hydrocarbon chains under high lipid concentration conditions. From cryo-TEM and DLS analyses under proper experimental conditions, the lipid−C_{70} interaction systems enabled the bicelle formation to lower the required lipid concentration and increased the diameter of the bicelle. It may be proposed that C_{70} encapsulated into the lipid membranes can assist them to retain a rigid and planar structure to stabilize the assembly in the bicelle structures. Fullerene-lipid self-assemblies show promising potential as nanoarchitecture in biological applications.

Recently we reported another type of C_{60} loading and delivery via supramolecular system in a pH-responsive drug-delivery application using aminocyclodextrin, in which a primary amine was displayed at the six positions of CD (ACD) (Fig. 11.8) [44]. A class of aminated CDs was already introduced more than 30 years ago, which can enhance the binding ability to the amphiphilic guests in which anionic molecules such as phosphates are encapsulated in the CD cavities via a combination of electrostatic and hydrophobic interactions [45,46]. Using C_{60} as a guest, the recognition mode becomes quite different. A new feature of ACD as a pH-responsive carrier of C_{60} could be accomplished by adjusting the stability of the

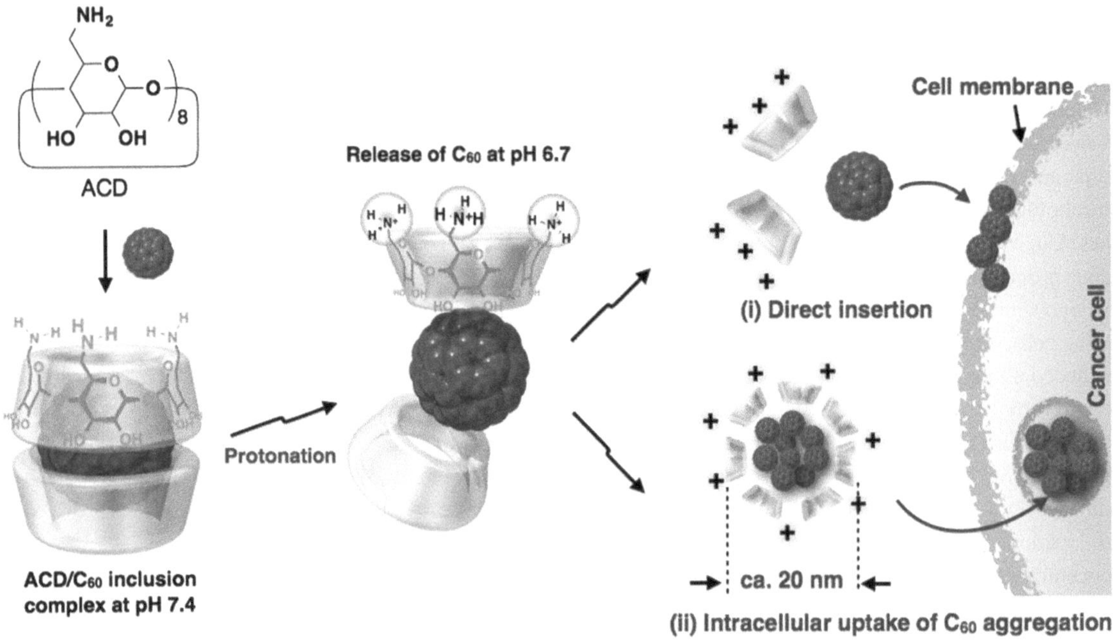

■ **FIGURE 11.8** Chemical structure of aminomodified CD (ACD), and schematic representation of 2:1 inclusion complex of ACD and C_{60} and proposed mechanism of C_{60} release and delivery in response to the protonation of amino groups at extracellular pH of cancer cell.

inclusion complex. ACD exhibited the same encapsulation property as unmodified CD, thus C_{60} was isolated as a single molecular state in neutral pH. On the other hand, protonation of ACD can act as a trigger for the deformation of ACD rings, leading to a unique release of C_{60} under slightly acid conditions. In a pH titration study, whereas amino groups of ACD appeared to gradually adopt protonation, C_{60} was drastically released with an inflection point at pH 6.7. Thus, a deformation energy gained from electrostatic repulsion among protonated amines can overcame the interactions between C_{60} and ACD. The released C_{60} in the aqueous environment settled down as an aggregate at a size of around 20 nm with water dispersibility. It suggests that the aggregate was stabilized by surrounding ACDs resulting in a micelle-like structure. From the practical viewpoint of the target delivery application, it is highly desirable to rationally design for the proper stability and controlled release of therapeutic applicants against physicochemical or biological signals. Methods to target intracellular/extracellular pH or pH gradients in cancer cells have been developed using water-soluble materials [47,48]. In our study, an in vitro experiment against HeLa cells at pH 6.4, whereby the extracellular pH condition of cancer cell was exploited, was performed, investigating the use of ACD/C_{60} complex. In all cases, dark toxicities were not observed, whereas ACD/C_{60} had a high level of photomortality for cancer cells compared with the neutral γCD/C_{60} complex. From TEM observation of the cancer cells after incubation of ACD/C_{60} at pH 6.4, it was revealed that C_{60} was located both in the cell membrane and within the intracellular vehicle. The results suggested two proposed mechanisms of C_{60} induction, (1) direct insertion of C_{60} to the cell membrane and (2) intracellular uptake of C_{60} aggregates probably through endocytosis (Fig. 11.8). The present method of ACD is a simple molecular design. However, considering the capability with variety functionalization to primary hydroxyl groups of CD, it has a great potential for efficient pH responsive drug delivery.

11.4 SUMMARY

In this chapter, we introduced concepts of CD-based host–guest interaction systems to realize the development of hydrosolubilization of poorly water-soluble molecules, and the manipulation of inclusion complex assembly on nanoarchitectures in therapeutic applications. Although the main focus of this chapter is in relation to its phototherapeutic applications in cancer therapy, the behavior of conventional CD/guest inclusion complexes demonstrates that they could be potential candidates for establishing novel and multifunctional nanomaterials, e.g., for the hierarchical assembly for enhancing the hydrosolubility and the efficacy of therapeutic agents, molecular assembly control in nanoarchitectures. Tailoring the chemistry of host molecules while considering their structural nature is expected to provide new insights into physiological stimuli-responsive molecular assembly and constituting nanomedicines. The CD-based supramolecular manipulation may provide new methods of designing sophisticated supramolecular nanoarchitectures to solve pressing needs in medical applications.

REFERENCES

[1] Moghimi SM, Hunter AC, Murray JC. Nanomedicine: current status and future prospects. FASEB J 2005;19(3):311–30.

[2] Rajewski RA, Stella VJ. Pharmaceutical applications of cyclodextrins. 2. In vivo drug delivery. J Pharm Sci 1996;85(11):1142–69.

[3] Hu QD, Tang GP, Chu PK. Cyclodextrin-based host–guest supramolecular nanoparticles for delivery: from design to applications. Acc Chem Res 2014;47(7): 2017–25.

[4] Davis ME, Brewster ME. Cyclodextrin-based pharmaceutics: past, present and future. Nat Rev Drug Discov 2004;3(12):1023–35.

[5] Li J, Loh XJ. Cyclodextrin-based supramolecular architectures: syntheses, structures, and applications for drug and gene delivery. Adv Drug Deliv Rev 2008;60(9): 1000–17.

[6] Freudenberg K, Cramer F, Plieninger H. Process for the preparation of inclusion compounds of physiologically active organic compounds. 1953 [Germany].
[7] Brewster ME, Loftsson T. Cyclodextrins as pharmaceutical solubilizers. Adv Drug Deliv Rev 2007;59(7):645–66.
[8] Uekama K, Hirayama F, Irie T. Cyclodextrin drug carrier systems. Chem Rev 1998; 98(5):2045–76.
[9] Alcaro S, et al. Preparation, characterization, molecular modeling and in vitro activity of paclitaxel-cyclodextrin complexes. Bioorg Med Chem Lett 2002;12(12): 1637–41.
[10] Pradines B, et al. Investigation of the complexation of albendazole with cyclodextrins for the design of new antiparasitic formulations. Carbohydr Res 2014;398:50–5.
[11] Loftsson T. Self-assembled cyclodextrin nanoparticles and drug delivery. J Inclusion Phenom Macrocycl Chem 2014;80(1–2):1–7.
[12] Pradines B, et al. The unexpected increase of clotrimazole apparent solubility using randomly methylated beta-cyclodextrin. J Mol Recognit 2015;28(2):96–102.
[13] Yallapu MM, Jaggi M, Chauhan SC. β-cyclodextrin-curcumin self-assembly enhances curcumin delivery in prostate cancer cells. Colloids Surf B Biointerfaces 2010;79(1):113–25.
[14] Sun T, et al. Strategy of directly employing paclitaxel to construct vesicles. J Phys Chem B 2012;116(50):14628–36.
[15] Li N, et al. Polysaccharide-gold nanocluster supramolecular conjugates as a versatile platform for the targeted delivery of anticancer drugs. Sci Rep 2014;4:4164.
[16] Bakry R, et al. Medicinal applications of fullerenes. Int J Nanomed 2007;2(4): 639–49.
[17] Dolmans DE, Fukumura D, Jain RK. Photodynamic therapy for cancer. Nat Rev Cancer 2003;3(5):380–7.
[18] Arbogast JW, et al. Photophysical properties of sixty atom carbon molecule (C60). J Phys Chem 1991;95(1):11–2.
[19] Kokubo K, et al. Facile synthesis of highly water-soluble fullerenes more than half-covered by hydroxyl groups. ACS Nano 2008;2(2):327–33.
[20] Oriana S, et al. Water-soluble C_{60}- and C_{70}-PVP polymers for biomaterials with efficient $^1O_2$ generation. Chem Commun 2013;49:9302–4.
[21] Wang M, et al. Synthesis and photodynamic effect of new highly photostable decacationically armed [60]- and [70]Fullerene decaiodide monoadducts to target pathogenic bacteria and cancer cells. J Med Chem 2012;55(9):4274–85.
[22] Wu H, et al. Solubilization of pristine fullerene by the unfolding mechanism of bovine serum albumin for cytotoxic application. Chem Commun 2011;47(38): 10659–61.
[23] Ikeda A. Water-soluble fullerenes using solubilizing agents, and their applications. J Inclusion Phenom Macrocycl Chem 2013;77(1–4):49–65.
[24] Nakanishi I, et al. DNA cleavage via superoxide anion formed in photoinduced electron transfer from NADH to γ-cyclodextrin-bicapped C_{60} in an oxygen-saturated aqueous solution. J Phys Chem B 2002;106(9):2372–80.
[25] Zhang W, et al. Water-soluble inclusion complex of fullerene with γ-cyclodextrin polymer for photodynamic therapy. J Mater Chem B 2014;2(31):5107–15.

[26] Andersson T, et al. C_{60} embedded in γ-cyclodextrin: a water-soluble fullerene. J Chem Soc Chem Commun 1992;8:604–6.

[27] Ikeda A, et al. Pseudorotaxane structure of a fullerene derivative–cyclodextrin 1:2 complex. Chem Commun 2011;47(48):12795–7.

[28] Liu Y, et al. Water-soluble supramolecular fullerene assembly mediated by metallobridged β-cyclodextrins. Angew Chem Int Ed 2004;43(20):2690–4.

[29] Murthy CN, Geckeler KE. The water-soluble β-cyclodextrin–[60]fullerene complex. Chem Commun 2001;13:1194–5.

[30] Liu Y, et al. Bundle-shaped cyclodextrin-Tb nano-supramolecular assembly mediated by C_{60}: intramolecular energy transfer. Nano Lett 2006;6(10):2196–200.

[31] Nobusawa K, Payra D, Naito M. Pyridyl-cyclodextrin for ultra-hydrosolubilization of [60]fullerene. Chem Commun 2014;50(61):8339–42.

[32] Yoshida ZI, et al. Molecular recognition of C_{60} with γ-cyclodextrin. Angew Chem Int Ed Engl 1994;33(15–16):1597–9.

[33] Nobusawa K, et al. Reversible solubilization and precipitation of carbon nanotubes through oxidation–reduction reactions of a solubilizing agent. Angew Chem Int Ed 2008;47(24):4577–80.

[34] Jeon IY, et al. Large-scale production of edge-selectively functionalized graphene nanoplatelets via ball milling and their use as metal-free electrocatalysts for oxygen reduction reaction. J Am Chem Soc 2013;135(4):1386–93.

[35] McCormack B, Gregoriadis G. Drugs-in-cyclodextrins-in liposomes: a novel concept in drug delivery. Int J Pharm 1994;112:249–58.

[36] Piel G, et al. Betamethasone-in-cyclodextrin-in-liposome: the effect of cyclodextrins on encapsulation efficiency and release kinetics. Int J Pharm 2006;312(1–2):75–82.

[37] Hungerbühler H, Guldi DM, Asmus KD. Incorporation of C_{60} into artificial lipid membranes. J Am Chem Soc 1993;115(8):3386–7.

[38] Bensasson RV, et al. C_{60} in model biological systems. A visible-UV absorption study of solvent-dependent parameters and solute aggregation. J Phys Chem 1994;98(13):3492–500.

[39] Ikeda A, et al. Efficient photocleavage of DNA utilising water-soluble lipid membrane-incorporated [60]fullerenes prepared using a [60]fullerene exchange method. Org Biomol Chem 2005;3(16):2907–9.

[40] Ikeda A, et al. An extremely effective DNA photocleavage utilizing functionalized liposomes with a fullerene-enriched lipid bilayer. J Am Chem Soc 2007;129(14):4140–1.

[41] Ikeda A, et al. Advantages and potential of lipid-membrane-incorporating fullerenes prepared by the fullerene-exchange method. Chem Asian J 2012;7(3):605–13.

[42] Ikeda A, et al. Location of [60]fullerene incorporation in lipid membranes. Chem Commun 2011;47(44):12095–7.

[43] Ikeda A, et al. [70]Fullerenes assist the formation of phospholipid bicelles at low lipid concentrations. Langmuir 2014;30(41):12315–20.

[44] Nobusawa K, et al. pH responsive smart carrier of [60] fullerene with 6-amino-cyclodextrin inclusion complex for photodynamic therapy. J Mater Chem 2012;22(42):22610–3.

[45] Boger J, Knowles JR. Symmetrical triamino-per-O-methyl-α-cyclodextrin: a host for phosphate esters exploiting both hydrophobic and electrostatic interactions in aqueous solution. J Am Chem Soc 1979;101(25):7631–3.

[46] Vizitiu D, Thatcher GRJ. Binding of phosphates to aminocyclodextrin biomimetics. J Org Chem 1999;64(17):6235−8.

[47] Lee ES, Gao Z, Bae YH. Recent progress in tumor pH targeting nanotechnology. J Control Release 2008;132(3):164−70.

[48] Chen Y-C, et al. Non-metallic nanomaterials in cancer theranostics: a review of silica- and carbon-based drug delivery systems. Sci Technol Adv Mater 2013;14(4):044407.

Nanoarchitectonics for Energy/ Environment

Nanoarchitectonics Towards Smart Chemical Sensing

S. Ishihara and J. Labuta
National Institute for Materials Science, Tsukuba, Japan

CHAPTER OUTLINE

12.1 INTRODUCTION

This section introduces the recent progress of chemical sensing realized by the combination and smart use of nanomaterials and nanodevices. Although there is no doubt that science and technology have contributed in making our society more affluent and convenient, the global development of industry and consumption of fossil fuel results in the emission of huge amounts of byproducts into the environment. In addition, serious accidents (e.g., explosions) in chemical plants, chemical storage or power plants may cause the release of fatal amounts of toxic substances into the environment.

Supra-materials Nanoarchitectonics. http://dx.doi.org/10.1016/B978-0-323-37829-1.00012-2

Table 12.1 Examples of Chemical Substances That Can Be Analyzed by Chemical Sensing

- Gases (e.g., O_2, CO, NO_2, H_2S)
- Volatile organic compounds (e.g., benzene, methanol, formaldehyde)
- Toxic cations/anions (e.g., Cd^{2+}, Hg^{2+}, Pb^{2+}, CN^-, F^-)
- Agricultural chemicals (e.g., methylviologen)
- Carcinogens (e.g., aromatic amine, dioxins)
- Explosives (e.g., trinitrotoluene)
- Chemical warfare agents (e.g., sarin, mustard gas)
- Radioactive substances (e.g., ^{137}Cs)
- Illegal drugs/pharmaceuticals (e.g., cocaine, purity and chirality of medicine)
- Biomolecules (e.g., antigens, viruses, glucose in blood)

Moreover, some chemical substances, such as agricultural chemicals and chemical warfare agents, are intentionally spread into our living environment. These chemicals can exhibit various long-term and/or acute toxic effects. For the purpose of improving our health, safety, and security it is important to develop smart ways to qualitatively and/or quantitatively estimate the presence of toxic substances potentially contained in the environment. In particular, toxic substances in gas, liquid, or solution show a wide range of impacts on the environment since they can migrate through air, water, plants, fish, animals, and food. Table 12.1 summarizes examples of chemical substances that can be analyzed by chemical sensing.

Fig. 12.1 shows a simplified principle of chemical sensors. Sensing nanomaterials have selective binding sites for a particular type of analyte (target). Surface functionalization is a convenient way to facilitate a selective and strong binding of analytes on nanomaterials. The interaction between a nanomaterial and an analyte leads to the change of physicochemical properties of nanomaterials (i.e., transduction). That change can be amplified and read out by nanodevices. In order to fabricate practical chemical sensors, it is important to design the aforementioned binding, transduction, and readout in a harmonized fashion from the comprehensive viewpoints of

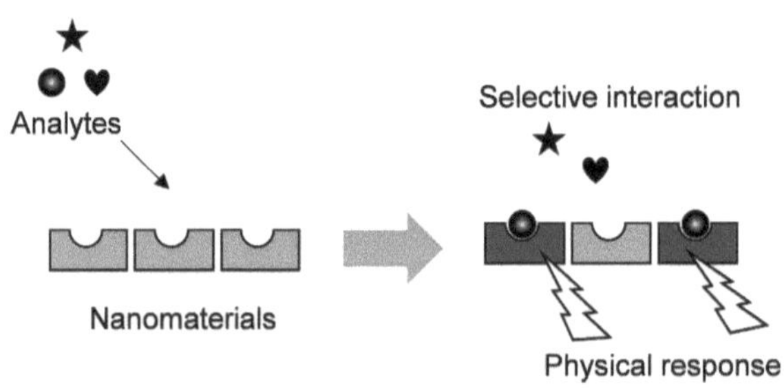

■ **FIGURE 12.1** Principle of chemical sensing.

Table 12.2 Examples of Chemical Sensing Discussed in This Section

Analyte	Nanomaterial	Detection Mode	References
Ammonia	Metal organic framework	Conductivity	[1,2]
	Carbon nanotube composite		
Mercury ion	DNA modified nanogap gold	Surface plasmon	[3]
Cesium ion	Fluorescent host molecule	Fluorescence	[4]
Methanol	Dye-clay composite	Visible color change	[5]
Aniline	Carbon nanocage	QCM	[6]
Explosives	Fluorescent polymer film	Fluorescence	[7]
Chemical warfare agents	Conjugate polymer	Visible color change	[8]
Chirality	DNA modified cantilever	Mechanical bending	[9]
	Achiral host molecule	NMR	[10,11]
Biomolecules, cocaine	DNA modified electrode	Electrochemistry	[12]

organic chemistry, inorganic chemistry, supramolecular chemistry, and physical chemistry.

As is seen in Table 12.2, which summarizes examples of sensors with an elegant design, various types of physical detection modes [photoabsorption spectroscopy, fluorescence spectroscopy, phosphorescence spectroscopy, surface plasmon resonance, quartz crystal microbalance (QCM), nuclear magnetic resonance spectroscopy (NMR), conductivity, field effect transistor, mass spectrometry, inductively coupled plasma atomic emission spectroscopy, elemental analysis, electrochemistry, mechanical response, chromatography, etc.] are available for chemical sensing.

12.2 RECENT EXAMPLES OF CHEMICAL SENSING

12.2.1 Ammonia Sensing

Although ammonia gas is highly toxic it is widely used in industry. In addition, ammonia is emitted from decaying food products. Therefore, the monitoring of ammonia gas is useful to improve industrial safety as well as agricultural management. Campbell et al. synthesized a two-dimensional metal-organic framework (2D-MOF) using 2,3,6,7,10,11-hexaiminotriphenylene (HITP) and copper ions (Fig. 12.2) [1]. Uniquely, the 2D-MOF demonstrated a high electrical conductivity of 0.2 Scm^{-1} (as a bulk pellet). Its conductivity increases in response to the binding of ammonia gas at the copper sites. Thus, the drop cast 2D-MOF works as a reversible chemiresistive sensor, by which the presence

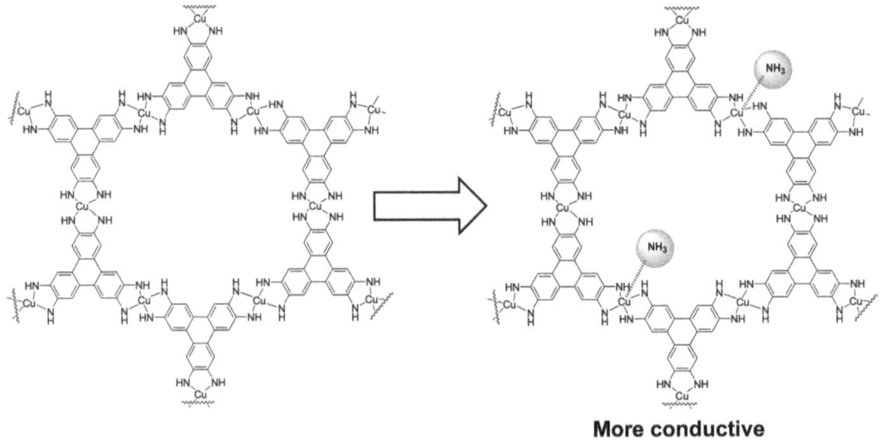

■ **FIGURE 12.2** Ammonia gas sensing based on a highly conductive two-dimensional metal organic framework.

of ammonia gas at sub-ppm level can be detected as a conductivity change using a simple potentiostat.

Azzarelli et al. fabricated a carbon nanotube (CNT) based chemiresistive sensor on the electronic circuitry of a small (45 × 45 mm^2 square) and commercially available near-field communication (NFC) tag (Fig. 12.3) [2]. The CNT materials selectively vary their conductivity in response to various chemical analytes (ammonia, hydrogen peroxide, water, and cyclohexanone) at ppt and ppm levels. The response can be monitored by smartphone. Thus, the sensing system is wireless, portable, cost-effective, and does not require a large energy supply.

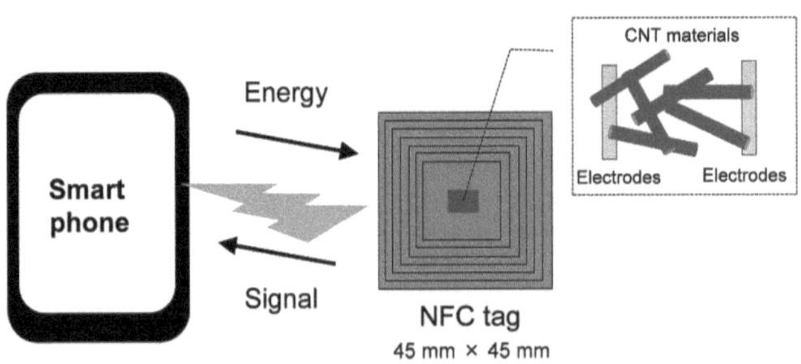

■ **FIGURE 12.3** Wireless ammonia gas sensing based on a chemiresistive sensor embedded on a near-field communication (NFC) tag.

12.2.2 Mercury Sensing

Mercury ions, potentially in industrial waste, can contaminate marine products, such as fish. A subsequent accumulation in humans via the food chain can lead to serious illness. Nagao and coworkers developed a highly sensitive plasmon sensor that can selectively detect ppt levels of mercury ions in water (Fig. 12.4) [3]. The surfaces of nanogapped gold arrays with a mid-infrared plasmonic structure were functionalized with mercury-binding DNA aptamers (single-stranded thiolated 15-base DNA oligonucleotides). The spectral variation of the vibration signal of the DNA aptamer upon binding with mercury ions can be monitored using a standard FT-IR spectrometer. The binding signal of mercury ions was successfully isolated from the binding of other biomolecules. A level of 37 ppt of mercury ions in lake water was readily detected. This sensor is promising in that the method is simple and enables the direct monitoring of metal ions in water without pretreating the sample. In addition, the method can be used more widely, since monitoring of other metal ions should be possible simply by altering the type of DNA aptamer.

12.2.3 Cesium Sensing

An accident in an atomic power plant could lead to an unexpected release of radioactive materials. Among such radioactive materials, ^{137}Cs remains in soil and water for extended periods, due to its long half-life (ca 30 years). For an efficient remediation of ^{137}Cs, a sensing method to detect ^{137}Cs contamination in the environment is in high demand. Mori et al. reported on an arm-shaped host material that fluoresces upon the binding of Cs ions (Fig. 12.5) [4]. The host material is only weakly fluorescent upon interaction with other alkali metal ions, such as sodium and potassium, with a strong and selective response to Cs ions. The mechanism is enabled by the selective inclusion of a Cs ion into an oligoethyleneglycol unit, leading to the alteration of the intramolecular energy transfer properties of the host molecule. By spraying a solution of the host material with the analyte

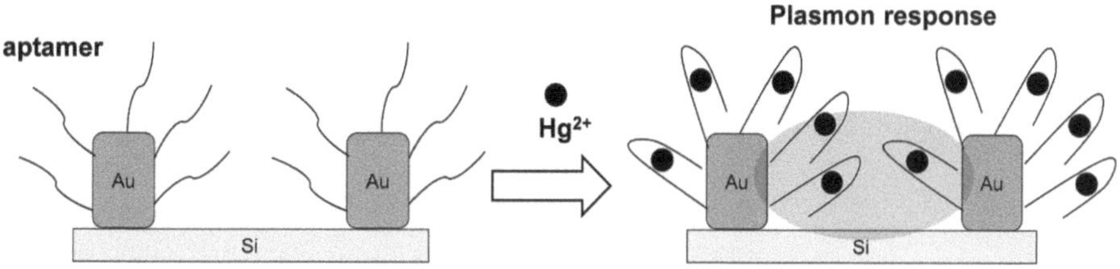

■ **FIGURE 12.4** Mercury ion sensing based on nanogapped gold arrays functionalized by mercury-selective DNA aptamer.

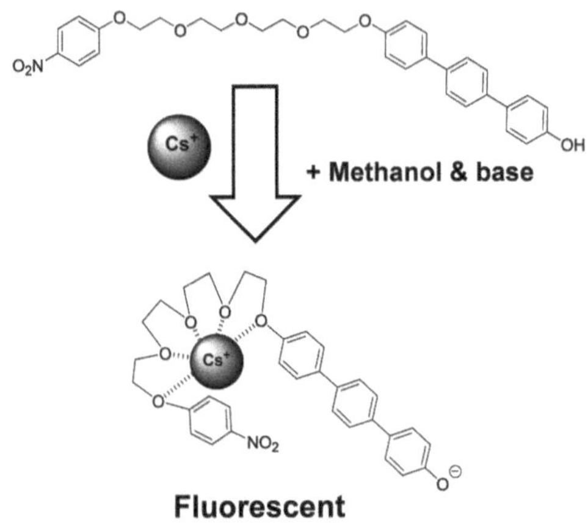

■ **FIGURE 12.5** Cs ion sensing based on arm-shaped fluorescent host material.

particles (e.g., onto soil) followed by irradiation with a UV lamp, it becomes possible to identify particles contaminated with Cs ions. This method enables a higher spatial resolution than existing radioscopes and gamma ray cameras. It could be a powerful tool for the efficient remediation of an environment polluted by radioactive Cs ions.

12.2.4 Methanol Sensing

Methanol and ethanol are drawing a great deal of attention as next-generation liquid fuels with a potential for wide use in the future. Since both methanol and ethanol are colorless liquids, it is difficult to differentiate methanol and ethanol with the naked eye. Methanol is highly toxic for the human metabolism. It can cause serious illnesses (damage to the eye) or even death when ingested. Therefore, a facile and cost-effective methodology for discriminating between methanol and ethanol is in demand.

Ishihara et al. reported that the composite film of layered double hydroxide (LDH) clay and oxoporphyrinogen (OxP) dye demonstrates visible color changes from red purple to blue purple upon exposure to methanol (Fig. 12.6) [5]. The composite film does not change color upon exposure to ethanol, so the composite film can be used similar to a "litmus test" for the detection of methanol. Although methanol and ethanol can be discriminated by analytical methods such as gas chromatography, NMR, mass spectrometry, and electrochemistry, all these methods require expensive instrumentation, an electricity supply, and a well-trained operator. In addition, these instruments are generally not portable, so are not suitable for common usage. The color change in OxP-LDH composite film results from

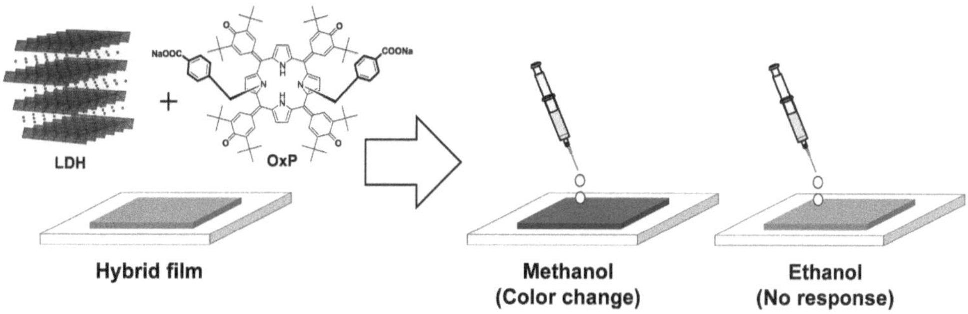

■ **FIGURE 12.6** Colorimetric methanol sensing based on layered double hydroxide (LDH) and oxoporphyrinogen (OxP) composite film. For actual color change, please refer to Ref. [5].

anions transferring from the LDH to the OxP, triggered only upon methanol exposure.

12.2.5 Carcinogen Sensing

Aromatic amines (such as aniline) found in cigarettes and industrial emissions are carcinogens that increase the risk for lung cancer. In fact, aromatic amines, such as aniline, are found in the breath and urine of most smokers. Although a complete recovery from lung cancer might be possible if found at an early stage, over one million people prematurely die from this disease every year. A cost-effective sensing procedure for aromatic amine would be useful to detect potential lung-cancer patients (due to daily exposure of aromatic amines) and a boon for cancer risk management in an industrial workplace. Kosaki et al. reported that a carbon nanocage (CNC), a porous carbon material with a large surface area (1600 $m^2\ g^{-1}$) and pore volume (2.10 $cm^3\ g^{-1}$), is a highly selective adsorbent of aniline (a representative aromatic amine). Theoretical calculation indicates that an aromatic moiety and remaining functional groups (e.g., carbonyl group) in CNC cooperatively interact with aniline. Carbon nanocages were incorporated into poly(methylmethacrylate) (PMMA) fiber film. Variations of mass upon exposure to aniline were observed using the QCM technique (Fig. 12.7) [6]. The QCM sensor responds to ppm levels of aniline but does not respond to other chemicals such as water, ammonia, or benzene. The QCM sensor demonstrates both good sensitivity and selectivity to aniline. Moreover, the QCM sensor quickly responds to aniline and can be used repetitively. This technique could be useful for high-throughput lung cancer risk management.

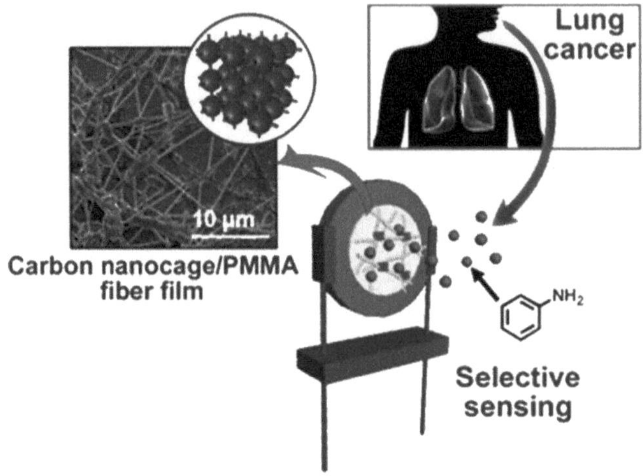

■ **FIGURE 12.7** Aromatic amine (aniline) sensing based on carbon nanocage/PMMA fiber film mounted on a QCM device. *Reprinted with permission from Kosaki Y, Izawa H, Ishihara S, Kawakami K, Sumita M, Tateyama Y, et al. ACS Appl Mater Interfaces 2013;5:2930–34. Copyright 2013 American Chemical Society.*

12.2.6 **Explosive Sensing**

Explosive materials are a threat to public safety and security (aviation safety, unexploded bombs, land mines, etc.). Therefore it is important to develop effective sensing materials and methods. Swager et al. reported on a highly sensitive fluorescent sensor for explosive substances (such as trinitrotoluene) using a porous π-conjugated polymer (Fig. 12.8) [7]. The polymer involves bulky iptycene moieties, preventing the π-π stacking of polymers as well as eximer formation and concentration quenching. Thus, the polymer is highly fluorescent even in a film state. The fluorescence is quenched by exposure to trace explosives such as trinitrotoluene.

12.2.7 **Chemical Warfare Agent Sensing**

Although chemical warfare agents are primarily used on the battlefield, they can also be misused by terrorists in urban areas (e.g., sarin gas attack in the Tokyo subway in 1995). Therefore, it is important to develop a portable, cost-effective sensor for chemical warfare agents that can be easily operated by nontrained users. Weis et al. reported on dithienobenzotropone-based conjugated alternating copolymers that change their color upon exposure to diethylchlorophosphate (DCP), a substance structurally related to nerve agents (Fig. 12.9) [8]. The copolymers have a reactive hydroxyl group that undergoes phosphorylation

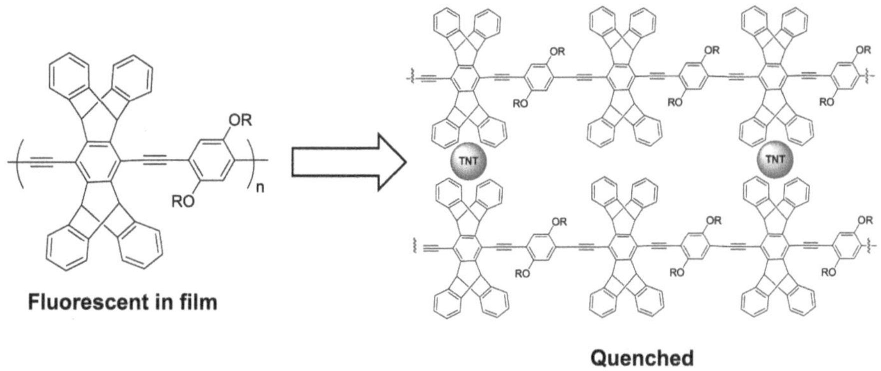

■ **FIGURE 12.8** Trinitrotoluene sensor based on fluorescent π-conjugated polymer.

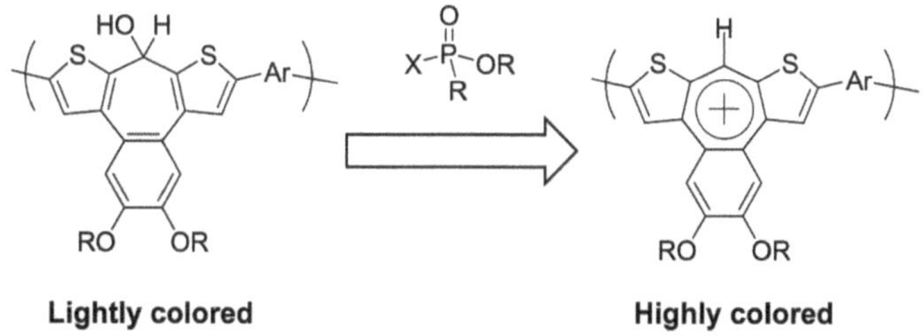

■ **FIGURE 12.9** Chemical warfare agent sensing based on dithienobenzotropone-based conjugated alternating copolymers.

with the subsequent ionization of its conjugated polymer backbone. The ionized copolymer is highly colored and enables colorimetric detection of DCP (in thin polymer film form).

12.2.8 Chiral Sensing

A chirality is a property of molecules, meaning the structure of a molecule is distinguishable from its mirror image. In nature, chiral molecules are abundant, especially in biological systems, such as proteins, sugars, amino acids, nucleotides, etc. Chiral molecules can be found or synthesized in both chiral forms, called enantiomers. Living organisms are built from chiral components that demonstrate a different response to both enantiomers, especially important for the administration of pharmaceuticals. Each enantiomer can have a different (sometimes adverse) pharmacological effect. Therefore, the determination of chiral properties (absolute configuration and/or enantiomeric purity) of chiral drugs is essential for their safe use. Although traditional methods (optical rotation, chiral column, NMR chiral solvating agent) for determining chiral properties exist,

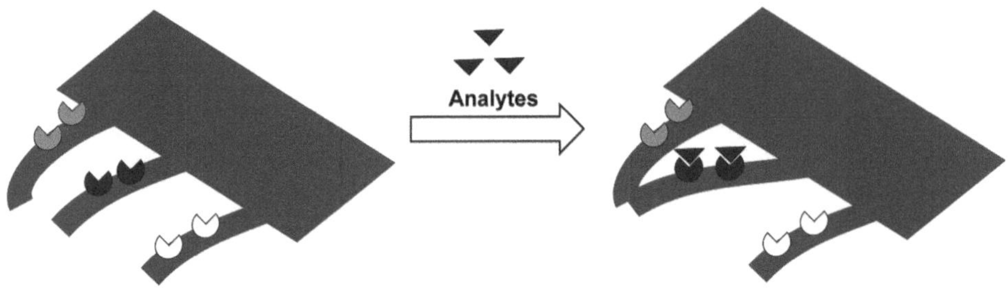

■ **FIGURE 12.10** Chiral sensing based on nanomechanical bending of antibody functionalized cantilever.

recent progress in nanotechnology and material science have shown some novel approaches to chiral sensing that may have some benefits over traditional methods.

Dutta et al. reported on a nanomechanical chiral sensor using microfabricated cantilevers (Fig. 12.10) [9]. The surface of the cantilever is functionalized with an antibody that selectively binds with one enantiomer of the amino acid. The binding of amino acid leads to the bending of the cantilever. The degree of bending is proportional to the concentration of the amino acid. Such a cantilever sensor can be fabricated at low cost using current silicon-based nanotechnology. Different antibodies can be embedded on each pin of a cantilever. This enables numerous arrays of sensors on a small chip.

In addition, Labuta et al. reported that some types of achiral host molecules work as NMR chiral solvating agents. The NMR signals of the host molecules split proportionally to the enantiomeric excess (*ee*) of chiral guests (Fig. 12.11) [10,11]. These achiral host molecules are found to be prochiral. Their structural symmetry is broken upon binding with a chiral guest. This results in the splitting of NMR signals of achiral host molecules. The kinetics of the binding equilibrium of a chiral guest/host complex is faster than the NMR chemical shift time scale. The NMR spectral splitting reflects an averaged chiral environment of host molecules. Consequently, the magnitude of NMR splitting is always linearly proportional to the *ee* of the chiral guest. Symmetrical achiral host molecules with these properties are called prochiral solvating agents (pro-CSA). The mechanism is in contrast to established traditional methods (e.g., chiral solvating agents, chiral derivatizing agents) that require that the NMR resolving agent must be chiral with the subsequent formation of diastereomeric complexes between chiral guest and chiral agent. The pro-CSA molecules established a novel type of NMR chirality sensing method.

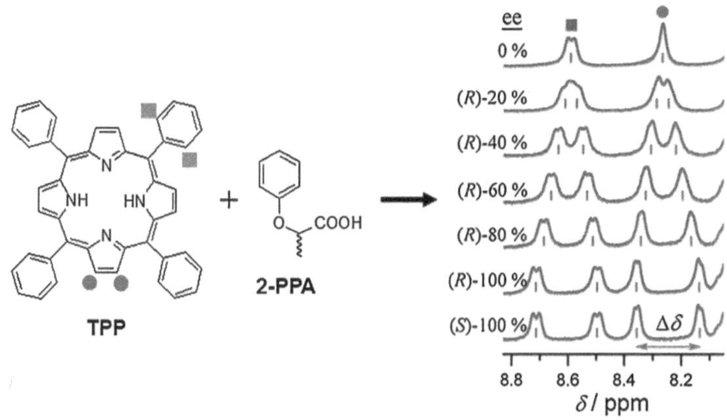

■ **FIGURE 12.11** An example of NMR chiral sensing method based on pro-CSA host (tetraphenylporphyrin, TPP) with chiral analyte (2-phenoxypropionic acid, 2-PPA). It can be seen that ^{1}H NMR spectral response ($\Delta\delta$) linearly depends on *ee*. *Adapted with permission from Labuta J, Ishihara S, Shundo A, Arai S, Takeoka S, Ariga K, et al. Chem Eur J 2011;17:3558–61 Copyright 2011 John Wiley and Sons.*

12.2.9 **Biological Molecule Sensing**

Das et al. reported highly sensitive and universal electrochemical sensors based on a DNA aptamer functionalized electrode (Fig. 12.12) [12]. The negatively charged DNA aptamer was initially bound with neutralizer, a conjugate of peptide nucleic acid, and cationic amino acids (such as lysine). The sequence of the neutralizer was intentionally designed to have some mismatch with the DNA aptamer so that the neutralizer can be easily

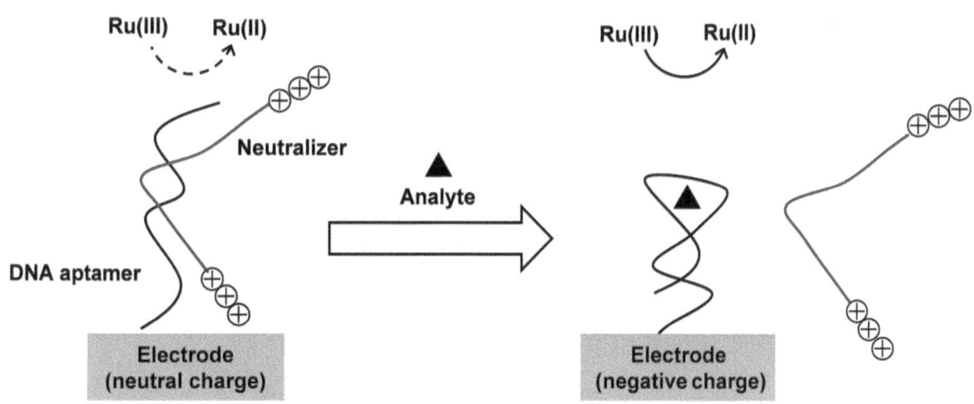

■ **FIGURE 12.12** Universal electrochemical sensing by using DNA aptamer and neutralizer displacement assay.

replaced with the target analyte. Consequently, the release of the neutralizer from the DNA aptamer causes a drastic change in the electrostatic charge of the electrode surface (i.e., from neutral to negative), which can be read out by the electrochemical reduction current of Ru(III) into Ru(II). The negatively charged surface formed by the release of neutralizer promotes the diffusion of Ru(III) ions to the electrode so that the reduction current of Ru(III) further increases. Thus, various types of analytes, such as adenosine triphosphate, cocaine, DNA, RNA, protein (thrombin) and bacteria, were quickly detected (within 1 min) with high sensitivity. This neutralizer displacement assay has a higher sensitivity than the conventional DNA aptamer method because the change of surface electrostatics is much larger (i.e., from neutral to negative) than the conventional method (from negative to negative).

12.3 CONCLUSIONS

This section introduced the significance and principles of recent examples of chemical sensing. These examples clearly show that ideas for chemical sensing are widespread in various research fields. Although there are many elegant examples of chemical sensors (as introduced here), many tasks remain to be overcome for practical applications. The ideal sensor should have high sensitivity, high selectivity, low detection limit, high speed, high stability, etc.

However, it may be difficult for individual sensing materials to achieve many of these goals at the same time. In order to overcome the problems and improve the sensing capability, "smart" nanomaterials should be incorporated in well-designed, fine-set and -tuned nanodevices. These systems can provide novel emerging functionalities that cannot be realized by individual molecules or components. This concept of "supramolecular nanoarchitectonics" should be key for the next generation of chemical sensing.

ACKNOWLEDGMENTS

This work was partly supported by JSPS fellowship, Kurata Grant from the Kurata Memorial Hitachi Science and Technology Foundation and International Center for Young Scientists (ICYS) fellowship, National Institute for Materials Science.

REFERENCES

[1] Campbell MG, Sheberla D, Liu SF, Swager TM, Dincă M. Angew Chem Int Ed 2015;54:4349–52.

[2] Azzarelli JM, Mirica KA, Ravnsbæk JB, Swager TM. Proc Natl Acad Sci USA 2014; 111:18162–6.

[3] Hoang CV, Oyama M, Saito O, Aono M, Nagao T. Sci Rep 2013;3:1175.
[4] Mori T, Akamatsu M, Okamoto K, Sumita M, Tateyama Y, Sakai H, et al. Sci Technol Adv Mater 2013;14:015002.
[5] Ishihara S, Iyi N, Labuta J, Deguchi K, Ohki S, Tansho M, et al. ACS Appl Mater Interfaces 2013;5:5927–30.
[6] Kosaki Y, Izawa H, Ishihara S, Kawakami K, Sumita M, Tateyama Y, et al. ACS Appl Mater Interfaces 2013;5:2930–4.
[7] Yang J-S, Swager TM. J Am Chem Soc 1998;120:5321–2.
[8] Weis JG, Swager TM. ACS Macro Lett 2015;4:138–42.
[9] Dutta P, Tipple CA, Lavrik NV, Datskos PG, Hofstetter H, Hofstetter O, et al. Anal Chem 2003;75:2342–8.
[10] Labuta J, Ishihara S, Shundo A, Arai S, Takeoka S, Ariga K, et al. Chem Eur J 2011; 17:3558–61.
[11] Labuta J, Hill JP, Ishihara S, Hanyková L, Ariga K. Acc Chem Res 2015;48:521–9.
[12] Das J, Cederquist KB, Zaragoza AA, Lee PE, Sargent EH, Kelley SO. Nat Chem 2014;4:642–8.

Nanoarchitectonics for Energy and Environment

Y. Manaka[1], T. Funaki[2], T.N. Murakami[2], N. Kameta[3] and Y. Hoshino[4]

[1]*Renewable Energy Research Center, National Institute of Advanced Industrial Science and Technology (AIST), Koriyama, Fukushima, Japan;* [2]*Research Center for Photovoltaics, National Institute of Advanced Industrial Science and Technology (AIST), Tsukuba, Ibaraki, Japan;* [3]*Research Institute for Sustainable Chemistry, National Institute of Advanced Industrial Science and Technology (AIST), Tsukuba, Ibaraki, Japan;* [4]*Kyushu University, Fukuoka, Japan*

CHAPTER OUTLINE

13.1 INTRODUCTION

In this chapter on nanoarchitectonics for energy and environment, leading-edge technologies, applied to both energy and the environment, are introduced with concrete examples. The section consists of four subsections: dye-sensitized

Supra-materials Nanoarchitectonics. http://dx.doi.org/10.1016/B978-0-323-37829-1.00013-4

solar cells (DSSCs), bioinspired catalysis, reversible H_2 storage, and energy efficient CO_2 absorbents (Sections 13.2–13.5, respectively). In all cases, the function of the materials is maximized by the tuning of molecules and/or the molecular-scale structure that are the main building and functional units in the materials, respectively. These are typically in the atomic-size scale. This "nanocreation" step is essential, and is thus usually studied and optimized prior to any of the other mentioned steps.

The first subsection is about DSSCs and describes the molecular design for efficient power conversion. To achieve an efficient power conversion there are two strategies, namely an increase in the photocurrent and an increase in the photovoltage. A bidentate ligand having a strong electron-donating ability, and/or an acceptor ligand with an extended π-conjugated field, can effectively increase the photocurrent. Surface structure consisting of dyes affects the photovoltage.

The next subsection is on bioinspired catalysis and describes rationally designed synthetic lipids in nanochannel reactions. These lipids selectively self-assembled into supramolecular nanotubes with a complete one-dimensional structure and a size-defined open nanochannel. Bioinspired catalysis was developed by the confinement or encapsulation of metal complexes or enzymes into the nanochannel, as well as individual modification of the inner/outer surfaces and membrane walls with functional groups.

The third subsection on reversible hydrogen storage illustrates a homogeneous catalyst design for an efficient H_2 conversion reaction. The formation of a proton relay pathway in nanoscale, between catalyst and H -like enzyme, greatly enhances this activity.

Finally, the fourth subsection is about an energy efficient CO_2 absorbent, and reviews how the design of the pK_a levels of the Brønsted acids in the absorbent is the most important step in this procedure. The preparation conditions and composition of nanosize gel particles have to be optimized in order to tune the pK_a of the functional groups in the particles. To accelerate absorption/desorption kinetics it is important to tune the size of the integral particles.

13.2 NANOARCHITECTONICS FOR DYE-SENSITIZED SOLAR CELLS

DSSCs have been drawing considerable attention as a new generation of cost-effective photovoltaics because of their fabrication from organic materials and their colorful nature [1–3]. For universal DSSC application, enhancement of the power conversion efficiency (PCE) is required. A DSSC consists of a dye-adsorbed porous semiconductor layer (commonly

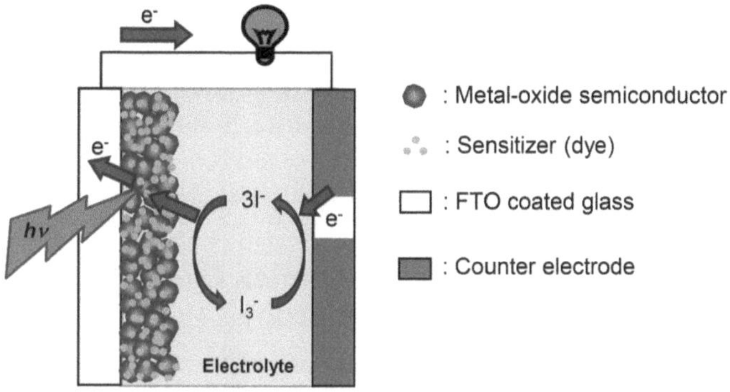

FIGURE 13.1 Schematic representation of a dye-sensitized solar cell.

comprising a 10- to 20-μm-thick porous layer of TiO_2 nanocrystalline particles with a diameter of 20 nm) on a transparent conductive substrate of F-doped tin oxide–coated glass (FTO-glass) as the photoelectrode, an electrolyte including redox couples, and a counter electrode with catalysts for the reduction of the redox system in the electrolyte (Fig. 13.1). The working principle of the DSSCs is briefly explained as follows. The dyes are excited via visible light absorption to inject electrons into the TiO_2 conduction band. The dyes become dye cations after injection. The electrons diffuse through the porous TiO_2 layer to the FTO, traveling via an external circuit to reach the counter electrode. At the counter electrodes, the oxidant of the redox couple is reduced to the reductant, and the reductant in turn reduces the dye cations back to the neutral dye. Thus, the dyes revert to their starting condition. The maximum photovoltage, represented as the open circuit voltage (V_{oc}) of the DSSC, is expressed as the energy difference between the Fermi level of TiO_2 and the redox potential of the electrolyte. For an increase in PCE, enhancement of the photocurrent and photovoltage is essential. To obtain a higher photocurrent, the dye should have a wider wavelength absorption. Dyes with smaller energy gaps between the highest occupied molecular orbital (HOMO) and the lowest unoccupied molecular orbital (LUMO) show wider photoabsorption. However, for electron injection from the exited dye to the TiO_2, the LUMO level of the dye should have a shallower position than the TiO_2 conduction band. Moreover, for dye regeneration from the dye cations to the neutral dye, the HOMO level of the dye should have a deeper position than the redox potential of the redox system in the electrolyte. On the other hand, to obtain a higher V_{oc}, the redox potential of the electrolyte should be at a deeper level and the Fermi level of the TiO_2 should be shallower. Therefore, there are two strategies to obtain a higher PCE; one is to increase the photocurrent with a dye having a wider photoabsorption, and the other is to increase the photovoltage with changing of the redox couples, owing to a deeper redox potential. The former strategy is described in the subsection pertaining to the development of a new Ru complex dye (2.1), while the latter is described in the development of an organic dye with a cobalt complex electrolyte (2.2). A future strategy would be a combination of these two technologies so that both the photocurrent and photovoltage could be improved simultaneously within the same technique.

13.2.1 Development of Ru Complex Dyes for Enhancing the Photocurrent

In the past decades, many different sensitizers have been designed and applied to DSSCs. Among the numerous sensitizers, ruthenium polypyridyl complexes have received significant attention owing to their superior performance in DSSCs. The complexes $[Ru(Hdcbpy)_2(NCS)_2](TBA)_2$ (dcbpy = 4,4′-dicarboxy-2,2′-bipyridine, NCS = isothiocyanato ligand, TBA = tetra-n-butylammonium) (N719) [4,5] and $[Ru(Htctpy)(NCS)_3](TBA)_3$ (tctpy = 4,4′,4″-tricarboxy-2,2′:6′,2″-terpyridine) (Black dye) [6] are two of the most successful sensitizers (Fig. 13.2). The DSSCs based on these complexes achieve 11% conversion efficiency from solar light to electricity. Generally, ruthenium dyes consist of a central ruthenium ion with donor and acceptor ligands having at least one anchoring group (Fig. 13.3). Light absorption in the visible region of the solar spectrum is mainly due to a metal-to-ligand charge transfer (MLCT) process. Many efforts have been made to change the ligands and optimize the ruthenium dyes to obtain sensitizers that have a wide spectral response, a high molecular extinction coefficient, and high chemical and thermal stabilities. For example, the absorption band can be extended to longer wavelengths by destabilizing the metal t_{2g} orbital using a strong donor ligand, and/or introducing a ligand with a low-lying π^*-level molecular orbital. To improve the conversion efficiency, an enhanced spectral response of the sensitizer in the longer-wavelength region is required, while maintaining sufficient thermodynamic driving forces for both electron injection and dye regeneration. Ruthenium terpyridyl complexes with strong donor ligands

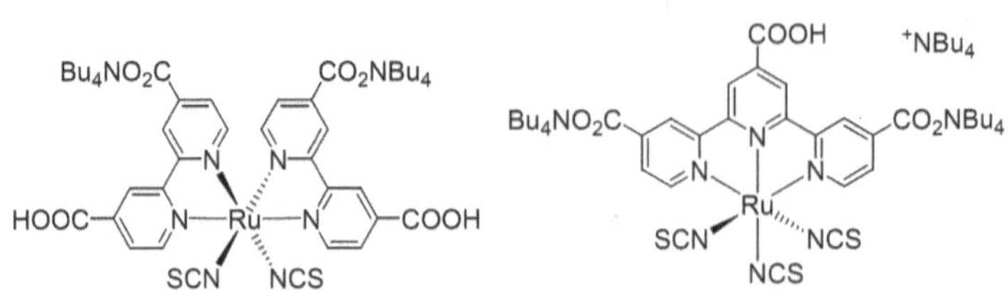

N719   Black dye.

■ **FIGURE 13.2** Structure of N719 and black dye.

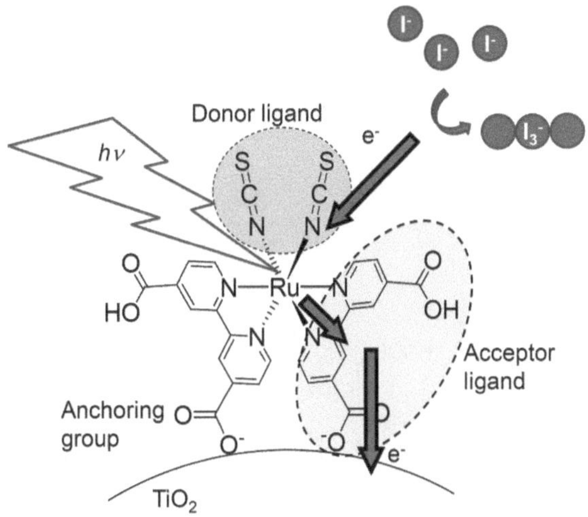

■ **FIGURE 13.3** Photoinduced electron transfer on the surface of TiO_2.

are suitable for the expansion of the absorption band in the longer-wavelength region [6,7]. Funaki et al. [8–13] and other groups [14–17] have examined ruthenium terpyridyl complexes of the type $[Ru(tctpy)(L1)(NCS)]^z$ (L1 = bidentate ligand, and z = 0 or +1) as sensitizers for DSSCs (Fig. 13.4). The introduction of bidentate ligands made it possible to tune the HOMO energy level (E_{HOMO}) of the resulting complexes while maintaining LUMO energy levels (E_{LUMO}) for efficient electron injection. The E_{HOMO} of the complexes was systematically tuned by changing the electron-donating ability of the bidentate ligands.

Cyclometalated ruthenium complexes show a superior light-harvesting property in the longer wavelength region because of the strong electron-donating ability of the cyclometalating ligand (Fig. 13.5) Ru(tctpy)(ppy)(NCS) (ppy = 2-phenylpyridinato) (FT3) shows a new absorption band above 700 nm [12]. Among the cyclometalated ruthenium dyes of the type Ru(tctpy)(L2)(NCS) (L2 = 2-phenylpyridinato and 2-phenylpyrimidinato), the DSSCs sensitized with Ru(tctpy)(btfm-ppym)(NCS) {btfm-ppym = 2-[2,4-bis(trifluoromethyl)phenyl]pyrimidinato} (FT89) exhibit the highest conversion efficiency (10.7%) [13]. Of note is that the broad incident photon-to-current conversion efficiency (IPCE) curve for FT89 covers the whole visible range extending into the near-IR region, and the action spectrum onset is near 920 nm (Fig. 13.6). Several ruthenium dyes exhibit remarkable light-harvesting properties in the red and near-IR region (Fig. 13.7) [18–23]. The DSSCs based on these dyes exhibit efficient panchromatic sensitization over the entire visible wavelength range

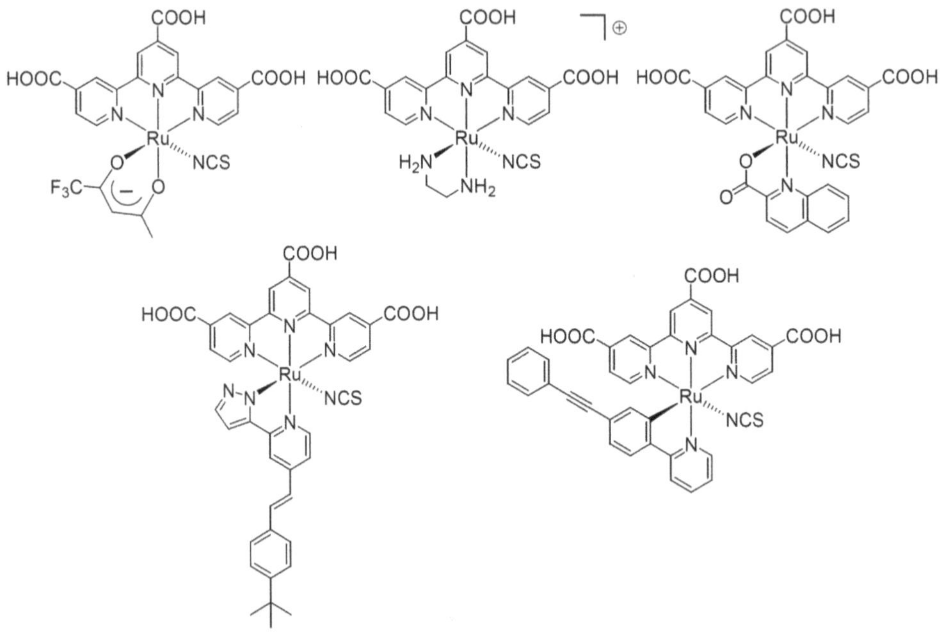

■ **FIGURE 13.4** Structures of ruthenium terpyridine complexes.

extending into the near-IR region, and the onset of the action spectrum is near 1000 nm (Fig. 13.8).

To realize high-performance DSSCs, an enhanced spectral response of the sensitizers in the lower energies is required, while maintaining sufficient thermodynamic driving forces for both electron injection and dye regeneration. The introduction of a bidentate ligand having a strong electron-donating ability and/or an acceptor ligand with an extended π-conjugated field improves light harvesting properties in the red and near-IR region about the ruthenium dyes.

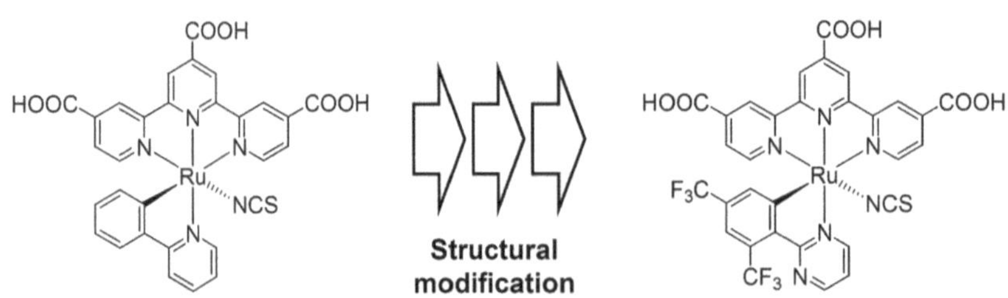

FT3 FT89

■ **FIGURE 13.5** Structures of cyclometalated ruthenium complexes.

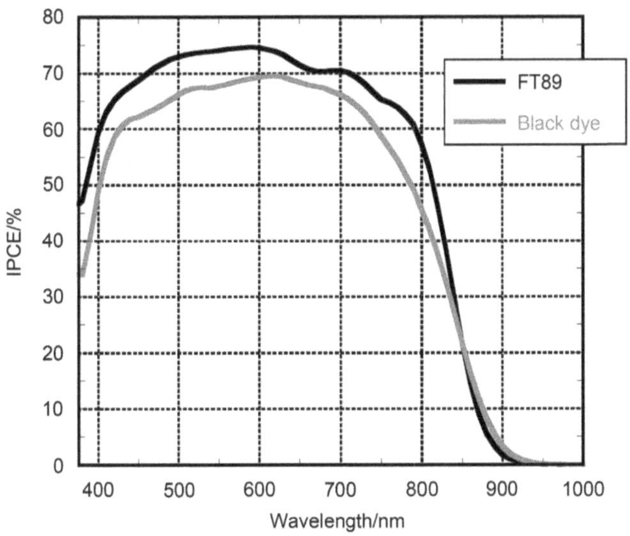

■ **FIGURE 13.6** Photocurrent action spectra of the cells based on FT89 and black dye.

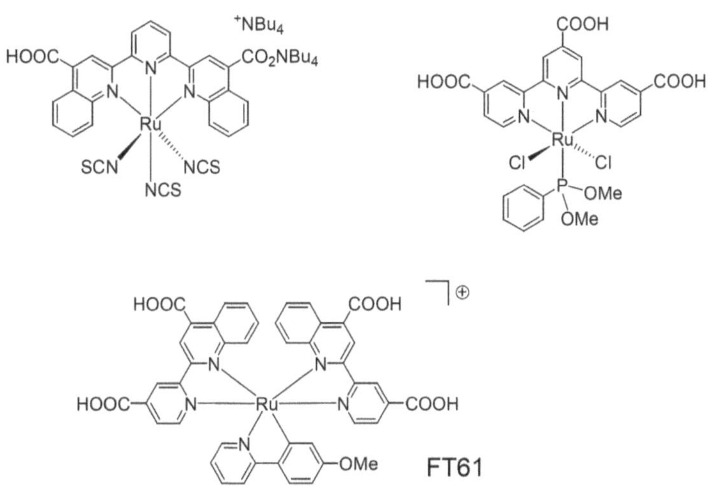

FT61

■ **FIGURE 13.7** Structure of the ruthenium complexes.

13.2.2 Searching for an Effective Dye Structure for High Photovoltage

DSSCs traditionally employ Ru complex dyes and the iodide/tri-iodide (I^-/I_3^-) redox system to achieve a high energy conversion efficiency [5]. However, the I^-/I_3^- redox system in conjunction with a Ru complex dye (N3 dye) loses 0.7 V in the regeneration of the dye, i.e., the difference

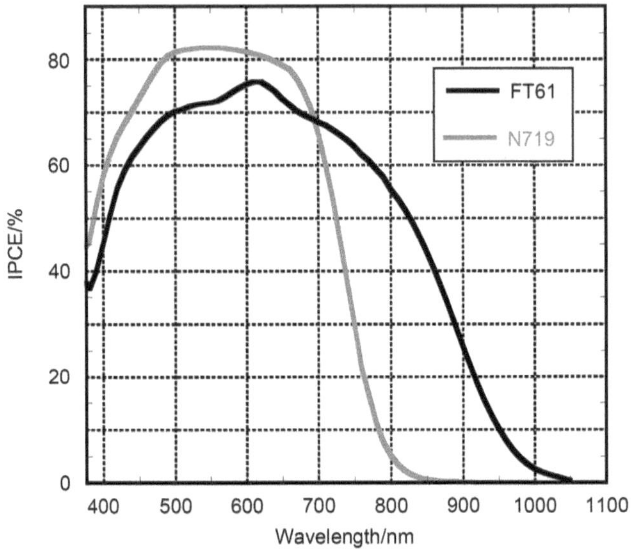

FIGURE 13.8 Photocurrent action spectra of the cells based on FT61 and N719.

between the I^-/I_3^- redox potential (0.42 V vs NHE, NHE = normal hydrogen electrode) and the HOMO potential of the N3 dye (1.1 V vs. NHE). Therefore, a redox system deeper than the I^-/I_3^- system is needed to obtain higher V_{oc} and PCE values. A cobalt complex (e.g., cobalt bipyridyl complex) redox (Co-redox) electrolyte, with a deeper redox potential, is one of the candidates for alternative redox couples needed to avoid the energy loss from dye regeneration. The initial application of Co-redox couples to DSSCs was carried out in conjunction with Ru complex dyes. At the time, the efficiency of the resulting DSSCs was much lower than that of a cell employing the conventional I^-/I_3^- redox couple [24–26]. Most cobalt complexes were limited in terms of the fast charge recombination between the injected electrons in TiO_2 and the cobalt complexes. This meant that the charge recombination reaction suppressed the electron accumulation in TiO_2 to decrease the V_{oc}, making the Fermi level shift deeper. Since 2008, however, a higher power conversion efficiency was achieved by employing bulky organic dyes, owing to the high V_{oc} of the cells [27–29]. The charge recombination may be effectively reduced by increasing the molecular size of the dye. Therefore, inhibition of the recombination reaction is essential for the employment of the Co-redox electrolyte, and the Co-redox couples perform well with dyes capable of blocking the approach of the redox species toward the surface of the TiO_2 electrode.

Murakami et al. developed efficient organic dyes for the Co-redox system and found possibilities of avoiding the recombination process and of adding a further blocking function on the design and modification of the dye structure [30–32]. Fig. 13.9 shows dyes with a steric barrier on the carbazole moiety or on the oligothiophene unit developed by Murakami et al. Higher V_{oc} and short-current density (J_{sc}) values were observed for the cell containing the MK-14 dye than for those having MK-1 and MK-20 dyes (Fig. 13.10 and Table 13.1). The large additional groups on the donor moiety inhibit the recombination because the approach of the cobalt complex redox coupled to the TiO_2 surface is avoided. MK-33 shows a higher V_{oc} and J_{sc} value than MK-14. The difference between MK-33 and MK-14 is the length of the side chain on the oligothiophene, and the adsorbed dye amount on the TiO_2 of MK-33 was 17% higher than that of MK-14. The shorter side alkyl chain increases the dye adsorption amount and the dye coverage on the TiO_2 surface, resulting in recombination inhibition. The high coverage of MK-33 on the TiO_2 substrate may enhance the blocking effect, thereby increasing the V_{oc}. Fig. 13.11 shows the open circuit voltage and electron lifetime as a function of the electron density of the cells. When recombination increases, the electron lifetime becomes shorter. MK-33 showed the longest electron lifetime. This was approximately 10 times longer than that of MK-1 because of the blocking effect of MK-33.

The donor type also affected the recombination reaction in DSSCs. Fig. 13.12 shows dyes with various donor groups. The DSSC I-V curves employing dyes MK-1, MK-31, and MK-88 are shown in Fig. 13.13, and their photovoltaic performance is described in Table 13.2. The free rotatable phenyl group of the triphenylamine donor acted as the steric hindrance necessary for recombination inhibition. In contrast, the partial charge in the coumarin donor reduced the V_{oc} and electron lifetime (Fig. 13.14). The recombination reaction in the DSSCs may be inhibited with the appropriate steric hindrance of the additional groups and donor structure.

13.3 NANOARCHITECTONICS FOR BIOINSPIRED CATALYSIS

13.3.1 Photocatalysts Based on Light-Harvesting Supramolecular Nanotubes

Biomimetic chemistry has allowed us to construct artificial light-harvesting antenna systems, based on rod-like nanostructures self-assembled from natural dyes such as chlorophyll and bacteriochlorophyll

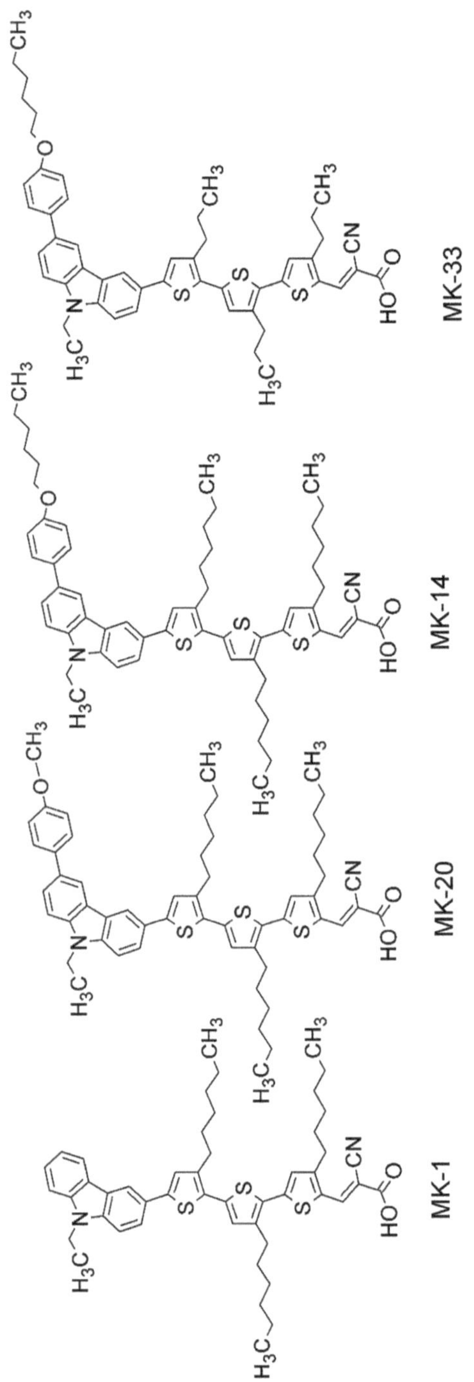

■ FIGURE 13.9 Dye structures with steric hindrance.

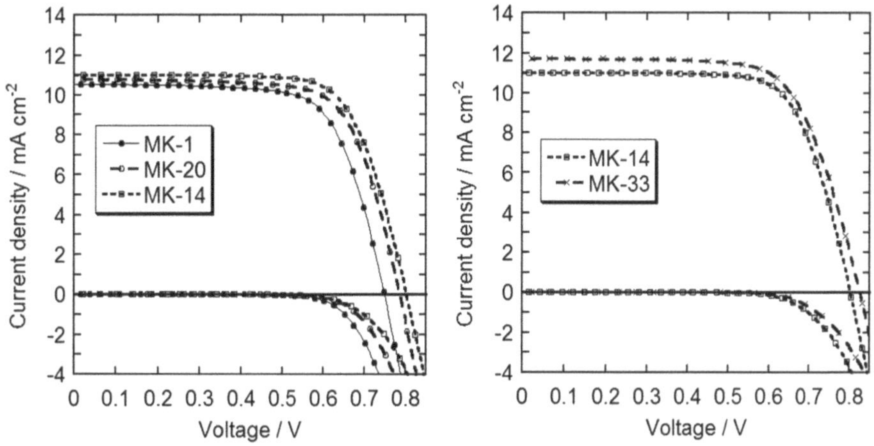

FIGURE 13.10 The I-V curves of the DSSCs with the dyes in Fig. 13.9.

Table 13.1 The Performance of the DSSCs in Fig. 13.10

Dye	J_{sc} (mA cm^{-2})	V_{oc} (mV)	FF (%)	Efficiency (%)
MK-1	10.5	748	70	5.5
MK-20	10.8	785	71	6.0
MK-14	11.0	802	71	6.3
MK-33	11.7	825	69	6.6

chromophores, as well as micelles, vesicles, nanodisks, nanofibers, and nanotubes self-assembled from synthetic amphiphiles, and also biomacromolecules integrated with dye moieties [33,34]. Nanotubes have a complete one-dimensional structure with well-defined size dimensions, and individually functionalized inner and outer surfaces [35] that are different from the branching structure of nanofibers. This unique morphology should lead to directional energy and charge transfers. In fact, the embedded dye moieties within the wall of supramolecular nanotubes and bionanotubes, including the tobacco mosaic virus, have been known to show excellent light-harvesting, photocatalytic, and electrical conduction abilities [36–46]. This subsection describes the rationally designed amphiphiles 1(R) (Scheme 13.1), quantitatively obtained from the dehydration reaction between a simple glycolipid and boronic acids, to selectively construct supramolecular nanotubes as a light-harvesting scaffold [47]. The light energy absorbed by the aromatic groups densely

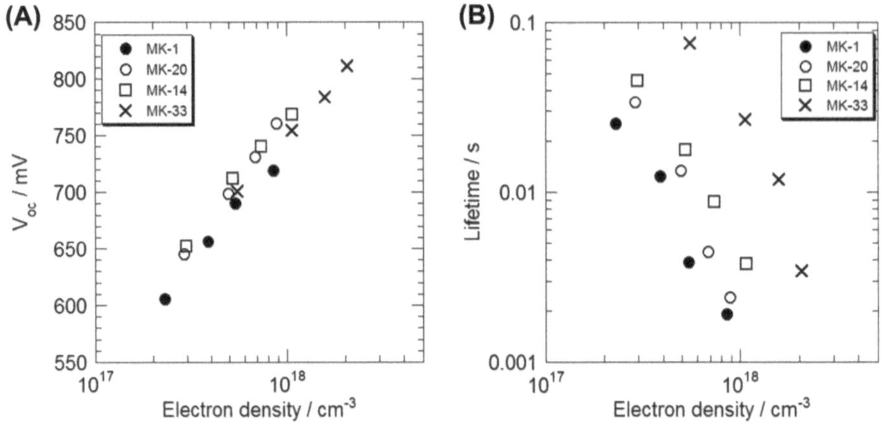

FIGURE 13.11 Relation between the open circuit voltage (A) and the electron lifetime (B) as a function of the electron density of the DSSCs.

organized within the bilayer membrane wall of the nanotube can be effectively transferred to acceptors encapsulated in the nanochannel. This results in a significant enhancement of the photocatalytic functions of a Re(I) complex as an acceptor [48]. Self-assembly of 1(R) in toluene selectively gave nanotubes consisting of bilayer membranes, in which 1(R) packs in an interdigitated fashion. The sizes of the inner diameter and the membrane thickness of the nanotubes depended on the aromatic groups in 1(R). The nanotube (denoted hereafter as N-nanotube) self-assembled from 1(R = naphthyl) had a uniform inner diameter of 15 nm and a

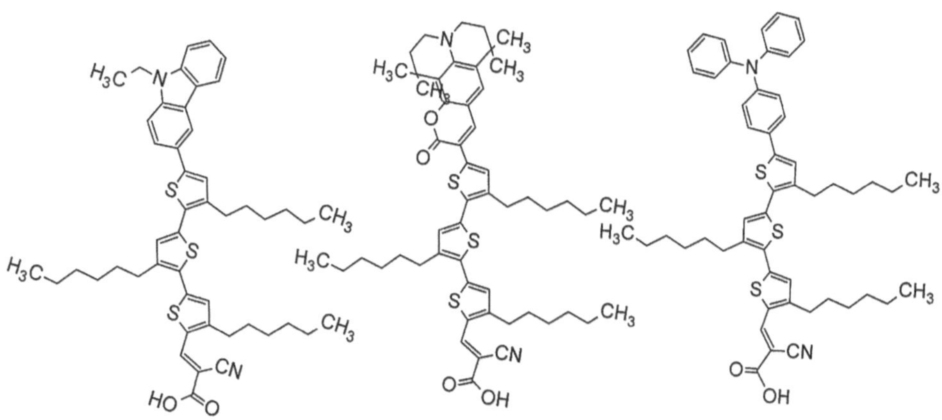

MK-1 MK-31 MK-88

FIGURE 13.12 Structure of dyes with various donors.

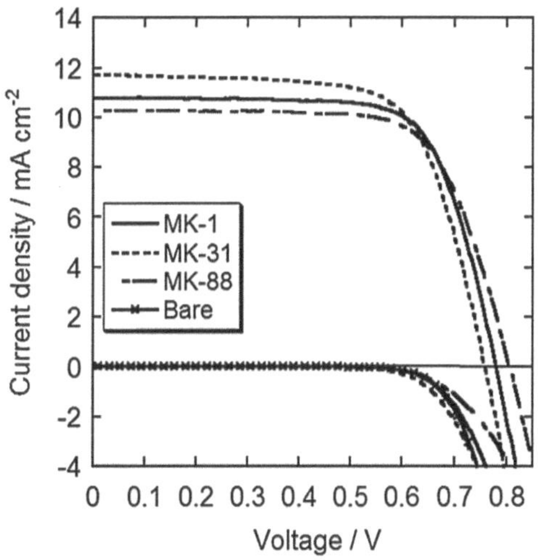

■ **FIGURE 13.13** The I-V curves of DSSCs with the dyes in Fig. 13.12.

membrane thickness of 18 nm. The latter is comparable to three times the value of the membrane stacking periodicity, $d = 5.90$ nm (Fig. 13.15). The absorption band ($\lambda_{max} = 280$ nm) for the N-nanotube was almost the same as that of 1(R = naphthyl) as a monomer. This means that few interactions are present between the naphthalene groups within the bilayer membranes of the N-nanotube in the ground state, despite the dense molecular packing. Upon excitation at 280 nm, the N-nanotube exhibited a relatively narrow fluorescence band at 350 nm, based on the monomer species of the naphthalene groups. The fluorescence quantum yield ($\phi = 0.39$) of the N-nanotube was comparable to that ($\phi = 0.33$) of the 1(R = naphthyl) monomer, although aggregates of dye monomers generally have lower ϕ values because of self-quenching based on the strong interactions in the excited state. The complex $[Re(4,4'\text{-dimethyl-}2,2'\text{-bipyridine})(CO)_3(PPh_3)]^+$ was used as a photocatalyst since the absorption band ($\lambda_{max} = 345$ nm), corresponding to the metal-to-ligand charge transfer of the Re(I) complex, overlapped well with the fluorescence band derived from the naphthalene

Table 13.2 Performance of the DSSCs in Fig. 13.13

Dye	J_{sc} (mA cm^{-2})	V_{oc} (mV)	FF (%)	Efficiency (%)
MK-1	11	0.779	0.72	6.0
MK-31	12	0.758	0.69	6.1
MK-88	10	0.806	0.70	5.8

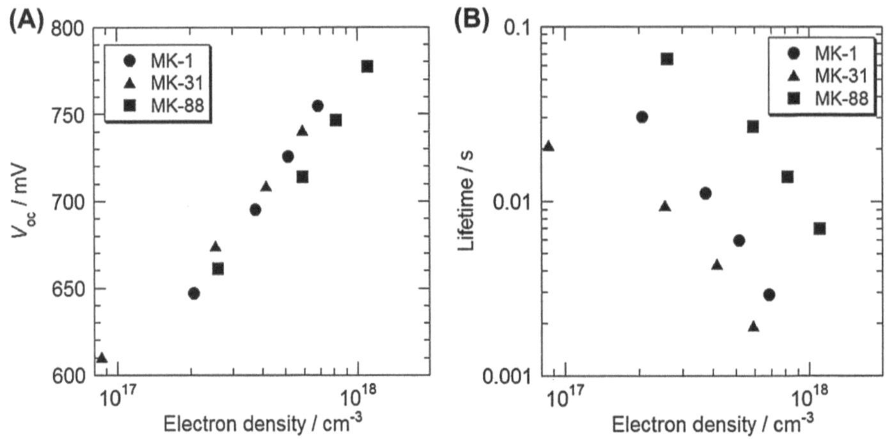

■ **FIGURE 13.14** Relation between the open circuit voltage (A) and the electron lifetime (B) as a function of the electron density of the DSSCs with the dyes in Fig. 13.12.

moieties in the N-nanotube. Namely, the combination of both chromophores suits the induction of the energy transfer. Fig. 13.16A shows the change in the fluorescence spectrum of the N-nanotube encapsulating the Re(I) complex in the nanochannel (10 mol% Re(I) complex/naphthalene

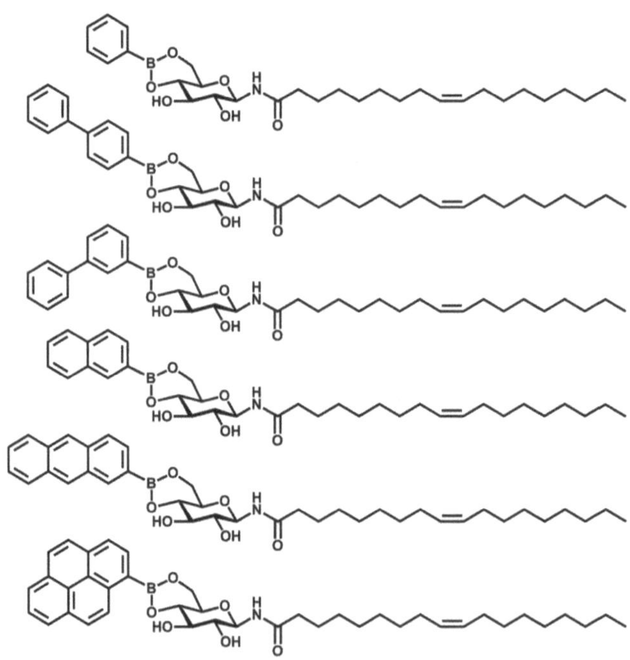

■ **SCHEME 13.1** Chemical structures of 1(R), R = phenyl, biphenyls, naphthyl, anthracenyl, and pyrenyl.

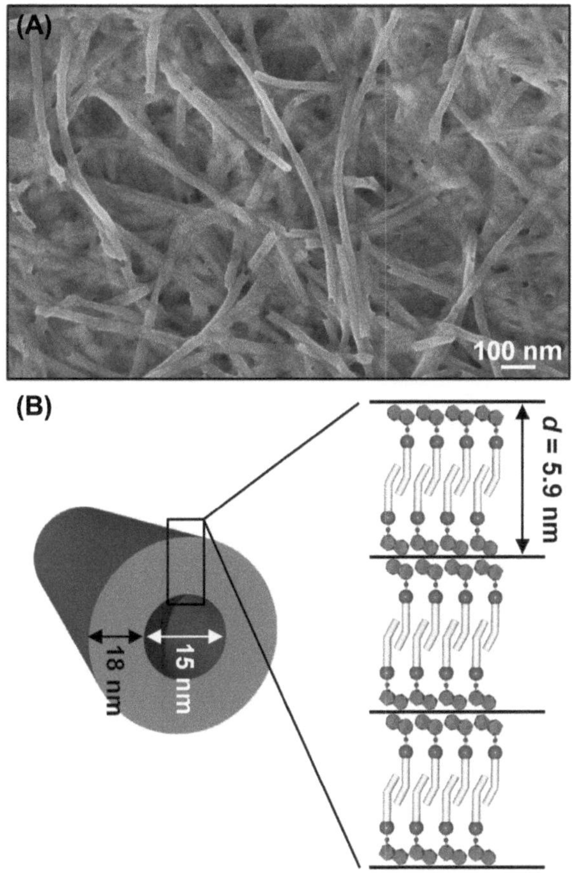

■ **FIGURE 13.15** (A) Scanning electron microscopic image of the N-nanotubes self-assembled from 1(R = naphthyl). (B) Schematic illustration of the N-nanotubes consisting of three bilayer membranes, in which 1(R = naphthyl) packs in an interdigitated fashion.

in the N-nanotube). The fluorescence band at 350 nm of the naphthalene groups in the N-nanotube membrane wall completely disappeared, while simultaneously, a broad fluorescence band of the Re(I) complex newly appeared at around 550 nm. This phenomenon was ascribable to the energy transfer from the naphthalene groups in the nanotube membrane wall to the encapsulated Re(I) complex in the nanotube channel (Fig. 13.16B), as emission upon direct excitation at 280 nm of the Re(I) complex was negligible (<0.5%). This was due to a lower concentration when compared to that of the naphthalene in the N-nanotube. The fluorescence of the encapsulated Re(I) complex, enhanced by the energy transfer under excitation at 280 nm, is much higher than that of the Re(I) complex by direct excitation at 345 nm. This indicates that the N-nanotube functions as an excellent

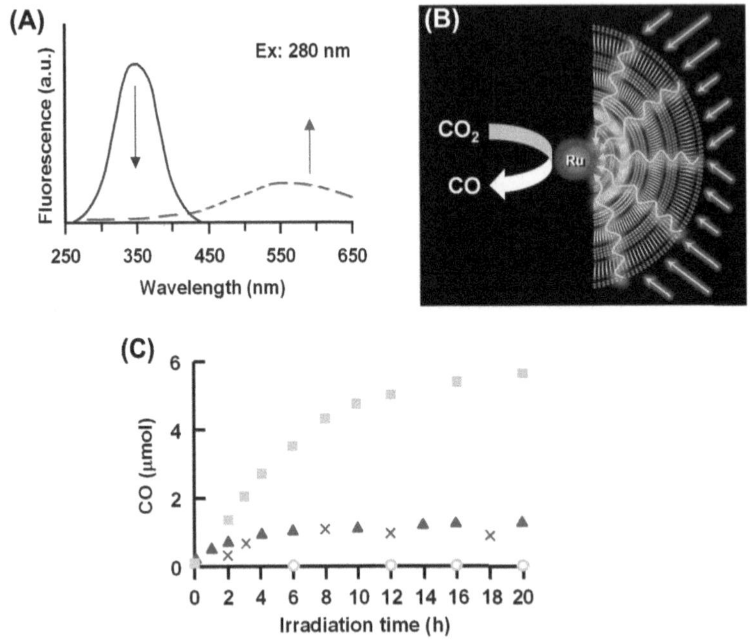

■ **FIGURE 13.16** (A) Fluorescence spectra of the N-nanotube encapsulating the Re(I) complex in the nanochannel (*a dotted line*) and the only N-nanotube (*a solid line*) in CH_3CN. (B) Schematic illustration of the enhancement of the photocatalytic activity of the encapsulated Re(I) complex in the N-nanotube acting as a light-harvesting antenna scaffold. (C) CO generation by photocatalysis for CO_2 reduction in the N-nanotube encapsulating the Re(I) complex under photoirradiation at 280 nm (■) and 345 nm (▲). CO generation by photocatalysis for CO_2 reduction in the bulk solution containing the Re(I) complex under photoirradiation at 345 nm (×). No CO generation from the N-nanotube without the Re(I) complex under photoirradiation at 280 nm (○).

light-harvesting antenna. The fluorescence decay of the naphthalene groups in the N-nanotube was remarkably shortened after encapsulation of the Re(I) complex, accompanied with the fast rise in the fluorescence decay profile of the Re(I) complex. Therefore, the quenching process is attributable to a resonance energy transfer without a radiation process. The energy transfer efficiency, estimated from the fluorescence quenching rate of the naphthalene groups in the nanotube membrane wall, reached almost 100% when the concentration of the encapsulated Re(I) complex was 10 mol%. The confinement of the Re(I) complex into a nanochannel with a 15 nm inner diameter by encapsulation should contribute to the high energy transfer efficiency. In fact, the energy transfer from the naphthalene groups in the nanotube membrane walls to the free Re(I) complex in the bulk solution hardly occurred, reflecting that the donor-acceptor distance (Förster distance) is the most important factor that affects efficiency. The photocatalytic reaction for CO_2 reduction via the encapsulated Re(I) complex in the N-nanotube was carried out in CH_3CN in the presence of triethanolamine as a sacrificial agent. Fig. 13.16C shows that CO evolution is strongly promoted by photoirradiation at 280 nm of the encapsulated Re(I) complex in the N-nanotube under a CO_2 atmosphere (closed square plots). However, the catalyst was gradually deactivated after a lapse of time. On the other hand, the same photoirradiation to the N-nanotube without the encapsulated Re(I) complex never produced CO (open circle plots). This supports the hypothesis that the above evolved CO does not originate from the organic components of the N-nanotube. Photoirradiation of the encapsulated Re(I) complex in the N-nanotube at 345 nm directly excites the Re(I) complex. At this wavelength, the free Re(I) complex in the bulk solution allowed us to observe smaller amounts of CO evolution (closed triangle plots and cross plots, respectively). The photocatalytic activity of the encapsulated Re(I) complex, enhanced by the energy transfer under excitation at 280 nm, is 3.5 times higher than that of the encapsulated Re(I) complex by direct excitation at 345 nm. This demonstrates that the N-nanotube acts as an excellent light-harvesting antenna (Fig. 13.16B). Furthermore, the conversion of the UV absorption energy to a lower excitation energy, due to such a light-harvesting antenna function of the N-nanotube, leads to the remarkable suppression of UV decomposition and ligand dissociation of the encapsulated Re(I) complex in the nanochannel. The energy transfer can be enhanced by appropriate selection and combination of aromatic groups in the nanotube membrane wall and guest acceptors in the nanochannel. Furthermore, control of the confinement effect, based on the encapsulation of guest acceptors in the nanochannel, will enable us to achieve a high energy transfer with 100% quantum efficiency. This can be seen in nature in the photosynthetic system. The precise design of amphiphilic monomers and the introduction of other boronic acids (Scheme 13.1), should contribute toward the control of the confinement effect as well as the expansion of the absorption band region acceptable to sunlight irradiation. Hierarchical organization into macroscale soft materials, such as gels and liquid crystals [49,50], will also open ways to develop optical and electronic soft materials applicable in photocatalysis and DSSCs.

13.3.2 Enzyme Reactors Based on Supramolecular Nanotubes With Tunable Diameters and Surfaces

Enzymes that are highly selective, and efficiently promote chemical reactions in the living body, have attracted much attention in fields in the chemical industry, such as green catalysis. Immobilization of enzymes in supports and carriers, such as polymer surfaces and gels [51,52], microchannels and capsules [53,54], and inorganic porous materials [55], have enabled us to construct enzyme reactors that produce target products from substrates. General immobilization methods are as follows: (1) Covalent bonding, ionic bonding, host−guest (affinity) interactions, and physical adsorption between the enzymes and the supports; (2) Cross-linking among enzymes using cross-linkers; (3) Encapsulations of the enzymes into capsules and the void spaces among the fiber networks of the gels. However, of concern in methods 1 and 2, is the deactivation of the enzymes due to their chemical and sterical changes, while in method 3, a decrease in the reaction rate due to the slow diffusion of the substrates may occur to the enzymes encapsulated in the restricted spaces. The latter part of this subsection describes the wedge-shaped bolaamphiphiles 2(n), 3(n), and 4(n) (Scheme 13.2), in which two hydrophilic headgroups of different sizes are connected to a hydrophobic spacer at each end [56], to form supramolecular nanotubes consisting of monolayer membranes as mesoscale hosts for the enzymes. The selectively encapsulated enzymes in the nanotubes exhibited high chemical and thermal stabilities, depending on the rational fitting of the size of the enzymes to the inner diameters of the nanotubes. As a result of this, the enzyme reactors were able to resolve the just-discussed problems, as described in the enzyme immobilization methods 1−3. If the wedge-shaped amphiphiles can only form a monolayer membrane with parallel molecular packing, the monolayer membrane is given a curvature based on the size difference between both headgroups, and forms a tubular morphology having different surfaces covered with each headgroup (Fig. 13.17A). However, the self-assembly of 2(n), having a relatively larger glucose headgroup and a smaller amino headgroup in water adjusted

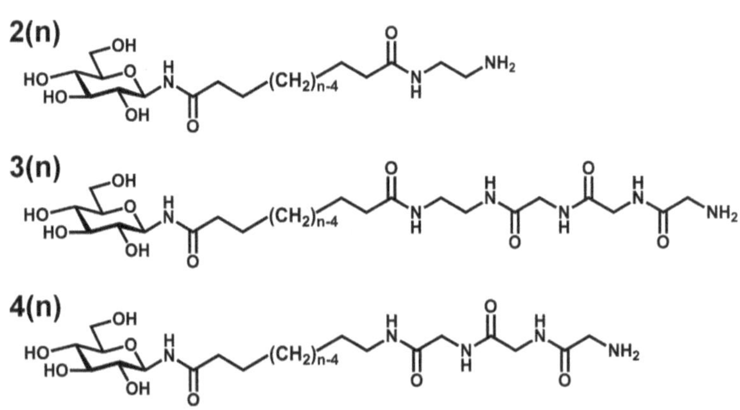

■ SCHEME 13.2 Chemical structures of wedge-shaped bolaamphiphiles, 2(n), 3(n), and 4(n).

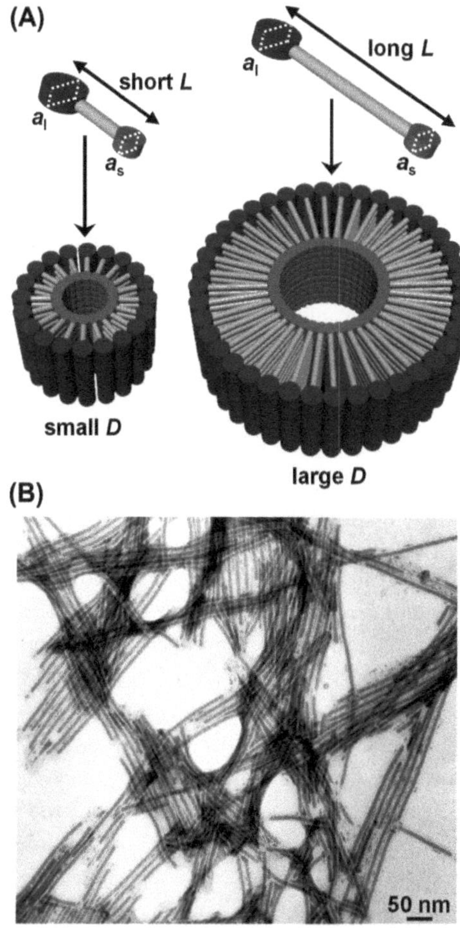

■ **FIGURE 13.17** (A) Schematic illustration of narrow and wide nanotubes, consisting of a monolayer membrane in which the wedge-shaped bolaamphiphiles with short and long oligomethylene spacers pack in parallel fashion, respectively. (B) Transmission electron microscopic image of the NT(10 nm i.d.) self-assembled from 3(16), negatively stained with phosphotungstate. The hollow cylinder space of the nanotubes can be visualized as being relatively darker than its surroundings.

to pH 10, gave tape-like structures in which the 2(n) bolaamphiphile packs in an antiparallel fashion within the monolayer membranes. Prior to the self-assembly process in water, the starting films were prepared by association of 2(n) with poly(thiopheneboronic acid) (PTB) to induce the parallel molecular packing of 2(n) in advance [57]. All the 2(n) moieties associated with PTB [2(n): the boronic acid moiety of PTB = 2:1] in the starting films that were prepared by dehydration reactions between the glucose head group of 2(n) and the boronic acid moiety of PTB, packed in a parallel fashion within the monolayer membranes. Dispersion of the obtained films in water caused the hydrolysis reaction of the boronate ester moieties. All the 2(n) moieties that dissociated from PTB kept the parallel molecular packing and selectively formed the nanotubes. The inner diameters were estimated to be 68, 69, 74, 79, and 80 nm for the nanotubes self-assembled from 2(12), 2(14), 2(16), 2(17), and 2(18), respectively. When the nanotubes consist of parallel molecular packing of the wedge-shaped amphiphiles, the inner diameters (D_{calc}) can be defined by the following equation: (1) $D_{calc} = 2a_sL/(a_l - a_s)$, where a_s and a_l are the cross-sectional areas of the small and large headgroups, respectively, and L is the extended molecular length (Fig. 13.17A). Based on Eq. (1), the elongation of the oligomethylene spacer enlarges the D_{calc} values, 1.5 nm steps per two methylenes. The experimental results almost reflect Eq. (1), although the incremental values become more irregular. The size of the inner diameter strongly depended on the pH conditions during the self-assembly procedure in water. At pH 6, 2(18) formed helical coils, while at pH 10 it formed a nanotube with an 80 nm inner diameter [abbreviated hereafter as NT(80 nm i.d.)] as described earlier. The helical coils transformed into nanotubes with a 20 nm inner diameter [abbreviated hereafter as NT(20 nm i.d.)] after aging for several weeks. The partial protonation of the amino group in 2(18) induced a large molecular tilt, and therefore resulted in the twisting and coiling of the monolayer membranes. The introduction of functional groups of specific intermolecular interaction abilities enabled wedge-shaped amphiphiles to regulate their parallel molecular packing. This resulted in the exclusive formation of nanotubes with different inner diameters [58]. Covalent modification of a triglycine group into 2(n), that is, 3(n) (n = 12, 14, 16, 18, and 20), brings about the selective formation of nanotubes that consist of a single monolayer lined by polyglycine-II-type hydrogen-bond networks among the triglycine moieties. All nanotubes have an inner diameter of about 10 nm [abbreviated hereafter as NT(10 nm i.d.)] that is independent of the oligomethylene spacer of 3(n) (Fig. 13.17B). The earlier prepared nanotubes have different inner and outer surfaces that are covered with amino and glucose headgroups, respectively. Therefore, the nanotube channels should have cationic and nonionic surfaces, based on the protonation and deprotonation of the amino headgroups on the inner surface, depending on the pH conditions. Green fluorescent protein (GFP, MW = 30,000, pI = 5.1) and carbonic anhydrase (MW = 30,000, pI = 5.9) were chosen as an anionic protein and an enzyme, respectively, since both proteins have similar sizes (3–4 nm) and surface charges at certain pH values. Chemical modification was carried out to visualize the dynamic encapsulation and transportation behavior of GFP in the nanotube channels by the fluorescence resonance energy transfer (FRET) technique. Alexa Fluor 546 (abbreviated hereafter as Alexa), an optical sensing probe for GFP, was covalently linked to the amino groups on the inner surface of NT(80 nm i.d.), NT(20 nm i.d.), and NT(10 nm i.d.). Fluorescence spectroscopy indicated that the percentage of Alexa immobilized on the amino groups on the inner surface was estimated to be about 10%. Fluorescence microscopy clearly displayed the nanotubes, proving that Alexa was dispersed on the inner surface of the membrane wall over the range of the displayed length. When the GFP approaches to within a few nanometers of the Alexa on the inner surface of the nanotube channels, Alexa acts as a fluorescence acceptor, and the GFP acts as a fluorescence donor in FRET. Time-lapse fluorescence microscopic observations showed that each open end of the Alexa-immobilized nanotube started to become visible, and as time elapsed, the two bright lines finally reached the center upon the addition of GFP at a pH of 6.8 (Fig. 13.18) [59]. This phenomenon, in which the GFP penetrates the

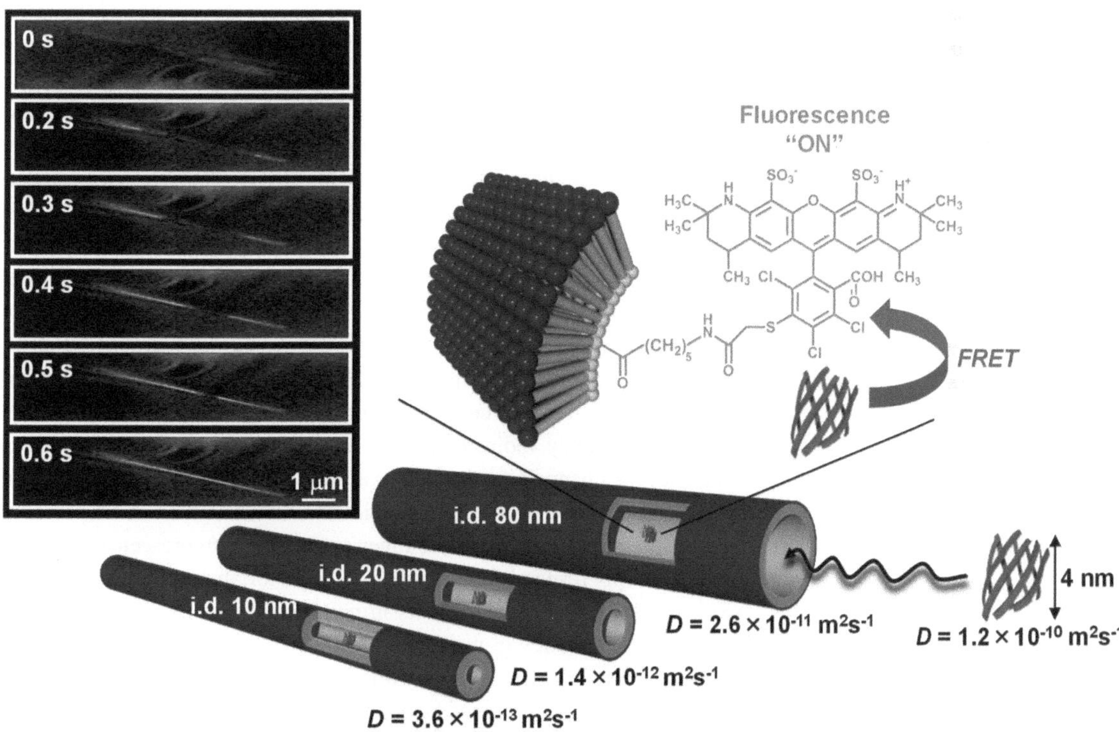

■ **FIGURE 13.18** (Left) Time-lapse fluorescence microscopic images of the Alexa-immobilized nanotubes with a 20 nm inner diameter upon addition of GFP. The elapsed time is indicated at the top of each image. (Right) Schematic illustration of the FRET from the encapsulated GFP to the interior Alexa on the inner surface of the nanotube. The *D* values of the encapsulated GFP in the nanotube channel compare with that of free GFP in the bulk solution, determined by dynamic light-scattering measurements.

Alexa-immobilized nanotube from both ends, and gradually excites the Alexa on the inner surface as it flows through the nanotube channel, is ascribable to FRET. The analysis of the obtained time-lapse images demonstrated that a diffusion phenomenon can simulate this GFP transportation feature well. The square of the GFP migration distance (X^2) was plotted against the time required for migration. Each obtained linear plot is clearly compatible with the diffusion process. In the present systems, GFP exists in high concentrations in the external solution, and diffuses from the open ends of the nanotubes ($X = 0$) to the center of the nanotube channel ($X > 0$). Therefore, the diffusion constant (D) can be defined as the equation, $X^2 = 4D\Delta t$. The D values are smaller than that of the free GFP in bulk that was independently determined by dynamic light scattering (Fig. 13.18). This depression in diffusion was also observed in the spherical protein ferritin with a 15 nm diameter in the NT(80 nm i.d.) nanochannel [60] and latex beads with a 30 nm diameter, in a fluid-state lipid nanotube with a 100 nm inner diameter composed of phospholipids [61]. Furthermore, the D values of the GFP decreased remarkably as the inner diameters of the nanotubes decreased. This is compatible with the theory that a strong reduction in diffusion is seen in restricted nanospaces. The storage and release behavior of the encapsulated GFP in the Alexa-immobilized nanotube channels was monitored by using this FRET system. The fluorescence spectra that represent the FRET from the encapsulated GFP to the interior Alexa hardly changed, even after 80 h at a pH of 6.8. This result indicates that the encapsulated GFP stays in the nanotube channels. The cationic inner surface of the nanotubes proved to effectively preserve the anionic GFP via electrostatic attraction, even though both open ends are exposed to the bulk solution. When the pH was increased to 8.9, the fluorescence intensity of the GFP increased, whereas the fluorescence intensity of Alexa emitted through FRET decreased. The reduction in FRET efficiency indicates the release of the encapsulated GFP from the nanotube channels to the bulk solution because a pH of 8.9 induced a dramatic decrease in the electrostatic attraction based on the deprotonation of the amino groups on the inner surface. The fluorescence intensity of GFP sharply responds to denaturation behavior, that is, native GFP fluoresces strongly, whereas it results in nonfluorescence upon denaturation. Elevated temperatures in the range of 20–90°C caused a decrease in the fluorescence intensity of the free GFP in bulk solution. Therefore, thermal denaturation of GFP proceeds on heating. Temperature elevation also caused a decrease in the fluorescence intensity of Alexa emitted through FRET from the encapsulated GFP to the interior Alexa of NT(80 nm i.d.) and NT(20 nm i.d.), even though both nanotubes showed no morphological changes or disassembly, based on a relatively high gel-to-liquid crystalline phase transition temperature, $T_{g\text{-}l} > 100°C$. The release of the encapsulated GFP to the bulk solution was not observed, suggesting that the fluorescence intensity of the encapsulated GFP itself decreases by thermal denaturation. On the other hand, the FRET efficiency from the encapsulated GFP to the Alexa of NT(10 nm i.d.) with $T_{g\text{-}l} > 100°C$ proved to be insensitive to the elevated temperatures, indicating that the encapsulated GFP shows no denaturation under the given conditions. The nanotube with a 10 nm inner diameter self-assembled from 4(n), composed of a short oligomethylene spacer, had relatively lower $T_{g\text{-}l}$ (=60°C) and was not able to stably store GFP at elevated temperatures above $T_{g\text{-}l}$ in which the solid monolayer membrane of the nanotube is converted to a fluid one [62,63]. This temperature dependence of the *endo*-sensing features by using the FRET system clearly shows that the thermal stability of GFP strongly depends on the inner diameters and the stabilities of the nanotubes. Encapsulation, transportation (diffusion), and release behavior and stability in the nanotube channels of carbonic anhydrase (CAB) are expected to be similar to those of GFP, when considering the size and surface charge resemblances of both proteins. The enzyme activity of the free CAB for the hydrolysis of *p*-nitrophenyl acetate in the bulk solution was completely lost after the temperature was elevated to a temperature range of 70–90°C for 1 h. On the other hand, over 90% of the enzyme activity of the encapsulated CAB in the NT(10 nm i.d.) nanochannel with $T_{g\text{-}l} > 100°C$ was retained under the same conditions (Fig. 13.19). Upon encapsulation, NT(80 nm i.d.) never exhibited the stabilization of the CAB. Instead, the thermal denaturation of, in this case, encapsulated CAB, was comparable to that of the free CAB in bulk solution, indicating that the tightly restricted 10 nm channel kinetically and thermodynamically prevented the denaturation of CAB. Interestingly, NT(20 nm i.d.) accelerated the denaturation of encapsulated CAB. Circular dichroism spectroscopy suggested that the nanochannel of NT(20 nm i.d.) induced the aggregation of the encapsulated CAB accompanied by the destruction of its α-helical structure, whereas the nanochannel of NT(10 nm i.d.) prevented aggregation [64]. This acceleration of the CAB aggregation was probably due to the loose geometry of the NT(20 nm i.d.) nanochannel relative to the size of CAB. Fluorescence lifetime measurements of a viscosity probe, 8-anilinonaphthalene-1-sulfonate, showed that the viscosity of the water in the NT(10 nm i.d.) nanochannel is twice as large as that of the bulk water, and is comparable to that of water at −10°C [65,66]. The polarity of the water in the NT(10 nm i.d.) nanochannel is about 15% lower than that of the bulk water, and is comparable to that of ethanol. Water having such physical properties, namely, the so-called "confined water," has been observable in cells shaped by the bilayer membranes of biological

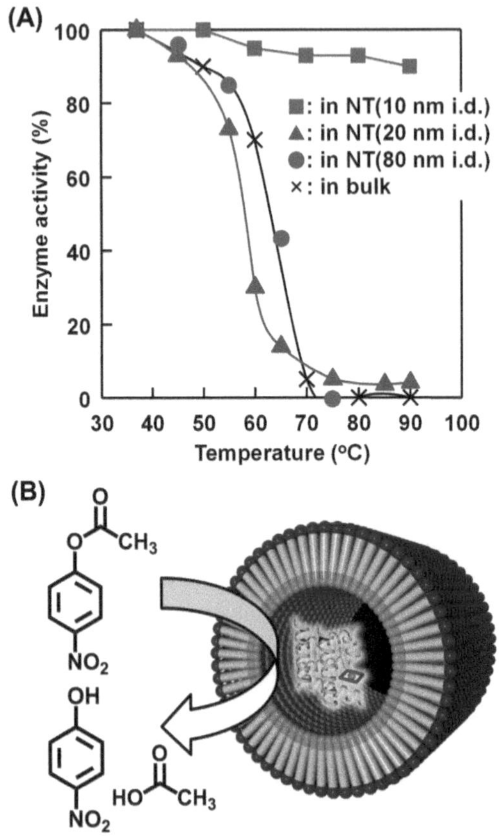

■ **FIGURE 13.19** (A) Enzyme activity of the encapsulated CAB in the nanotube channels and the free CAB in the bulk solution under different temperature conditions. (B) Phenol generation by enzymatic catalysis with high thermal stability for hydrolysis of the phenol acetate in the nanotube channel encapsulating CAB.

phospholipids. Therefore, the confined water in the nanotube channel, shaped by the monolayer membrane of the synthetic lipids, should contribute to stabilize the hydration structure of CAB. This is an important factor for keeping the native structure and the folded conformation.

The present enzyme reactor that performs well at high temperature conditions is able to prevent a decrease in the reaction rate, due to the slow diffusion of the substrates from the bulk solution to the narrow nanotube channel existing enzymes. Rational design of the synthetic amphiphiles and diversification of self-assembly processes enable us to utilize various unstable enzymes, and also to select harsh reaction conditions. Furthermore, cooperation of the catalytic abilities of the enzymes with supramolecular chirality [67] of the nanotubes, based on the amplification of the original chirality in the amphiphiles, will lead to enantiomer synthesis in high yield and high selectivity.

13.4 NANOARCHITECTONICS FOR REVERSIBLE HYDROGEN STORAGE

The concept of nanoarchitectonics can be applied to the chemistry of hydrogen storage. Today, H_2 has attracted a great deal of attention in energy-related chemical materials. H_2 is well known as a clean and green energy source because it produces only water molecules without carbon dioxide (CO_2) as waste during power generation. In addition, it can be supplied by using renewable energy via the electrolysis of water that exists in abundance. H_2-related technologies, such as fuel cell technology and H_2 power generation with turbine engines, are developing at an amazingly accelerated pace [68–70]. However, H_2 is a gas molecule under ambient temperature and pressure, making it difficult to handle, store, and transport. To solve these problems, some physical technologies, such as high-pressure cylinders to compress the H_2 gas and convert it into liquid H_2 at low temperatures, are currently being studied. On the other hand, recent studies have shown that it is possible to store and produce H_2 molecules via chemical conversion pathways. For example, hydrogenation and dehydrogenation of molecules using catalysts are well-studied basic chemical reactions. By using these reactions for H_2 storage and production, many molecules can be regarded as H_2 carriers [71–73]. The interconversion cycle between CO_2/H_2 and formic acid is one of the most exceptionally interesting reversible H_2 carrier systems because formic acid is a nontoxic liquid under ambient temperature and contains 4.3 wt.% H_2. Furthermore, the energy required for the conversion is low, e.g., hydrogenation of CO_2 to formic acid in the gas phase is $\Delta G^{\circ}_{298} = +32.9\ \text{kJ mol}^{-1}$ (endergonic) and hydrogenation of CO_2 to formic acid in the aqueous phase is $\Delta G^{\circ}_{298} = -4.0\ \text{kJ mol}^{-1}$ (exergonic) (Fig. 13.20) [74–76]. Not only can this system act as a possible H_2 carrier candidate but it can also allow the utilization of CO_2 from the point of view of environmental chemistry.

$$H_2(g) + CO_2(g) \longrightarrow HCO_2H(l)$$

$$\Delta G^0 = +32.9\ \text{kJ mol}^{-1}$$
$$\Delta H^0 = -31.5\ \text{kJ mol}^{-1}$$
$$\Delta S^0 = -215\ \text{J mol}^{-1}\ \text{K}^{-1}$$

FIGURE 13.20 Hydrogenation of CO_2 in the gas phase.

Hydrogenation of CO_2 into formic acid using homogeneous catalysts has been studied since the 1970s [77]. Compared to heterogeneous catalysts, homogeneous catalysts offer higher activity and greater reaction selectivity. However, to achieve high hydrogenation reaction efficiencies, high-pressure H_2 and CO_2 gases are required because the activity of the catalyst is insufficient to carry out the reaction. Moreover, organic solvents and/or organic additives are often used, even if these are not suitable for green chemistry. Nevertheless, many reports still employ such reaction conditions [78–80].

However, Himeda et al. reported that hydrogenation of CO_2 can be achieved without any organic solvents and organic additives by using homogeneous half-sandwich metal complex catalysts [81–85]. This holds especially for catalysts with a dihydroxyl group on the ortho position of the bipyridine ligand (6DHBP) (Fig. 13.21). These catalysts exhibited a much higher hydrogenation ability than any other reported homogeneous catalysts at room temperature and ambient pressure in alkaline solutions (Turnover frequency = 27 h^{-1}) [86]. A strategy to achieve CO_2 hydrogenation under these mild conditions was to mimic the active site of the enzyme hydrogenase that catalyzes the reduction/oxidation of H_2 in nature, in the catalyst design. Some hydrogenase groups have a base residue close to their active metal cluster to assist the heterolysis of H_2 (Fig. 13.22) [87–90]. The position of the base residue is like a pendant against the active center, so the base is called a pendant base, and the assisted effect of the base is called the pendant base effect [91,92]. In the 6DHBP catalyst, the dihydroxyl groups on the ortho position of the bipyridine ligand mimic the pendant base to generate a hydride complex via the heterolysis of H_2. In fact, the dihydroxy groups were deprotonated during the reaction, and the dioxyanion was generated because the hydrogenation of CO_2 took place in an alkaline aqueous solution (Fig. 13.21). The hydride complex is assumed to react with CO_2 (or carbonate, or bicarbonate) to produce the formate through an insertion reaction. Although the high catalytic activity observed with 6DHBP suggests that it is due to the pedant base effect, it is in fact the

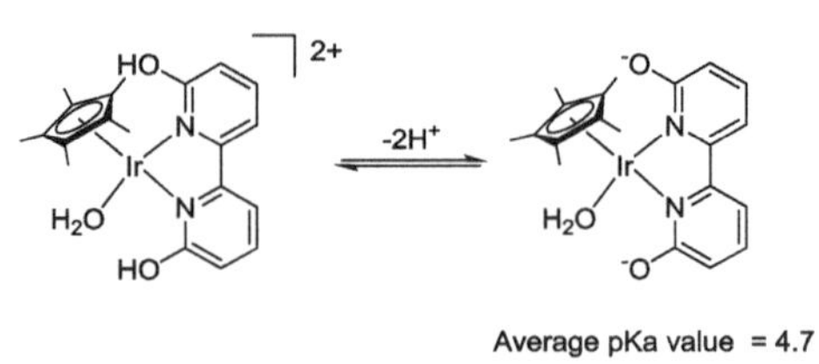

■ **FIGURE 13.21** Homogeneous bipyridine-based catalyst (6DHBP) for CO_2 hydrogenation in aqueous media.

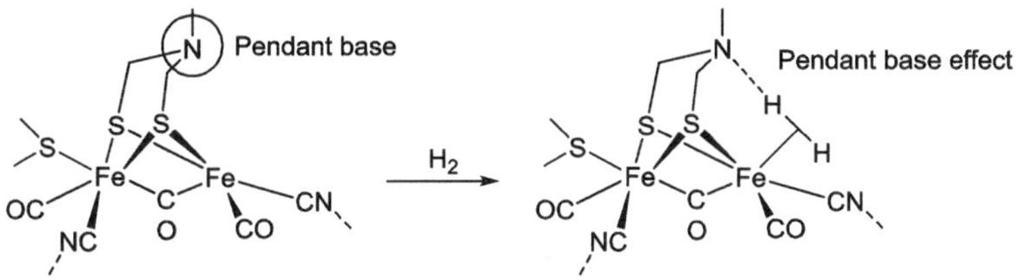

■ **FIGURE 13.22** Schematic representation of the heterolysis of H_2 via support of a pendant base on the active site of hydrogenase.

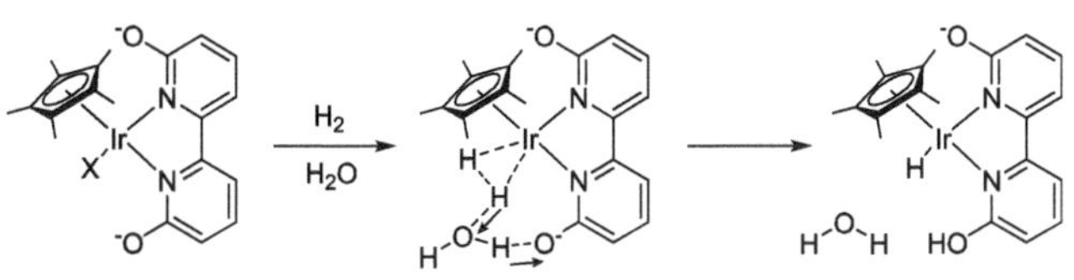

■ **FIGURE 13.23** Proposed mechanism of heterolysis of H_2 assisted by the oxyanion (pendant-base) and water via nano-order proton relay in an alkaline solution with a pH of 8.6. Arrows show the possible trajectory of the proton relay.

nano-order chemical effect that is responsible for this enhancement. Using DFT calculations to simulate the reaction intermediate, Himeda reported that a water molecule, existing between the center metal and the hydroxyl anion, assists the heterolysis of H_2 through proton relay (Fig. 13.23) [93]. This is supported by the results from the kinetic isotope effect experiment with D_2O that also suggest the assistance of the water molecule during heterolysis of H_2 [86]. Other groups have also reported this kind of water-assisted effect on heterolysis using computational simulations [94–96]. Although the effect of the nano-order space filling by the substituent of the catalyst and water is considered as a second coordination sphere effect [97], this effect also deserves to be regarded as an example of nanoarchitectonics.

The catalysts Himeda developed can dehydrogenate formic acid to supply hydrogen [98–100]. Interestingly, only a change in pH is required to reverse the reaction. Under the alkaline pH conditions mentioned previously, the catalyst converts H_2 into the formate. On the other hand, formic acid is dehydrogenated into H_2 and CO_2 under acidic conditions. Formic acid dehydrogenation generally proceeds via three steps (Fig. 13.24). First, the catalyst reacts with the formate anion to form the formato complex. Next, CO_2 is released by β-hydrogen elimination, and a metal-hydride complex is formed. Finally, the hydride complex reacts with a proton to generate H_2 gas.

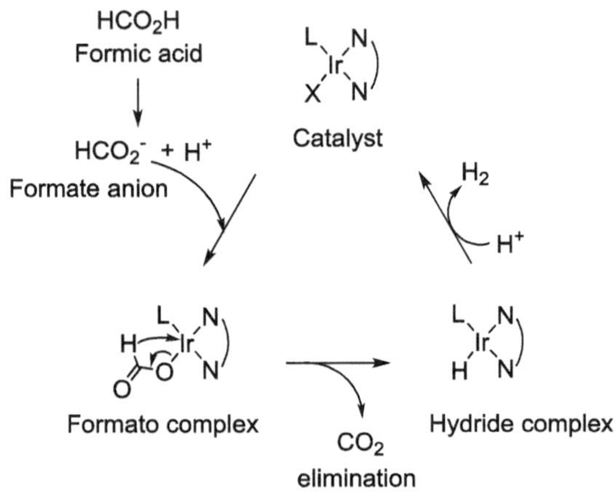

■ **FIGURE 13.24** Proposed reaction mechanism of the dehydrogenation of formic acid with homogeneous catalysts in aqueous media. L = ligand, N^N = bidentate ligand, X = a vacant site.

The pendant base can also assist via a proton relay during the reaction between the hydride complex and the proton [101]. The proton relay stabilizes the catalyst−H_2 transition state and lowers the energy barrier for H_2 generation (Fig. 13.25). Himeda experimentally demonstrated that the β-hydrogen elimination step is the rate-determining step for 6DHBP in the dehydrogenation reaction of formic acid using the kinetic isotope effect. This is because the final step is accelerated by the proton relay [101]. Moreover, the catalytic activity was also studied as a function of pH. This study revealed that the catalytic activity of 6DHBP was at a maximum at a pH of 4. This pH value is close to the pK_a of the hydroxyl group of the ligand, therefore, it was concluded that deprotonation affects the reaction rate. In other words, the formation of a proton relay pathway with a hydronium ion and a strong base such as an oxyanion in nano order is effective for the efficient dehydrogenation reaction of formic acid.

A reversible hydrogen storage system with homogeneous catalysts is still being developed. A catalyst design using a nanoarchitectonics approach

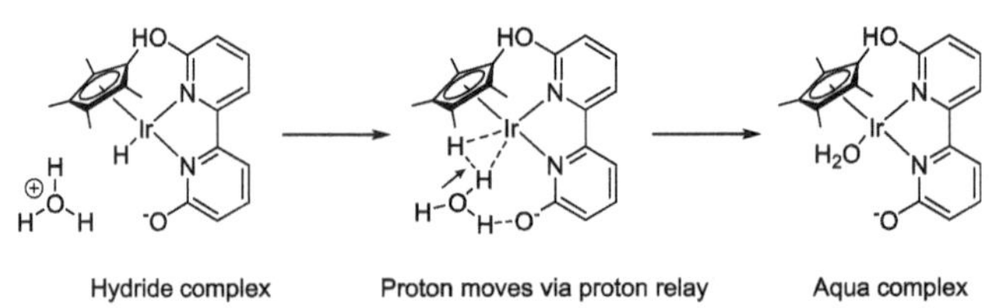

■ **FIGURE 13.25** Hydrogen generation assisted by the oxyanion and water (in the form of hydronium) via nano-order proton relay in an acidic solution with a pH of 3.5.

is one of the most promising methods for achieving highly efficient reactions. This subsection only discusses the application of the nanoarchitectonic approach to homogeneous catalysts. However, this approach can also be applied to various other types of catalysts, e.g., heterogeneous catalysts, MOFs, and immobilized catalysts.

13.5 DEVELOPMENT OF TEMPERATURE-RESPONSIVE HYDROGEL NANOPARTICLES FOR ENERGY-EFFICIENT CO_2 SEPARATION

The development of energy-efficient and scalable CO_2-capture processes is essential for the future use of carbon-based energy [102,103]. The capture of CO_2 from high-humidity exhaust gases of coal-fired power plants, using aqueous solutions of amines (such as ethanol amine) as absorbents, has been considered to be the most feasible process [104,105]. The amine solutions absorbed CO_2 at a low temperature ($\sim$40°C) by an exothermic reaction, and desorbed CO_2 upon heating (>140°C) (Fig. 13.26) [106,107]. Though these aqueous amine processes exhibited a high capacity for CO_2 capture, a high-energy consumption—equivalent to 20–40% of a typical power plant output—prevented the use of this process in these plants [106–110]. Furthermore, amines and their oxidized products are easily volatilized during the high-temperature regeneration process, and need to be recovered in order to avoid air pollution [111]. Thus, nonvolatile sorbents that desorb CO_2 at low temperatures (<100°C) needed to be developed [108–110].

The dynamic and reversible transition of the pK_a values of Brønsted acid and base groups in hemoglobin plays an important role in the energy efficient CO_2 transfer through the blood (Bohr effect) [112,113]. In the lungs,

■ **FIGURE 13.26** Schematic illustration of the traditional CO_2 capture process using amine solution.

the pK_a value of the ammonium and imidazolium cations in/on the hemoglobin decreases due to the drastic conformational change of the polypeptides, caused by oxygen binding to heme (Fig. 13.27). The pK_a shift triggers the release of H^+ ions into the blood, lowers the pH of the blood, and drives the CO_2 efficiently away from the lungs without heating the system. When the hemoglobin reaches a capillary where the concentration of oxygen is low, dissociation of oxygen from the heme triggers a change in conformation back to the original structure. Subsequently, the pK_a value of the ammonium and imidazolium cations in/on the hemoglobin increases back to its original value. The high pK_a value of the ammonium ions on the hemoglobin enables the efficient removal of CO_2 from the tissues around the capillaries in the form of ammonium and imidazolium bicarbonate.

Inspired by reversible ion capture triggered by the conformational change of biomacromolecules, Shinkai et al. developed photoresponsive synthetic receptors for the reversible recognition of specific ions [114–116]. It has also been reported that temperature-responsive poly(N-isopropylacrylamide) (pNIPAm) hydrogel films, nanoparticles (NPs), and fibers comprising

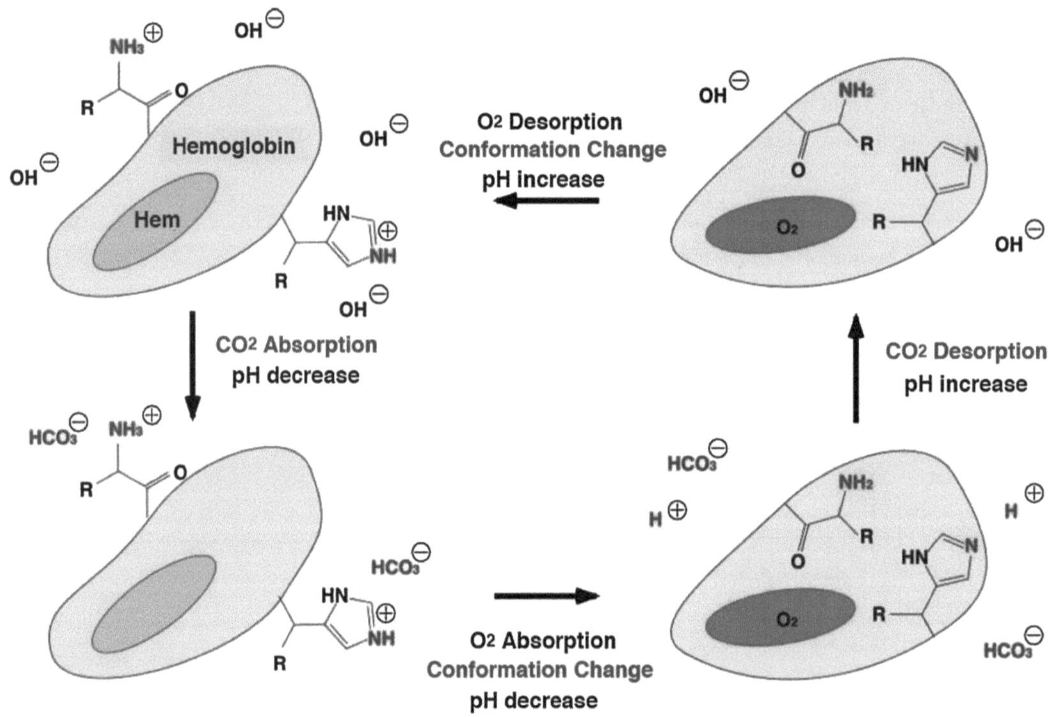

■ **FIGURE 13.27** Molecular mechanism of reversible CO_2 absorption by hemoglobin triggered by dissociation and binding of oxygen to the heme.

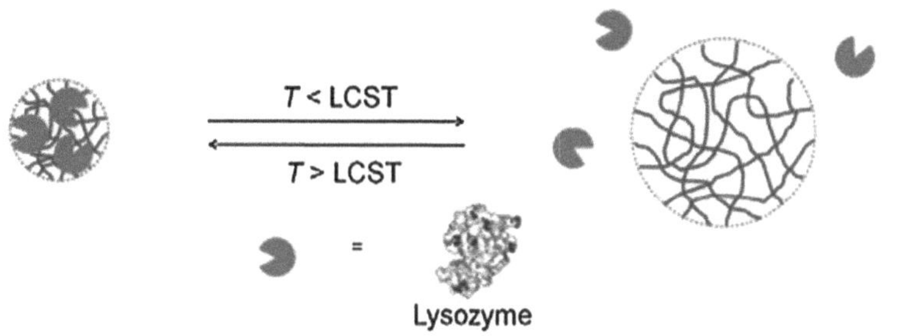

■ **FIGURE 13.28** Illustration of temperature-responsive catch and release of proteins by functionalized pNIPAm hydrogel particles [132].

functional groups reversibly captured targets such as ionic dyes [117,118], drugs [119,120], peptides [121], proteins [122], nucleotides [123], and cells [124,125] under heating and cooling cycles (Fig. 13.28). The binding affinity of pNIPAm NPs to capture target ions [126], peptides [127], and proteins [128] can be further enhanced by polymerizing NPs in the presence of target molecules that serve as a template. In this polymerization procedure of molecular/ion imprinting, the three-dimensional shape of the target molecule is transferred into the polymer network, leaving complementary binding sites in the network after removal of the target [129–135]. The imprinted NPs showed a greater affinity shift in response to the phase transition than the nonimprinted NPs [128].

Recently, it was reported that the pK_a value of Brønsted acids and bases incorporated in pNIPAm shifted reversibly in response to the phase transitions of polymers induced by cooling–heating cycles [123,136–139]. Thus, pNIPAm-based NPs that show large and reversible pK_a shifts can further be prepared by the "proton imprinting" [140] and "microenvironment imprinting" [141] strategy: When acidic monomers such as acrylic acids were protonated in the polymerization process, the protonated acids were easily incorporated into the less-polar domains of growing collapsed-pNIPAm NPs, and were subsequently stabilized by a cross-linker. As a result, high pK_a acids were imprinted in the NPs. When the degree of cross-linking was low, the structure of the high-pK_a acidic sites was reversibly denatured by temperature-induced conformational changes of the polymer chains without forming aggregates of NPs, resulting in a large reversible pK_a shift as proteins (Fig. 13.29A). The greatest pK_a variation range observed in the proton-imprinted NPs was significantly larger than that of the nonimprinted particles (Fig. 13.29B) and enzymes that transport protons in the blood [112]. The pK_f shift enabled the reversible capture and release of the target ions [123,138–141].

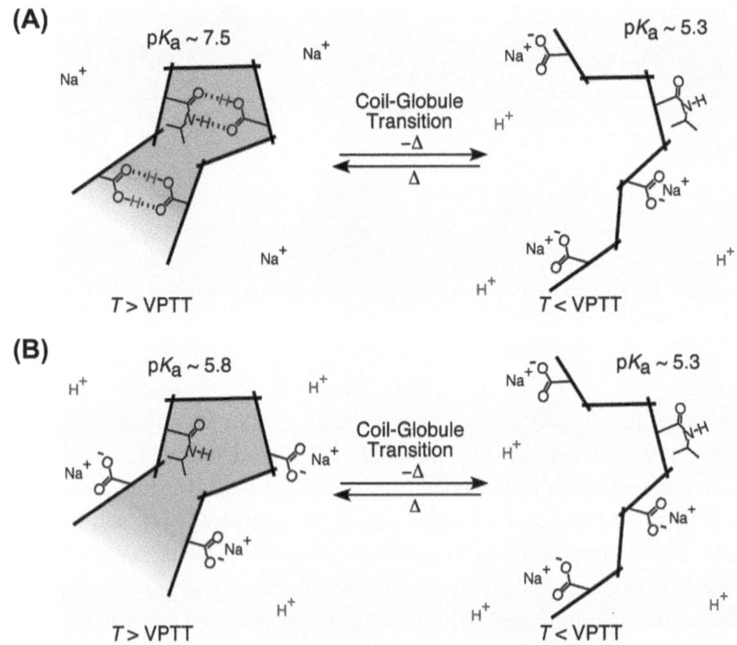

■ **FIGURE 13.29** Mechanism of temperature-responsive pK_a shift of (A) proton-imprinted and (B) nonimprinted NPs triggered by the phase transition [141].

Taking advantage of the reversible pK_a shift of the pNIPAm NPs, materials to reversibly absorb and desorb CO_2 via a phase transition, induced by small temperature interval cooling and heating cycles, can be synthesized [138,139,141]. Amine-containing monomers, such as *N*-[3-(dimethylamino)propyl]methacrylamide (DMAPM), are incorporated in the NPs as functional units to capture CO_2. Below the volume phase transition temperature (VPTT), most amines in the swollen NPs are capable of forming ion pairs with absorbed bicarbonate ions. However, above the VPTT, shrinkage of NPs lowers the pK_a and the number of amine groups exposed to water, thereby resulting in an almost complete desorption of CO_2 (Fig. 13.30). The NPs can reversibly absorb more than the DMAPM monomer and polymer without NIPAm. This indicates the importance of the temperature-responsive phase transition of polymers in determining the degree of absorption.

The capacity for CO_2 absorption per amine in the NPs in solution was either comparable, or superior to, the traditional CO_2 absorbents comprising ethanolamine [138]. However, the absorption capacity per weight of the solution was much smaller than that of the aqueous solution of ethanol amines

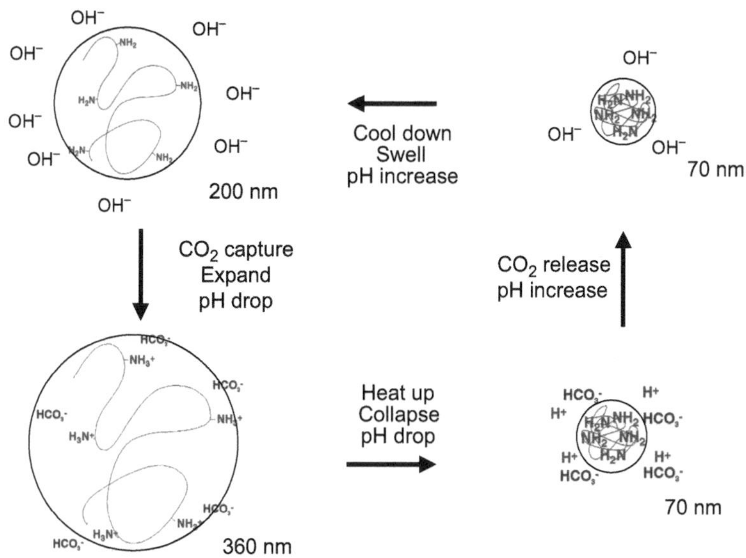

■ **FIGURE 13.30** Mechanism of reversible CO_2 absorption by pNIPAm NPs functionalized by DMAPM triggered by the temperature-induced volume phase transition [139].

because the concentration of NPs in the solution (~0.1 wt.%) was much lower than that of the amine solution (20–30 wt.%) [138]. As a result, the energy consumed by the solution of NPs to sequester CO_2 was significantly larger than the amine solution, in spite of the use of a low temperature for the regeneration process. In order to improve the energy efficiency, the concentration of NPs had to be dramatically increased.

Hydrogel films, consisting of temperature-responsive NP (NP films), were developed in order to achieve energy efficient CO_2 sorbents with high-absorption capacity per weight [139,141]. The films were prepared by simply casting the solution of temperature-responsive NPs onto the surface of a solid substrate (Fig. 13.31). The films brought together the best of the high stoichiometric absorption efficiency of the NP solutions with the large absorption capacity per weight of the solid sorbents. Furthermore, it was revealed that the temperature-responsive hydrogel films consisting of NPs respond much faster to external stimuli when compared to monolithic hydrogel films (Fig. 13.32) [139,141–143].

The reversible absorption capacity of the hydrogel film can be improved by incorporating a larger amount of DMAPM. However, NPs with a larger amount of DMAPM do not always release CO_2 completely when heating to 75°C [141].

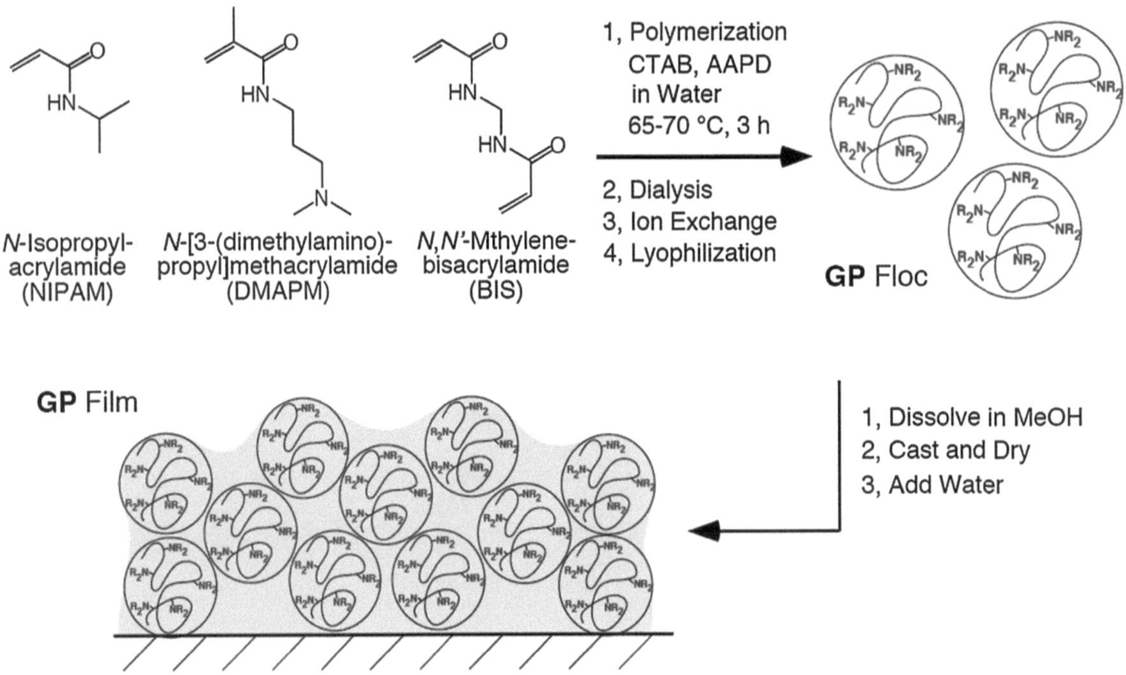

■ **FIGURE 13.31** Schematic illustration of the procedure to prepare temperature-responsible microgel films [140].

In order to prepare thermally responsive NP films that reversibly capture and release large amounts of CO_2 over a narrow temperature range (30–75°C), the pK_a value of the ammonium ions in the NPs at both the CO_2 capture temperature (30°C) and the release temperature (75°C) has to be the carefully optimized [141]. At 30°C, a high pK_a value, above the pK_{a1} of H_2CO_3, is required for efficient CO_2 capture. Conversely, a low pK_a value at 75°C, below the pK_{a1} of H_2CO_3, is required for the efficient release of CO_2.

The pK_a value of the NPs can be readily adjusted to the desired level via four methods [141]: (1) Controlling the VPTT of the GPs above and close to 30°C—the phase transition induces a large pK_a transition over a wide temperature range above the VPTT; (2) Controlling the size of the GPs—smaller GPs show higher pK_a values; (3) Controlling the swelling ratio that influences the degree of variation of the pK_a value, and consequently the pK_a value of the GPs; (4) Controlling the imprinted microenvironment of the GPs—GPs synthesized in the presence of a large amount of protons exhibit higher pK_a values.

The optimal NPs, obtained according to the design rationale, showed a high capture capacity (68 mL CO_2/g dry NPs, 3.0 mmol CO_2/g dry NPs), although

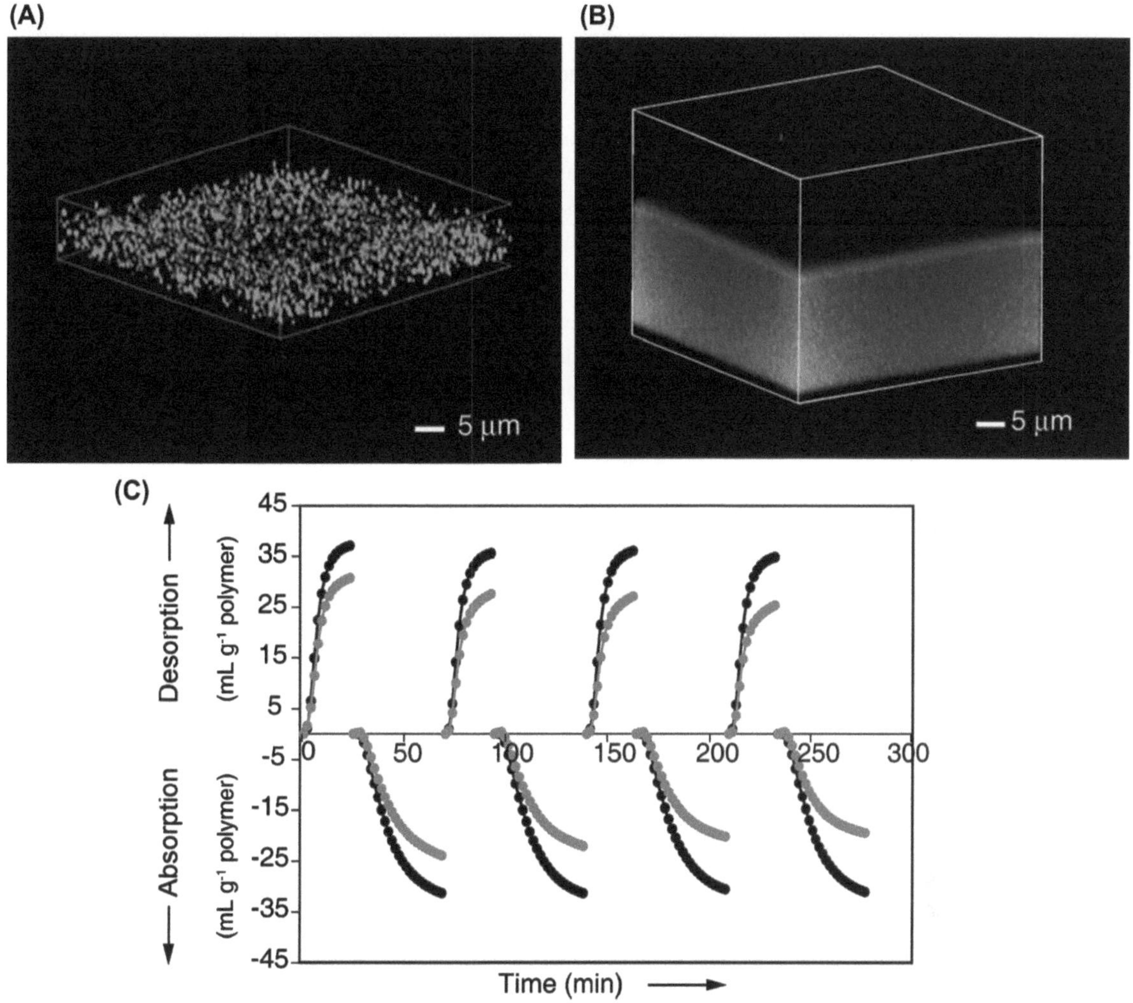

■ **FIGURE 13.32** Confocal fluorescence images of (A) a gel particle film and (B) a monolithic gel film. (C) Temperature-responsive absorption (30°C) and desorption (75°C) of CO_2 from a gel particle film (black) and a monolithic gel film (gray) over four cycles [140].

the regeneration temperature was as low as 75°C. Hydrogel films, consisting of NPs with a hemoglobin mimic function, are expected to become the new class of CO_2 absorbents used to sequester CO_2 at low energy cost.

13.6 CONCLUSIONS

This section introduces recent research on efficient materials for application to both energy and the environment. In most cases, researchers put a lot of effort in the molecular design (nanocreation) of the main material components. In the development of DSSCs, dye molecules are designed to achieve enhanced photocurrent and photovoltage in the wide wavelength region. In the research of bioinspired catalysis, synthetic lipids are designed to prepare nanochannels as enzyme reactors. For reversible hydrogen storage, bioinspired catalysts are designed for efficient and reversible conversion reactions. In the development of an energy efficient CO_2 absorbent, the pK_a of the ammonium anion in the hydrogel nanoparticle is tuned.

In order to achieve the desired functional materials using the optimized functional units described in this chapter, it would be equally important to organize the building/functional unit as the final product. Depending on the order and organization of the nanoscale functional unit in materials and/or structure of the substrate on which the functional units are immobilized (nano organization), the maximized function observed in the nanocreation step may disappear. In the case of DSSCs, a higher photocurrent might be achieved through immobilization of the dye to base material, with a surface area that is large enough and has interconnecting pores that are large enough for the ions to diffuse. To accelerate reversible H_2 storage and the CO_2 capture process, it would be important to tune the design of the porous substrate.

By tuning the materials through optimization of both the molecular design (nanocreation step) and the organized structure (nano organization step), highly efficient materials for energy and environmental application should be realized.

REFERENCES

[1] O'Regan B, Grätzel M. A low-cost, high-efficiency solar cell based on dye-sensitized colloidal TiO_2 films. Nature 1991;353:737–40.

[2] Grätzel M. Recent advances in sensitized mesoscopic solar cells. Acc Chem Res 2009;42:1788–98.

[3] Hagfeldt A, Boschloo G, Sun L, Kloo L, Pettersson H. Dye-sensitized solar cells. Chem Rev 2010;110:6595–663.

[4] Nazeeruddin MK, Key A, Rodicio I, Humphry-Baker R, Müller E, Liska P, et al. Conversion of light to electricity by cis-X2bis(2,2′-bipyridyl-4,4′-dicarboxylate) ruthenium(II) charge-transfer sensitizers ($X = Cl^-$, Br^-, I^-, CN^-, and SCN^-) on nanocrystalline titanium dioxide electrodes. J Am Chem Soc 1993;115:6382–90.

[5] Nazeeruddin MK, De Angelis F, Fantacci S, Selloni A, Viscardi G, Liska P, et al. Combined experimental and DFT-TDDFT computational study of photoelectrochemical cell ruthenium sensitizers. J Am Chem Soc 2005;127: 16835–47.

[6] Nazeeruddin MK, Péchy P, Renouard T, Zakeeruddin SM, Humphry-Baker R, Comte P, et al. Engineering of efficient panchromatic sensitizers for nanocrystalline TiO_2-based solar cells. J Am Chem Soc 2001;123:1613–24.

[7] Robson KCD, Koivisto BD, Yella A, Sporinova B, Nazeeruddin MK, Baumgartner T, et al. Design and development of functionalized cyclometalated ruthenium chromophores for light-harvesting applications. Inorg Chem 2011;50: 5494–508.

[8] Islam A, Sugihara H, Yanagida M, Hara K, Fujihashi G, Tachibana Y, et al. Efficient panchromatic sensitization of nanocrystalline TiO_2 films by β-diketonato ruthenium polypyridyl complexes. New J Chem 2002;26:966–8.

[9] Yamaguchi T, Yanagida M, Katoh R, Sugihara H, Arakawa H. Synthesis and application of ruthenium(II) tricarboxyterpyridyl complex with a nitrogen chelete ligand for solar cells based on nanocrystalline TiO_2 films. Chem Lett 2004;33: 986–7.

[10] Funaki T, Yanagida M, Onozawa-Komatsuzaki N, Kawanishi Y, Kasuga K, Sugihara H. A 2-quinolinecarboxylate-substituted ruthenium(II) complex as a new type of sensitizer for dye-sensitized solar cells. Inorg Chim Acta 2009;362: 2519–22.

[11] Funaki T, Yanagida M, Onozawa-Komatsuzaki N, Kasuga K, Kawanishi Y, Sugihara H. Efficient panchromatic sensitization of nanocrystalline TiO_2-based solar cells using 2-pyridinecarboxylate-substituted ruthenium(II) complexes. Chem Lett 2009;38:62–3.

[12] Funaki T, Yanagida M, Onozawa-Komatsuzaki N, Kasuga K, Kawanishi Y, Sayama K, et al. Synthesis of a new class of cyclometallated ruthenium(II) complexes and their application in dye-sensitized solar cells. Inorg Chem Commun 2009;12:842–5.

[13] Funaki T, Funakoshi H, Onozawa-Komatsuzaki N, Kasuga K, Sayama K, Sugihara H. Cyclometalated ruthenium(II) complexes as near-IR sensitizers for high efficiency dye-sensitized solar cells. Angew Chem Int Ed 2012;51: 7528–31.

[14] Islam A, Chowdhury FA, Chiba Y, Komiya R, Fuke N, Ikeda N, et al. Synthesis and characterization of new efficient tricarboxyterpyridyl (β-diketonato) ruthenium(II) sensitizers and their applications in dye-sensitized solar cells. Chem Mater 2006;18:5178–85.

[15] Chen BS, Chen K, Hong YH, Liu WH, Li TH, Lai CH, et al. Neutral, panchromatic Ru(II) terpyridine sensitizers bearing pyridine pyrazolate chelates with superior DSSC performance. Chem Commun 2009:5844–66.

[16] Islam A, Singh SP, Han L. Synthesis and application of new ruthenium complexes containing β-diketonato ligands as sensitizers for nanocrystalline TiO_2 solar cells. Int J Photoenergy 2011;2011:204639–47.

[17] Li G, Yella A, Brown DG, Gorelsky SI, Nazeeruddin MK, Grätzel M, et al. Near-IR photoresponse of ruthenium dipyrrinate terpyridine sensitizers in the dye-sensitized solar cells. Inorg Chem 2014;53:5417–9.

[18] Onozawa-Komatsuzaki N, Yanagida M, Funaki T, Kasuga K, Sayama K, Sugihara H. Near-IR sensitization of nanocrystalline TiO_2 with a new ruthenium complex having a 2,6-bis(4-carboxyquinolin-2-yl)pyridine ligand. Inorg Chem Commun 2009;12:1212–5.

[19] Onozawa-Komatsuzaki N, Yanagida M, Funaki T, Kasuga K, Sayama K, Sugihara H. Near-IR dye-sensitized solar cells using a new type of ruthenium complexes having 2,6-bis(quinolin-2-yl)pyridine derivatives. Sol Energy Mater Sol Cells 2011;95:310–4.

[20] Onozawa-Komatsuzaki N, Funaki T, Kasuga K, Nakazawa Y, Sayama K, Sugihara H. Synthesis and electrochemical properties of 2,6-bis(quinoline-2-yl) pyridyl ruthenium complexes as near-infrared sensitizers for dye-sensitized solar cells. Jpn J Appl Phys 2012;51:10NE11.

[21] Kinoshita T, Dy JT, Uchida S, Kubo T, Segawa H. Wideband dye-sensitized solar cells employing a phosphine-coordinated ruthenium sensitizer. Nat Photonics 2013;7:535–9.

[22] Funaki T, Onozawa-Komatsuzaki N, Kasuga K, Sayama K, Sugihara H. New class of thiocyanate-free cyclometalated ruthenium(II) complexes having a pyridylquinoline derivative for near-infrared sensitization of dye-sensitized solar cells. Inorg Chem Commun 2013;35:281–3.

[23] Funaki T, Kusama H, Onozawa-Komatsuzaki N, Kasuga K, Sayama K, Sugihara H. Near-IR sensitization of dye-sensitized solar cells using thiocyanate-free cyclometalated ruthenium(II) complexes having a pyridylquinoline ligand. Eur J Inorg Chem 2014;2014:1303–11.

[24] Nusbaumer H, Moser JE, Zakeeruddin SM, Nazeeruddin MK, Grätzel M. Co^{II} $(dbbip)_2^{2+}$ complex rivals tri-iodide/iodide redox mediator in dye-sensitized photovoltaic cells. J Phys Chem B 2001;105:10461–4.

[25] Sapp SA, Elliott CM, Contado C, Caramori S, Bignozzi CA. Substituted polypyridine complexes of cobalt(II/III) as efficient electron-transfer mediators in dye-sensitized solar cells. J Am Chem Soc 2002;124:11215–22.

[26] Nusbaumer H, Zakeeruddin SM, Moser JE, Grätzel M. An alternative efficient redox couple for the dye-sensitized solar cell system. Chem Eur J 2003;9:3756–63.

[27] Feldt SM, Gibson EA, Gabrielsson E, Sun L, Boschloo G, Hagfeldt A. Design of organic dyes and cobalt polypyridine redox mediators for high-efficiency dye-sensitized solar cells. J Am Chem Soc 2010;132:16714–24.

[28] Feldt SM, Wang G, Boschloo G, Hagfeldt A. Effects of driving forces for recombination and regeneration on the photovoltaic performance of dye-sensitized solar cells using cobalt polypyridine redox couples. J Phys Chem C 2011;115:21500–7.

[29] Yella A, Lee H, Tsao HN, Yi C, Chandiran AK, Nazeeruddin MK, et al. Porphyrin-sensitized solar cells with cobalt (II/III)-based redox electrolyte exceed 12 percent efficiency. Science 2011;334:629–34.

[30] Murakami TN, Koumura N, Uchiyama T, Uemura Y, Obuchi K, Masaki N, et al. Recombination inhibitive structure of organic dyes for cobalt complex redox electrolytes in dye-sensitised solar cells. J Mat Chem A 2013;1:792–8.

[31] Uchiyama T, Murakami TN, Yoshii N, Uemura Y, Koumura N, Masaki N, et al. An increase in energy conversion efficiency by decreasing cobalt redox electrolyte diffusion resistance in dye-sensitized solar cells. Chem Lett 2013;42:453–4.

[32] Murakami TN, Koumura N, Kimura M, Mori S. Structural effect of donor in organic dye on recombination in dye-sensitized solar cells with cobalt complex electrolyte. Langmuir 2014;30:2274–9.

[33] Frischmann PD, Mahata K, Wurthner F. Powering the future of molecular artificial photosynthesis with light-harvesting metallosupramolecular dye assemblies. Chem Soc Rev 2013;42:1847–70.

[34] Babu SS, Praveen VK, Ajayaghosh A. Functional π-gelators and their applications. Chem Rev 2014;114:1973–2129.

[35] Shimizu T, Minamikawa H, Kogiso M, Aoyagi M, Kameta N, Ding W, et al. Self-organized nanotube materials and their application in bioengineering. Polym J 2014;46:831–58.

[36] Yamamoto Y, Fukushima T, Suna Y, Ishii N, Saeki A, Seki S, et al. Photoconductive coaxial nanotubes of molecularly connected electron donor and acceptor layers. Science 2006;314:1761–4.

[37] Roger C, Miloslavina Y, Brunner D, Holzwarth A, Wurthner F. Self-assembled zinc chlorin rod antennae powered by peripheral light-harvesting chromophores. J Am Chem Soc 2008;130:5929–39.

[38] Shao H, Seifert J, Romano NC, Gao M, Helmus JJ, Jaroniec CP, et al. Amphiphilic self-assembly of an n-type nanotube. Angew Chem Int Ed 2010;49:7688–91.

[39] Eisele DM, Berlepsch HV, Böttcher C, Stevenson KJ, Vanden Bout DA, Kirstein S, et al. Photoinitiated growth of sub-7 nm silver nanowires within a chemically active organic nanotubular template. J Am Chem Soc 2010;132:2104–5.

[40] Miller RA, Stephanopoulos N, McFarland JM, Rosko AS, Geissler PL, Francis MB. Impact of assembly state on the defect tolerance of TMV-based light harvesting arrays. J Am Chem Soc 2010;132:6068–74.

[41] de la Escosura A, Janssen PGA, Schenning APHJ, Nolte RJM, Cornelissen JJLM. Encapsulation of DNA-templated chromophore assemblies within virus protein nanotubes. Angew Chem Int Ed 2010;49:5335–8.

[42] Kholkin A, Amdursky N, Bdikin I, Gazit E, Rosenman G. Strong piezoelectricity in bioinspired peptide nanotubes. ACS Nano 2010;4:610–4.

[43] Walker EK, Vanden Bout DA, Stevenson KJ. Aqueous electrogenerated chemiluminescence of self-assembled double walled tubular J-aggregates of amphiphilic cyanine dyes. J Phys Chem C 2011;115:2470–5.

[44] Zhang W, Jin W, Fukushima T, Saeki A, Seki S, Aida T. Supramolecular linear heterojunction composed of graphite-like semiconducting nanotubular segments. Science 2011;334:340–3.

[45] Eisele DM, Cone CW, Bloemsma EA, Vlaming SM, van der Kwaak CGF, Silbey RJ, et al. Utilizing redox-chemistry to elucidate the nature of exciton transitions in supramolecular dye nanotubes. Nat Chem 2012;4:655–62.

[46] Kim JH, Lee M, Lee JS, Park CB. Self-assembled light harvesting peptide nanotubes for mimicking natural photosynthesis. Angew Chem Int Ed 2012;51:517–20.

[47] Kameta N, Ishikawa K, Masuda M, Asakawa M, Shimizu T. Soft nanotubes acting as a light-harvesting antenna system. Chem Mater 2012;24:209–14.

[48] Kameta N, Masuda M, Shimizu T, Shiroishi H. [in preparation].

[49] Kameta N, Masuda M, Shimizu T. Two-step naked-eye detection of lectin by hierarchical organization of soft nanotubes into liquid crystal and gel phases. Chem Commun 2015;51:6816–9.

[50] Kameta N, Masuda M, Shimizu T. Qualitative/chiral sensing of amino acids by naked-eye fluorescence change based on morphological transformation and hierarchizing in supramolecular assemblies of pyrene-conjugated glycolipids. Chem Commun 2015;51:11104–7.
[51] Goddard JM, Hotchkiss JH. Polymer surface modification for the attachment of bioactive compounds. Prog Polym Sci 2007;32:698–725.
[52] Komatsu T. Protein-based nanotubes for biomedical applications. Nanoscale 2012; 4:1910–8.
[53] Miyazaki M, Maeda H. Microchannel enzyme reactors and their applications for processing. Trends Biotechnol 2006;24:463–70.
[54] Yu LT, Banerjee IA, Gao XY, Nuraje N, Matsui H. Fabrication and application of enzyme-incorporated peptide nanotubes. Bioconjugate Chem 2005;16:1484–7.
[55] Hudson S, Cooney J, Magner E. Proteins in mesoporous silicates. Angew Chem Int Ed 2008;47:8582–94.
[56] Kameta N, Minamikawa H, Masuda M. Supramolecular organic nanotubes: how to utilize the inner nanospace and the outer space. Soft Matter 2011;7:4539–61.
[57] Kameta N, Ishikawa K, Masuda M, Shimizu T. Control of self-assembled morphology and molecular packing of asymmetric glycolipids by association/dissociation with poly(thiopheneboronic acid). Langmuir 2013;29:13291–8.
[58] Kameta N, Yoshida K, Masuda M, Shimizu T. Supramolecular nanotube hydrogels: remarkable resistance effect of confined proteins to denaturants. Chem Mater 2009;21:5892–8.
[59] Kameta N, Minamikawa H, Someya Y, Yui H, Masuda M, Shimizu T. Confinement effect of organic nanotubes toward green fluorescent protein (GFP) depending on the inner diameter size. Chem Eur J 2010;16:4217–23.
[60] Kameta N, Masuda M, Minamikawa H, Mishima Y, Yamashita I, Shimizu T. Functionalizable organic nanochannels based on lipid nanotubes: encapsulation and nanofluidic behavior of biomacromolecules. Chem Mater 2007;19:3553–60.
[61] Karlsson R, Karlsson M, Karlsson A, Cans AS, Bergenholtz J, Akerman B, et al. Moving-wall-driven flows in nanofluidic systems. Langmuir 2002;18:4186–90.
[62] Kameta N, Masuda M, Mizuno G, Morii N, Shimizu T. Supramolecular nanotube endo sensing for a guest protein. Small 2008;4:561–5.
[63] Kameta N, Minamikawa H, Masuda M, Mizuno G, Shimizu T. Controllable biomolecule release from self-assembled organic nanotubes with asymmetric surfaces: pH and temperature dependence. Soft Matter 2008;4:1681–7.
[64] Kameta N, Masuda M, Shimizu T. Soft nanotube hydrogels functioning as artificial chaperones. ACS Nano 2012;6:5249–58.
[65] Yui H, Guo Y, Koyama K, Sawada T, John G, Yang B, et al. Local environment and property of water inside the hollow cylinder of a lipid nanotube. Langmuir 2005;21:721–7.
[66] Kameta N. Soft nanotube hosts for capsulation and release of molecules, macromolecules, and nanomaterials. J Incl Phenom Macrocycl Chem 2014;79: 1–22.
[67] Kameta N, Masuda M, Shimizu T. Soft nanotubes acting as confinement effecters and chirality induces for achiral polythiophenes. Chem Commun 2016;52:1346–9.
[68] Carrette L, Friedrich KA, Stimming U. Fuel cells: principles, types, fuels, and applications. ChemPhysChem 2000;1:162–93.

[69] Aricò AS, Bruce P, Scrosati B, Tarascon JM, Schalkwijk WV. Nanostructured materials for advanced energy conversion and storage devices. Nat Mat 2005;4:366–77.

[70] Wright IG, Gibbons TB. Recent developments in gas turbine materials and technology and their implications for syngas firing. Int J Hydrogen Energ 2007;32:3610–21.

[71] Hamilton CW, Baker RT, Staubitz A, Manners I. B-N compounds for chemical hydrogen storage. Chem Soc Rev 2009;38:279–93.

[72] Johnson TC, Morris DJ, Wills M. Hydrogen generation from formic acid and alcohols using homogeneous catalysts. Chem Soc Rev 2010;39:81–8.

[73] Yadav M, Xu Q. Liquid-phase chemical hydrogen storage materials. Energy Environ Sci 2012;5:9698–725.

[74] Jessop PG, Ikariya T, Noyori R. Homogeneous hydrogenation of carbon dioxide. Chem Rev 1995;95:259–72.

[75] Grasemann M, Laurenczy G. Formic acid as a hydrogen source - recent developments and future trends. Energy Environ Sci 2012;5:8171–81.

[76] Wang WH, Himeda Y, Muckerman JT, Manbeck GF, Fujita E. CO_2 hydrogenation to formate and methanol as an alternative to photo- and electrochemical CO_2 reduction. Chem Rev 2015. http://dx.doi.org/10.1021/acs.chemrev.5b00197.

[77] Inoue Y, Izumida H, Sasaki Y, Hashimoto H. Catalytic fixation of carbon dioxide to formic acid by transition-metal complexes under mild conditions. Chem Lett 1976;5:863–4.

[78] Leitner W. Carbon dioxide as a raw material: the synthesis of formic acid and its derivatives from CO_2. Angew Chem Int Ed 1995;34:2207–21.

[79] Jessop PG, Joo F, Tai CC. Recent advances in the homogeneous hydrogenation of carbon dioxide. Coord Chem Rev 2004;248:2425–42.

[80] Wang W, Wang S, Ma X, Gong J. Recent advances in catalytic hydrogenation of carbon dioxide. Chem Soc Rev 2011;40:3703–27.

[81] Himeda Y, Onozawa-Komatsuzaki N, Sugihara H, Arakawa H, Kasuga K. Half-sandwich complexes with 4,7-dihydroxy-1,10-phenanthroline: water-soluble, highly efficient catalysts for hydrogenation of bicarbonate attributable to the generation of an oxyanion on the catalyst ligand. Organometallics 2004;23:1480–3.

[82] Himeda Y, Onozawa-Komatsuzaki N, Sugihara H, Kasuga K. Recyclable catalyst for conversion of carbon dioxide into formate attributable to an oxyanion on the catalyst ligand. J Am Chem Soc 2005;127:13118–9.

[83] Himeda Y, Onozawa-Komatsuzaki N, Sugihara H, Kasuga K. Highly efficient conversion of carbon dioxide catalyzed by half-sandwich complexes with pyridinol ligand: the electronic effect of oxyanion. J Photochem Photobiol A Chem 2006;182:306–9.

[84] Himeda Y, Onozawa-Komatsuzaki N, Sugihara H, Kasuga K. Simultaneous tuning of activity and water solubility of complex catalysts by acid-base equilibrium of ligands for conversion of carbon dioxide. Organometallics 2007;26:702–12.

[85] Suna Y, Ertem MZ, Wang WH, Kambayashi H, Manaka Y, Muckerman JT, et al. Positional effects of hydroxy groups on catalytic activity of proton-responsive half-sandwich Cp* iridium(III) complexes. Organometallics 2014;33:6519–30.

[86] Wang WH, Hull JF, Muckerman JT, Fujita E, Himeda Y. Second-coordination-sphere and electronic effects enhance iridium (III)-catalyzed homogeneous hydrogenation of carbon dioxide in water near ambient temperature and pressure. Energy Environ Sci 2012;5:7923–6.

[87] Peters JW, Lanzilotta WN, Lemon BJ, Seefeldt LC. X-ray crystal structure of the Fe-only hyrogenase (CpI) from clostridium pasteurianum to 1.8 angstrom resolution. Science 1998;282:1853–8.

[88] Pandey AS, Harris TV, Giles LJ, Peters JW, Szilagyi RK. Dithiomethylether as a ligand in the hydrogenase H-cluster. J Am Chem Soc 2008;130:4533–40.

[89] Silakov A, Wenk B, Reijerse E, Lubitz W. ^{14}N HYSCORE investigation of the H-cluster of [FeFe] hydrogenase: evidence for a nitrogen in the dithiol bridge. Phys Chem Chem Phys 2009;11:6592–9.

[90] Lubitz W, Ogata H, Rüdiger O, Reijerse E. Hydrogenases. Chem Rev 2014;114: 4081–148.

[91] Yang JY, Bullock RM, Shaw WJ, Twamley B, Fraze K, DuBois MR, et al. Mechanistic insights into catalytic H_2 oxidation by Ni complexes containing a diphosphine ligand with a positioned amine base. J Am Chem Soc 2009;131: 5935–45.

[92] DuBois DL, Bullock RM. Molecular electrocatalysts for the oxidation of hydrogen and the production of hydrogen - the role of pendant amines as proton relays. Eur J Inorg Chem 2011;7:1017–27.

[93] Wang WH, Muckerman JT, Fujita E, Himeda Y. Mechanistic insight through factors controlling effective hydrogenation of CO_2 catalyzed by bioinspired proton-responsive iridium (III) complexes. ACS Catal 2013;3:856–60.

[94] Rossin A, Gonsalvi L, Phillips AD, Maresca O, Lledós A, Peruzzini M. Water-assisted H-H bond splitting mediated by $[CpRu(PTA)_2Cl]$ (PTA = 1,3,5-triaza-7-phosphaadamantane). A DFT analysis. Organometallics 2007;26:3289–96.

[95] Friedrich A, Drees M, Günne JS, Schneider S. Highly stereoselective proton/hydride exchange: assistance of hydrogen bonding for the heterolytic splitting of H_2. J Am Chem Soc 2009;131:17552–3.

[96] Kovacs G, Rossin A, Gonsalvi L, Lledos A, Peruzzini M. Comparative DFT analysis of ligand and solvent effects on the mechanism of H_2 activation in water mediated by half-sandwich complexes $[Cp'Ru(PTA)_2Cl]$ ($Cp' = C_5H_5$, C_5Me_5; PTA = 1,3,5-triaza-7-phosphaadamantane). Organometallics 2010;29:5121–31.

[97] DuBois MR, DuBois DL. The roles of the first and second coordination spheres in the design of molecular catalysts for H_2 production and oxidation. Chem Soc Rev 2009;38:62–72.

[98] Himeda Y, Miyazawa S, Hirose T. Interconversion between formic acid and H_2/CO_2 using rhodium and ruthenium catalysts for CO_2 fixation and H_2 storage. ChemSusChem 2011;4:487–93.

[99] Manaka Y, Wang WH, Suna Y, Kambayashi H, Muckerman JT, Fujita E, et al. Efficient H_2 generation from formic acid using azole complexes in water. Catal Sci Technol 2014;4:34–7.

[100] Wang WH, Ertem MZ, Xu S, Onishi N, Manaka Y, Suna Y, et al. Highly robust hydrogen generation by bioinspired Ir complexes for dehydrogenation of formic acid in water: experimental and theoretical mechanistic investigations at different pH. ACS Catal 2015;5:5496–504.

[101] Wang WH, Xu S, Manaka Y, Suna Y, Kambayashi H, Muckerman JT, et al. Formic acid dehydrogenation with bioinspired iridium complexes: a kinetic isotope effect study and mechanistic insight. ChemSusChem 2014;7:1976–83.

[102] Olah GA, Goeppert A, Prakash GKS. Beyond oil and gas: the methanol economy. 2nd ed. Weinheim: Wiley-VCH; 2009.

[103] Olah GA, Goeppert A, Prakash GKS. Chemical recycling of carbon dioxide to methanol and dimethyl ether: from greenhouse gas to renewable, environmentally carbon neutral fuels and synthetic hydrocarbons. J Org Chem 2009;74:487–98.

[104] Rochelle GT. Amine scrubbing for CO_2 capture. Science 2009;325:1652–4.

[105] Chowdhury FA, Okabe H, Yamada H, Onoda M, Fujioka Y. Synthesis and selection of hindered new amine absorbents for CO_2 capture. Energy Proc 2011; 4:201–8.

[106] Fout TE, Murphy JT. DOE/NETL's carbon capture R&D program for existing coal-fired power plants. DOE/NETL 2009/1356. Pittsburgh, PA: National Energy Technology Laboratory; 2009.

[107] Aaron D, Tsouris C. Separation of CO_2 from flue gas: a review. Sep Sci Technol 2005;40:321–48.

[108] D'Alessandro DM, Smit B, Long JR. Carbon dioxide capture: prospects for new materials. Angew Chem Int Ed 2010;49:6058–82.

[109] Park Y, Shin D, Jang YN, Park AHA. CO_2 capture capacity and swelling measurements of liquid-like nanoparticle organic hybrid materials via attenuated total reflectance Fourier transform infrared spectroscopy. J Chem Eng Data 2012;57:40–5.

[110] Huang HY, Yang RT, Chinn D, Munson CL. Amine-grafted MCM-48 and silica xerogel as superior sorbents for acidic gas removal from natural gas. Ind Eng Chem Res 2003;42:2427–33.

[111] Pirngruber GD, Cassiano-Gaspar S, Louret S, Chaumonnot A, Delfort B. Amines immobilized on a solid support for postcombustion CO_2 capture—A preliminary analysis of the performance in a VSA or TSA process based on the adsorption isotherms and kinetic data. Energy Proc 2009;1:1335–42.

[112] Kilmartin JV, Breen JJ, Roberts GCK, Ho C. Direct measurement of the pK values of an alkaline Bohr group in human hemoglobin. Proc Natl Acad Sci 1973;70: 1246–9.

[113] Voet D, Voet JG. Biochemistry. 4th ed. New York, USA: Wiley; 1995.

[114] Shinkai S, Ogawa T, Nakaji T, Kusano Y, Nanabe O. Photocontrolled extraction ability of azobenzene-bridged azacrown ether. Tetrahedron Lett 1979;20: 4569–72.

[115] Shinkai S, Nakaji T, Nishida Y, Ogawa T, Manabe O. Photoresponsive crown ethers. 1. Cis-trans isomerism of azobenzene as a tool to enforce conformational changes of crown ethers and polymers. J Am Chem Soc 1980;102:5860–5.

[116] Shinkai S, Nakaji T, Ogawa T, Shigematsu K, Manabe O. Photoresponsive crown ethers. 2. Photocontrol of ion extraction and ion transport by a bis(crown ether) with a butterfly-like motion. J Am Chem Soc 1981;103:111–5.

[117] Oya T, Enoki T, Grosberg AY, Masamune S, Sakiyama T, Takeoka Y, et al. Reversible molecular adsorption based on multiple-point interaction by shrinkable gels. Science 1999;286:1543–5.

[118] Lynch I, de Gregorio P, Dawson KA. Simultaneous release of hydrophobic and cationic solutes from thin-film "plum-pudding" gels: A multifunctional platform for surface drug delivery? J Phys Chem B 2005;109:6257–61.

[119] Serpe MJ, Yarmey KA, Nolan CM, Lyon LA. Doxorubicin uptake and release from microgel thin films. Biomacromolecules 2005;6:408–13.

[120] Hoare T, Pelton R. Impact of microgel morphology on functionalized microgel–drug interactions. Langmuir 2008;24:1005–12.

[121] Hoshino Y, Haberaecker III WW, Kodama T, Zeng Z, Okahata Y, Shea KJ. Affinity purification of multifunctional polymer nanoparticles. J Am Chem Soc 2010;132:13648–50.

[122] Yoshimatsu K, Lesel BK, Yonamine Y, Beierle JM, Hoshino Y, Shea KJ. Temperature-responsive "catch and release" of proteins by using multifunctional polymer-based nanoparticles. Angew Chem Int Ed 2012;51:2405–8.

[123] Nagase K, Kobayashi J, Kikuchi A, Akiyama Y, Kanazawa J, Okano T. Preparation of thermoresponsive cationic copolymer brush surfaces and application of the surface to separation of biomolecules. Biomacromolecules 2008;9:1340–7.

[124] Akiyama Y, Kikuchi A, Yamato M, Okano T. Ultrathin poly(*N*-isopropylacrylamide) grafted layer on polystyrene surfaces for cell adhesion/detachment control. Langmuir 2004;20:5506–11.

[125] Kim YJ, Ebara M, Aoyagi T. A smart nanofiber web that captures and releases cells. Angew Chem Int Ed 2012;51:10537–41.

[126] Kanazawa R, Yoshida T, Gotoh T, Sakohara S. Preparation of molecular imprinted thermosensitive gel adsorbents and adsorption/desorption properties of heavy metal ions by temperature swing. J Chem Eng Jpn 2004;37:59–66.

[127] Hoshino Y, Kodama T, Okahata Y, Shea KJ. Peptide imprinted polymer nanoparticles: a plastic antibody. J Am Chem Soc 2008;130:15242–3.

[128] Pan G, Guo Q, Cao C, Yang H, Li B. Thermo-responsive molecularly imprinted nanogels for specific recognition and controlled release of proteins. Soft Matter 2013;9:3840–50.

[129] Wulff G. Molecular imprinting in cross-linked materials with the aid of molecular templates—A way towards artificial antibodies. Angew Chem Int Ed 1995;34:1812–32.

[130] Vlatakis G, Andersson LI, Müller R, Mosbach K. Drug assay using antibody mimics made by molecular imprinting. Nature 1993;361:645–7.

[131] Yan M, Ramstrom O, editors. Molecularly imprinted materials: science and technology. Boca Raton, FL, USA: CRC Press; 2004.

[132] Sellergren B. Molecularly imprinted polymers. Amsterdam: Elsevier; 2001.

[133] Shea KJ. Trends Polym Sci 1994;2:166.

[134] Haupt K. Molecularly imprinted polymers: the next generation. Anal Chem 2003;75:376A–83A.

[135] Çakir P, Cutivet A, Resmini M, Bui BT, Haupt K. Protein-size molecularly imprinted polymer nanogels as synthetic antibodies, by localized polymerization with multi-initiators. Adv Mater 2013;25:1048–51.

[136] Feil H, Bae YH, Feijen J, Kim SW. Mutual influence of pH and temperature on the swelling of ionizable and thermosensitive hydrogels. Macromolecules 1992;25:5528–30.

[137] Yang Y, Mijalis AJ, Fu H, Agosto C, Tan KJ, Batteas JD, et al. Reversible changes in solution pH resulting from changes in thermoresponsive polymer solubility. J Am Chem Soc 2012;134:7378–83.

[138] Hoshino Y, Imamura K, Yue M, Inoue G, Miura Y. Reversible absorption of CO_2 triggered by phase transition of amine-containing micro- and nanogel particles. J Am Chem Soc 2012;134:18177–80.

[139] Yue M, Hoshino Y, Ohshiro Y, Imamura K, Miura Y. Temperature-responsive microgel films as reversible carbon dioxide absorbents in wet environment. Angew Chem Int Ed 2014;53:2654–7.

[140] Hoshino Y, Ohashi RC, Miura Y. Rational design of synthetic nanoparticles with a large reversible shift of acid dissociation constants: proton imprinting in stimuli responsive nanogel particles. Adv Mater 2014;26:3718–23.

[141] Yue M, Hoshino Y, Miura Y. Design rationale of thermally responsive microgel particle films that reversibly absorb large amounts of CO_2: fine tuning the pK_a of ammonium ions in the particles. Chem Sci 2015;6:6112–23.

[142] Lynch I, Dawson KA. Synthesis and characterization of an extremely versatile structural motif called the "plum-pudding" gel. J Phys Chem B 2003;107:9629–37.

[143] Nayak S, Debord SB, Lyon LA. Investigations into the deswelling dynamics and thermodynamics of thermoresponsive microgel composite films. Langmuir 2003;19:7374–9.

Index

'*Note*: Page numbers followed by "f" indicate figures and "t" indicate tables.'

O

P

FSC
www.fsc.org
MIX
Papier | Fördert
gute Waldnutzung
FSC® C083411

Kolibri 360 GmbH
Ferdinand-Jühlke-Straße 7
99095 Erfurt, Deutschland
produktsicherheit@kolibri360.de